QUORUM SENSING
MOLECULAR MECHANISM AND BIOTECHNOLOGICAL APPLICATION

群体感应
分子机制和生物技术应用

[意] 朱塞皮娜·托莫纳罗（Giuseppina Tommonaro） 主编

刘 宁 等 译著

中国农业科学技术出版社

图书在版编目（CIP）数据

群体感应：分子机制和生物技术应用／（意）朱塞皮娜·托莫纳罗（Giuseppina Tommonaro）主编；刘宁等译著. --北京：中国农业科学技术出版社，2024.8
书名原文：Quorum Sensing：Molecular Mechanism and Biotechnological Application
ISBN 978-7-5116-6520-1

Ⅰ.①群… Ⅱ.①朱…②刘… Ⅲ.①微生物群落-研究 Ⅳ.①Q938.1

中国国家版本馆 CIP 数据核字（2023）第 208762 号

责任编辑　张诗瑶
责任校对　李向荣
责任印制　姜义伟　王思文

出 版 者	中国农业科学技术出版社 北京市中关村南大街 12 号　　邮编：100081
电　　话	（010）82106625（编辑室）　　（010）82106624（发行部） （010）82109709（读者服务部）
网　　址	https://castp.caas.cn
经 销 者	各地新华书店
印 刷 者	北京建宏印刷有限公司
开　　本	185 mm×260 mm　1/16
印　　张	15
字　　数	404 千字
版　　次	2024 年 8 月第 1 版　2024 年 8 月第 1 次印刷
定　　价	140.00 元

◀◀◀ 版权所有·翻印必究 ▶▶▶

Quorum Sensing: Molecular Mechanism and Biotechnological Application
Giuseppina Tommonaro
ISBN: 978-0-12-814905-8

Copyright © 2019 Giuseppina Tommonaro. Published by Elsevier Inc. All rights reserved.
Authorized Chinese translation published by China Agricultural Science and Technology Press Ltd.

《群体感应：分子机制和生物技术应用》（刘宁等译著）
ISBN: 978-7-5116-6520-1

Copyright © Elsevier Inc. and China Agricultural Science and Technology Press Ltd. All rights reserved.
No part of this publication may be reproduced or transmitted in any form or by any means, electronic or mechanical, including photocopying, recording, or any information storage and retrieval system, without permission in writing from Elsevier Inc. Details on how to seek permission, further information about the Elsevier's permissions policies and arrangements with organizations such as the Copyright Clearance Center and the Copyright Licensing Agency, can be found at our website: www.elsevier.com/permissions.

This book and the individual contributions contained in it are protected under copyright by Elsevier Inc. and China Agricultural Science and Technology Press Ltd. (other than as may be noted herein) .

This edition of Quorum Sensing: Molecular Mechanism and Biotechnological Application is published by China Agricultural Science and Technology Press Ltd. under arrangement with ELSEVIER INC.
This edition is authorized for sale in China only, excluding Hong Kong, Macau and Taiwan. Unauthorized export of this edition is a violation of the Copyright Act. Violation of this Law is subject to Civil and Criminal Penalties.

本版由 ELSEVIER INC. 授权中国农业科学技术出版社在中国大陆地区（不包括香港、澳门以及台湾地区）出版发行。
本版仅限在中国大陆地区（不包括香港、澳门以及台湾地区）出版及标价销售。未经许可之出口，视为违反著作权法，将受民事及刑事法律之制裁。
本书封底贴有 Elsevier 防伪标签，无标签者不得销售。

注　意

本书涉及领域的知识和实践标准在不断变化。新的研究和经验拓展我们的理解，因此须对研究方法、专业实践或医疗方法作出调整。从业者和研究人员必须始终依靠自身经验和知识来评估和使用本书中提到的所有信息、方法、化合物或本书中描述的实验。在使用这些信息或方法时，他们应注意自身和他人的安全，包括注意他们负有专业责任的当事人的安全。在法律允许的最大范围内，爱思唯尔、译文的原文作者、原文编辑及原文内容提供者均不对因产品责任、疏忽或其他人身或财产伤害及/或损失承担责任，亦不对由于使用或操作文中提到的方法、产品、说明或思想而导致的人身或财产伤害及/或损失承担责任。

《群体感应：分子机制和生物技术应用》译著者名单

刘　宁　中国农业科学院特产研究所

张亚玉　成都大学

刘守安　吉林大学

张　悦　中国农业科学院特产研究所

丁冠中　中国农业科学院特产研究所

穆　朋　中国科学院东北地理与农业生态研究所

梁池嘉　中国农业科学院特产研究所

周　帆　中国农业科学院特产研究所

王　岩　中国农业科学院特产研究所

译者前言

在现代农业研究中，提升植物生长和抗逆性已成为学术界和产业界共同追求的目标。人参（Panax ginseng）作为一种重要的中药材，其栽培技术的不断创新和提升不仅关乎其经济价值，更为传统医学的可持续发展提供了新的可能。近年来，研究者们逐渐认识到，土壤微生物在调控人参生长、改善土壤健康及提升人参品质方面发挥着至关重要的作用。

近年来，本团队通过宏基因组学、代谢组学和合成生物学等前沿技术，深入探讨了人参根际微生物群落的组成、结构与功能。研究表明，林下人参和园参根际参与氮、磷等养分循环的微生物存在显著差异，这些差异不仅影响了人参的生长、发育和品质，同时对土壤健康和生态平衡也产生了重要影响。利用合成生物学构建合成菌群不仅能改善人参栽培土壤的理化性质，还能通过分泌植物生长调节物质和其他生物活性物质，促进人参生长，增强人参抗逆性。因此，探索如何优化根际微生物群落组成、结构与功能，成为推动人参产业绿色可持续发展的必由之路。

群体感应（quorum sensing，QS）作为细菌之间的一种通信机制，在微生物领域被广泛关注。细菌通过群体感应调节自身的生理活动，协调群体行为，这一机制在植物与微生物互作中发挥重要作用。土壤微生物通过释放群体感应信号分子调控其趋化和定殖能力，从而影响植物生长和土壤健康。深入研究微生物群体的调控机制，不仅为微生物在农业中的应用提供了理论基础，也为人参的高效绿色栽培开辟了新的研究方向。

此译著的完成得到了中国农业科学院科技创新工程项目［CAAS-ASTIP-ISAPS-2024（No008yc.）］和吉林省科技发展计划项目创新平台（基地）和人才专项（20230508030RC）的资助。在翻译此书的过程中，衷心感谢所有参与者的辛勤付出、不懈努力以及坚定的信念，感谢出版方的鼓励、帮助、支持与信任，才使这部著作顺利完成。同时，我们深知，由于语言和文化的差异，翻译过程中难免存在一些不足之处，诚恳阅读本书的专家、学者、老师和同学们能给予包容和理解，并积极提出宝贵的意见和建议，以帮助我们在后续的学习工作中不断提升和改进。愿本书能够为相关研究领域贡献绵薄之力，促进人参、西洋参及其他药用植物栽培领域的绿色、可持续和高质量发展。

刘 宁

2024 年 6 月

前 言

群体感应（quorum sensing，QS）是细菌通过合作行为进行的过程，影响基因调控。细胞间通信系统根据细胞密度和生长阶段，涉及信号分子的产生与传递。致病性，即侵染宿主并引起疾病的能力，其受细胞通信信号分子的调控。本书探讨了 QS 作为发现新抗生素的一种潜在途径，同时也有利于增进对微生物世界的了解。

本书提供了关于群体感应机制及其在生物材料设计、药物输送、海洋生物膜等领域的生物技术应用的多种研究方法。一方面，有助于理解调控这一机制的分子基础；另一方面，介绍了应用领域的新发现。此外，本书也可以作为一本实用手册，对于从事微生物学、环境和食品科学、海洋生物学和化学领域的研究人员极具价值；同时，也帮助初学者更深入地了解这一令人兴奋的领域。

本书分为 4 个部分（1. 概论：化学和微生物学；2. 跨界通信；3. 群体感应抑制；4. 应用），内容涵盖信号分子的化学机制，极端微生物的最新研究成果，植物与人类的跨界交流，不同分子作为群体感应抑制剂的作用，以及若干生物技术研究实例。

首先，感谢所有作者接受我的邀请并合作完成本项目，感谢他们的帮助，正是由于我们的共同努力，才能将本书打造成为一本高质量且内容很全面的著作；其次，要特别感谢爱思唯尔员工的敬业精神和热情支持；最后，我想感谢我的导师 Salvatore De Rosa 博士（化学家，1946—2014）和我不可或缺的实验室助理 Carmine Iodice 先生，感谢他们对我研究工作的信任与支持。

感谢！

献给我生命中最重要的人：Michela，Serena 和 Sara。

主编　Giuseppina Tommonaro

目　　录

第1部分　概况：化学和微生物学

第1章　革兰氏阴性细菌的化学语言 2
  1　前言 2
  2　革兰氏阴性细菌中群体感应系统构建 3
　　2.1　基于AHLs的群体感应 3
　　2.2　基于AI-2的群体感应 5
　　2.3　使用其他自诱导物的QS系统 5
  3　自诱导物的化学性质 6
　　3.1　N-酰基高丝氨酸内酯 7
　　3.2　环呋喃酮化合物 8
　　3.3　其他 9
　　3.4　海洋环境中的群体感应 9
  4　群体淬灭分子 10
　　4.1　AHL受体抑制剂的研究：LasR拮抗剂 10
　　4.2　QS酶的抑制作用 14
  5　结论 16
  术语表 16
  缩写词 17
  参考文献 17

第2章　群体感应(QS)分子的鉴定分析方法 23
  1　前言 23
  2　样品的制备 25
　　2.1　液液萃取（Liquid-Liquid Extraction，LLE） 25
　　2.2　固相萃取（SPE） 25
  3　QS分子的检测、鉴定和定量 26
　　3.1　AHL检测的分析方法 26
　　3.2　AHL鉴定和定量分析方法 28
　　3.3　其他QS分子鉴定和定量分析方法 33
  4　结论 35
  术语表 35
  缩写词 36

参考文献 · 37

第3章 海洋生物膜和环境中的群体感应 · 43
1 前言 · 43
2 海洋环境中群体感应的发现和日益增长的研究兴趣 · 43
2.1 20世纪70年代：在海洋环境中发现群体感应——弧菌-乌贼模型 · 43
2.2 20世纪90年代至21世纪10年代：对海洋环境中群体感应研究兴趣的日益增长 · 44
3 概述海洋环境和生物膜中参与群体感应的原核生物多样性和化合物 · 45
3.1 海洋环境中AIs特征分析的试验方法 · 45
3.2 AHLs或自诱导物-1（AI-1） · 46
3.3 自诱导物-2（AI-2） · 49
3.4 其他AIs · 49
3.5 AIs在海洋环境中的扩散 · 50
3.6 哈维弧菌（和 Harveyi 支系）是研究海洋细菌群体感应的模型 · 51
4 表面生物膜和共生功能体：微生物栖息地的多样性使海洋环境中的群体感应成为可能 · 52
4.1 （微）藻际和海洋植物的叶际 · 52
4.2 下沉的海洋雪颗粒周围的生物膜 · 53
4.3 海洋微生物垫和其他类型的亚潮带生物膜 · 53
4.4 珊瑚和其他刺胞动物相关的群落 · 54
4.5 海绵相关群落 · 55
5 在海洋环境中由群体感应调控的原核生物功能多样性 · 55
5.1 发光和色素的产生 · 55
5.2 运动性 · 56
5.3 生物膜形成、成熟和散布的开始 · 56
5.4 毒力 · 57
5.5 定殖因子的分泌 · 58
5.6 水平基因转移 · 58
5.7 营养获取 · 58
5.8 抗菌化合物的产生 · 59
5.9 细胞群体内表型异质性的诱导和维持 · 59
6 在海洋环境中寻找和应用群体感应抑制剂：水产养殖保护与绿色防污技术创新 · 60
6.1 一个模型故事：红色海洋藻类 Delisea pulchra 中发现防污化合物 · 60
6.2 从海洋生物中分离的群体感应淬灭化合物的多样性 · 61
6.3 海洋群体淬灭酶的多样性 · 62
6.4 在海洋水下表面生物污垢的应用 · 63
6.5 在水产养殖中的应用 · 63
7 总结 · 64
术语表 · 65
缩写词 · 65

致谢	65
参考文献	65

第4章 极端微生物中的群体感应

1 极端微生物和极端生态系统简介	77
2 极端微生物的群体感应	79
2.1 嗜盐菌	79
2.2 嗜热菌	81
2.3 嗜酸菌	83
2.4 古菌	83
2.5 其他极端细菌	85
3 极端微生物生物分子合成：QS 调控	86
3.1 极端酶	86
3.2 胞外多糖	86
4 极端生物分子：潜在的生物技术应用	87
5 结论	89
术语表	90
缩写词	91
致谢	91
参考文献	92

第2部分 跨界通信

第5章 群体感应在植物致病性上的作用

1 前言	100
2 群体感应现象	102
2.1 植物相关的微生物群体感应	102
3 细菌群体感应对植物免疫和生理的影响	105
3.1 植物先天免疫	105
3.2 群体感应分子与植物免疫	106
3.3 群体感应分子影响植物生理	107
4 群体淬灭在植物病理中的作用	108
4.1 植物群体感应抑制剂	108
4.2 抗植物致病细菌的细菌群体感应抑制剂	109
5 植物-微生物相互作用中糖和糖的信号转导	110
6 总结和展望	111
术语表	111
缩写词	111
参考文献	112
拓展阅读	118

第6章　群体感应和肠道微生物组 ··· 119
1　前言 ··· 119
2　自诱导物 ··· 119
3　细菌中的儿茶酚胺信号传导 ··· 122
4　肠道中的营养信号 ··· 123
5　结论 ··· 128
术语表 ··· 128
缩写词 ··· 128
参考文献 ··· 129

第3部分　群体感应抑制

第7章　生物膜中酶促群体淬灭 ··· 136
1　前言 ··· 136
2　表层上的生命：细菌生物膜 ··· 136
3　生物膜和抗生素耐药性 ··· 137
　3.1　多物种生物膜 ··· 138
　3.2　铜绿假单胞菌生物膜的交流 ··· 139
　3.3　群体感应信号 ··· 139
　3.4　群体淬灭酶 ··· 140
　3.5　淬灭喹诺酮依赖的QS信号 ··· 142
　3.6　群体淬灭酶的应用 ··· 143
　3.7　表面的功能化 ··· 143
　3.8　膜生物反应器 ··· 145
　3.9　吸入式乳酸酶类 ··· 145
4　讨论和展望 ··· 145
术语表 ··· 146
缩写词 ··· 146
致谢 ··· 147
参考文献 ··· 147

第8章　多酚对微生物细胞间通信的影响 ··· 152
1　什么是群体感应？ ··· 152
2　多酚 ··· 154
　2.1　结构与分布 ··· 154
　2.2　多酚的抗菌活性 ··· 157
　2.3　多酚作为群体淬灭剂 ··· 157
3　从海洋生物中提取的具有抗菌活性和QQ活性的多酚 ··· 160
　3.1　褐藻多酚的结构、表征和生物活性 ··· 161
　3.2　褐藻多酚具有抗真菌和QQ活性 ··· 163

4 地衣次生代谢物：一般性、抗菌和抗真菌活性 163
    4.1 地衣多酚的 QS 和抗生物膜活性 165
5 结论 166
术语表 166
缩写词 167
参考文献 168

第4部分 应用

第9章 铜绿假单胞菌群体感应和生物膜抑制 176
1 前言 176
2 群体感应和群体淬灭 177
    2.1 铜绿假单胞菌群体感应系统和生物膜 178
    2.2 群体感应抑制剂的筛选 179
    2.3 天然群体感应抑制剂 180
    2.4 合成群体感应调节器 189
3 结论与意见 191
术语表 191
缩写词 192
致谢 193
参考文献 193

第10章 神经性疾病中的环肽：cyclo(His-pro)的情况 201
1 前言 201
2 环肽 202
    2.1 环肽类 203
    2.2 QS 系统中的 CDPs 206
3 CHP 在神经系统疾病中的作用 208
    3.1 背景 208
    3.2 当前的理解 209
    3.3 展望 212
4 结论 213
术语表 213
缩写词 214
参考文献 215
推荐阅读 225

原著贡献者 226

第1部分
概况:化学和微生物学

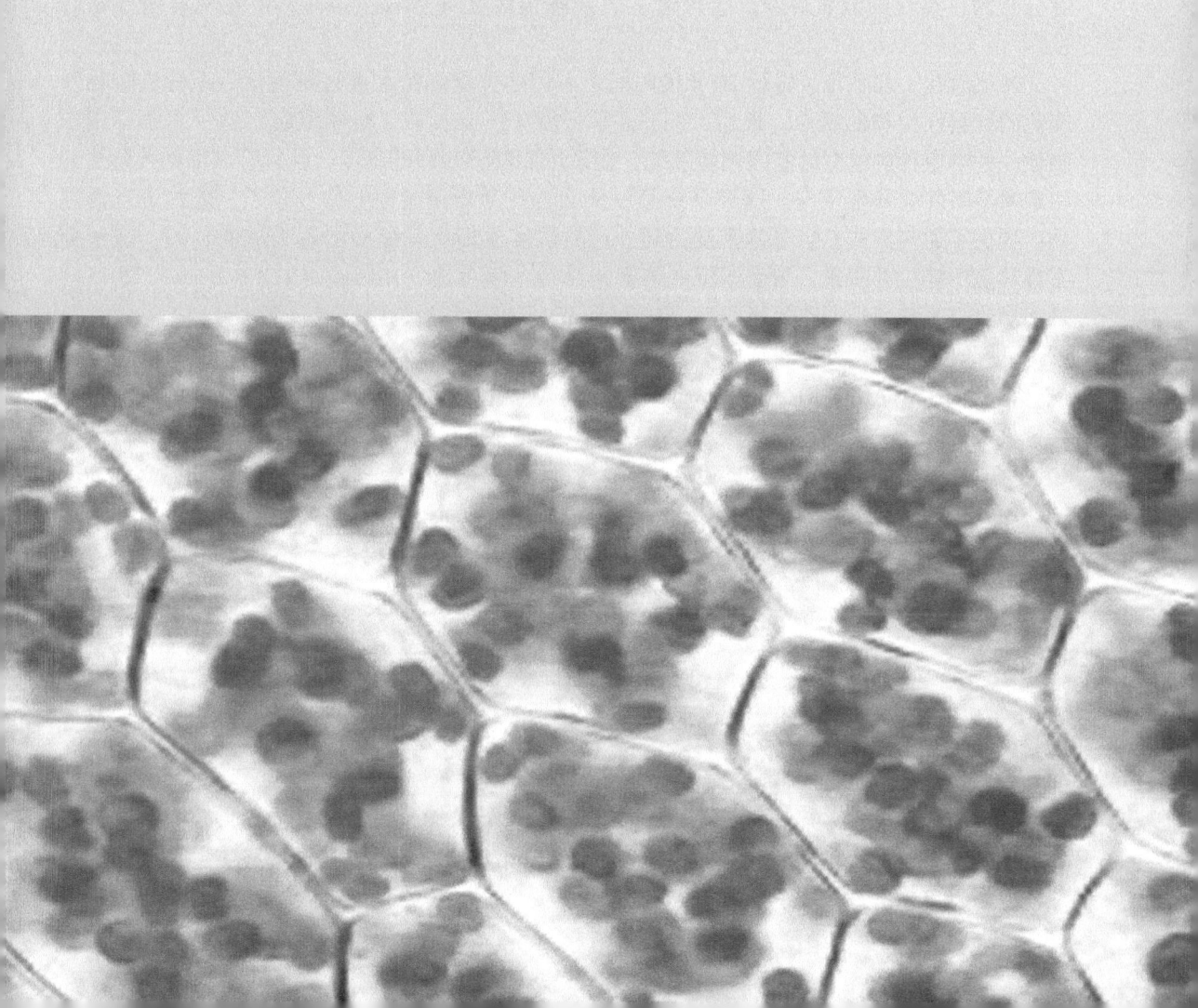

第1章 革兰氏阴性细菌的化学语言

Gerardo Della Sala[*], Roberta Teta[†],
Germana Esposito[†], Valeria Costantino[†]

[*]Laboratory of Pre-Clinical and Translational Research, IRCCS-CROB, Referral Cancer Center of Basilicata, Rionero in Vulture, Italy, [†]TheBlueChemistryLab, Department of Pharmacy, University of Naples Federico II, Napoli, Italy

1 前言

大约40年前，两篇论文报道了同一重大发现：细菌是社会生物，彼此通过化学语言交流来协调群体活动。1965年，Tomasz在《自然》杂志上发表了一篇文章，首次报道了肺炎球菌（*Pneumococcus*）通过化学因子（即感受态因子）进行调节的机制。几年后，Nealson等（1970）在《细菌学杂志》上报道了"在费氏弧菌（*Vibrio fisheri*）中发现自诱导物活性"。

从那时起，几乎花了将近30年的时间，人们才接受细菌是通过细胞信号系统进行相互交流的社会性生物的观点。最后，Fuqua等（1994）引入了"群体感应"这一术语，用于描述一种响应种群密度的基因调控系统，该系统能够感知种群密度，且只有在细胞数量达到一定水平时才会发生反应。这个系统是以什么方式起作用的呢？信号分子，被称为自诱导物，提供了通信方式。当群体密度达到一定的浓度（即化学信号的适当浓度）时，这些分子在环境中产生并积累，与受体蛋白结合并激活基因表达（Henseand and Schuster，2015），从而导致一系列信号通路的级联反应被激活，包括毒力和生物膜形成（图1-1）。

从那时起，许多文章报道了不同细菌种类中的这种信号系统；现在通常将其称为群体感应（quorum sensing, QS）（Fuqua and Greenberg，2002）。如今，QS被认为是细菌一个普遍特征，无论是革兰氏阴性细菌，还是革兰氏阳性细菌，其独特的化学信号结构的差异，使细菌能够以一种模仿多细胞生物的方式来协调它们的集体行为。这种细胞间通信的概念作为细菌间的社会活动，被称为社会微生物学。Parsek和Greenberg（2005）引入这一术语来描述包括生物膜形成和毒力因子控制在内的社会现象。

许多革兰氏阴性细菌合成 *N*-酰基高丝氨酸内酯（AHLs）作为化学信号，然后利用LuxR家族蛋白来感知环境中AHLs的浓度，并激活LuxI合酶，即其配对蛋白。第一个详细描述革兰氏阴性细菌的QS系统是费氏弧菌，其定殖在夏威夷短尾乌贼（*Euprymna scolopes*）的发光器官中（Ruby and Lee，1998）。最近的研究表明，一些细菌表现出没有同源自诱导合酶LuxI的LuxR同源物。它们被称为LuxR"单体"或"孤体"（Brameyer et al.，2014），其在细菌的种间和界间通信中发挥作用。

除AHLs外，也被称为自诱导物-1，过去15年的研究中揭示了其他类型信号分子的存在，因此，也存在着新的QS途径，这些途径可使几种细菌共存，从而构成了一个复杂的网

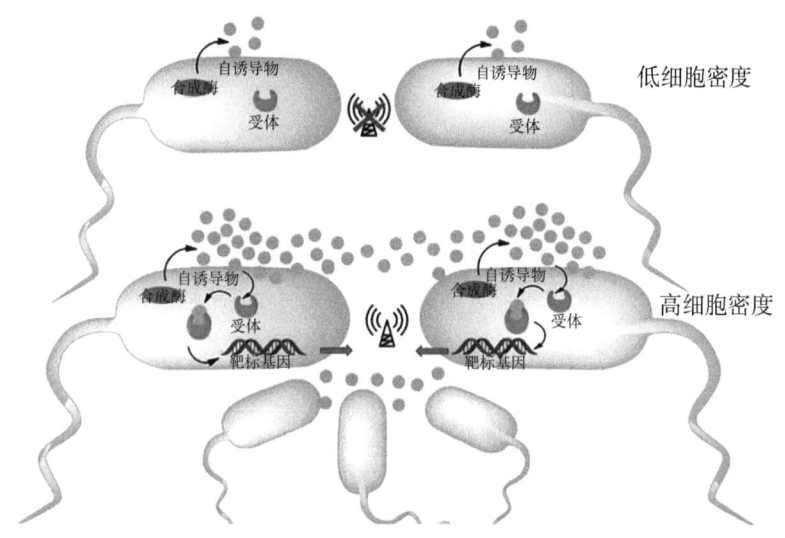

图 1-1 革兰氏阴性细菌中的群体感应系统

注：QS 依赖于自诱导信号的合成，这些信号以群体密度依赖的方式产生；当达到阈值浓度（群体）时，这些分子与转录调控因子相互作用，允许在高细胞密度下表达特定的基因。

络，控制着细菌的毒力产生和生理过程。

2 革兰氏阴性细菌中群体感应系统构建

革兰氏阴性细菌主要利用两种自诱导物：AHLs（自诱导物-1）和自诱导物 2（AI-2，也被革兰氏阳性细菌使用）。随着群体感应（QS）研究的不断深入，发现新的 QS 信号，如自诱导物-3（AI-3）、喹诺酮类信号（PQS）和可扩散信号因子（DSF）。

因此，基于不同自诱导物的 QS 系统进行简单描述，以便全面了解细菌病原体是如何通过 QS 依赖的机制调控毒力。

2.1 基于 AHLs 的群体感应

QS 系统最常见的信号分子是 AHLs，其仅介导革兰氏阴性细菌以及革兰氏阴性细菌与宿主之间的通信（Britstein et al., 2018；Parsek and Greenberg, 2000）。AHLs 是一种小分子，由脂肪酰基链与高丝氨酸内酯环相连，其长度、β 位氧化状态和饱和度都可发生变化。通常，AHLs（除少数例外）是由 LuxI 型 AHL 合酶通过催化 S-腺苷甲硫氨酸（SAM）和酰化酰基载体蛋白（acyl-ACP）之间的缩合反应合成（La Sarre and Federle, 2013）。

在高细胞密度下，当自诱导物 AHL 达到临界浓度（群体感应密度）时，AHL 会与细胞质中的 LuxR 型受体结合，形成一个转录调控复合物。之后，LuxR-AHL 复合物可以结合到 QS 调控基因上游特定启动子序列（lux-boxes）上，从而影响基因的表达。

紫色色素杆菌（*Chromobacterium violaceum*）携带一个典型的 LuxI/LuxR QS 系统，该系统包括 CviI（LuxI 型 AHL 合酶），CviR（LuxR 型 AHL 受体），以及 AHL（*N*-己酰-*L*-高丝氨酸内酯，C6-HSL），用于调控紫色杆菌素的产生。

膜集成的感应激酶可以替代细胞质中的 LuxR 受体，膜集成的感应激酶可与 AHL 配体相互作用：在哈维氏弧菌（V. harvey）中，LuxN 感应激酶可以在高细胞密度下识别 N-（3-羟基-丁酰）-L-高丝氨酸内酯（3-OH-C4-HSL），并催化 LuxU 去磷酸化，LuxU 是一个能调节 LuxO 转录响应的磷酸化蛋白。

值得注意的是，每个 AHL 受体在识别和招募 AHL 时都表现出一定程度的选择性，这与 AHL 脂肪酰基链的长度、氧化程度和饱和度相关。通常，一种细菌拥有一对同源合酶/受体对，能够产生并响应特定的 AHL 信号。然而，一些研究也发现某些物种超出了普遍接受的规则。例如，在铜绿假单胞菌（Pseudomonas aeruginosa）的 QS 谱分析中，可推断出某些物种使用多个合成酶/受体对，同时负责不同化学信号的生物合成和转导。此外，在细菌中也存在与其同源 LuxI 合成酶"单独"的孤立 LuxR-类型受体，如铜绿假单胞菌的 QscR 和大肠杆菌（Escherichia coli）的 SidA。

由于病原体铜绿假单胞菌对人类健康具有重大影响，经过多年来深入研究发现，其利用一个复杂的 QS 网络。在铜绿假单胞菌中，QS 显示出至少三种不同的信号/受体对（图 1-2），它们以等级的方式进行协调。该病原体表达两种 LuxI/LuxR 型的 QS 系统，即 LasI/LasR 和 RhlI/RhlR，分别产生和感知小分子：N-（3-氧-十二酰）-L-高丝氨酸内酯（3-O-C12-HSL）和 N-丁酰-L-高丝氨酸内酯（C4-HSL）。在铜绿假单胞菌中，Las 系统调节两个 QS 系统。首先，识别出正确的 AHL，LasR 受体驱动 LasI 合酶和 RhlR 受体同时表达。然后，RhlR/C4-HSL 复合物自诱导 rhlI 基因的转录。Las 和 Rhl 系统都参与了基于 PQS（假单胞菌喹诺酮信号，见本章第 2.3 节）的第三个 QS 系统的调控。尤其，LasR 和 RhlR 分别以正向和负向方式调节 PQS QS 系统中相关基因的表达。PQS 自诱导其自身的合成，并激活 RhlR 的表达，从而通过负反馈机制自我限制其产生。此外，每个 QS 系统都对各种毒力因子编码的基因具有直接调控作用。在这个 QS 网络中，"单体" LuxR 型蛋白 QscR 发挥了重要作用：QscR 识别 3-O-C12-HSL（LasR 配体），并负责调控 Las 和 Rhl QS 系统，从而关闭这个复杂的自调节回路。

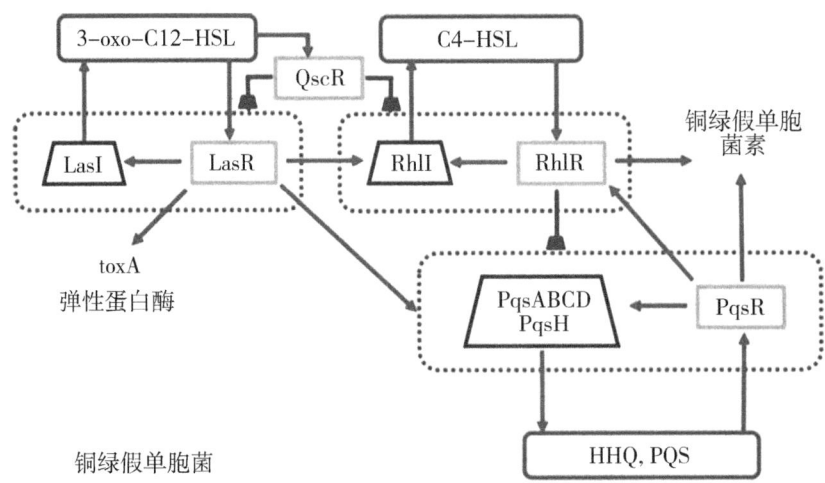

图 1-2　铜绿假单胞菌的群体感应网络

注：铜绿假单胞菌群体感应系统存在 3 个不同的信号/受体对：LasI/LasR、RhlI/RhlR 和 PQS/PqsR，以及"单体" LuxR 型蛋白 QscR。

2.2 基于 AI-2 的群体感应

自诱导物-2（AI-2）是由环状呋喃酮化合物为代表的一类 QS 信号家族（Guo et al., 2013）。哈维弧菌中 AI-2 的化学型结构已被确定为 (2R,4S)-2-甲基-2,3,3,4-四羟基四氢呋喃的硼酸酯，而在肠道沙门菌血清型鼠伤寒杆菌（*Salmonella enterica* serovar Typhimurium）中却发现 AI-2 缺乏硼酸酯。

一般而言，AI-2 型分子的生物合成主要通过两个步骤进行：首先，核苷酶 MTAN（也称为 Pfs）裂解为 S-腺苷-L-高半胱氨酸，通过去除腺嘌呤生成 S-核糖基-L-同型半胱氨酸（SRH）；随后，SRH 由金属酶 LuxS 转化为同型半胱氨酸和 AI-2 前体 4,5-二羟基-2,3-戊二酮（DPD）。DPD 很不稳定，经过自发环化和重排，从而生成不同的环形呋喃酮化合物，代表 AI-2 型 QS 信号家族。

在哈维弧菌和霍乱弧菌（*V. cholerae*）中，当 AI-2 达到高浓度时，可扩散到细胞膜中，并与 LuxP/LuxQ 受体/感受器激酶复合物发生相互作用，进而整合到细菌的细胞膜中（图 1-3）。AI-2/LuxPQ 复合体使 LuxU 去磷酸化，而 LuxU 反过来也使 LuxO 去磷酸并失活，从而抑制调节小 RNA 的表达，进而开启毒力因子的表达。

在大肠杆菌和鼠伤寒沙门菌（*S. typhimurium*）中，AI-2 介导的 QS 通路揭示了一种不同的途径。当达到群体感应阈值时，AI-2 通过一种转运蛋白 LsrB 内化进入细胞中，然后被激酶 LsrK 磷酸化；磷酸化的 AI-2 可以结合抑制因子 LsrR 来解除 lsr 操纵子的抑制并激活靶基因，而其中一些基因负责致病性（如涉及大肠杆菌生物膜形成的基因）。

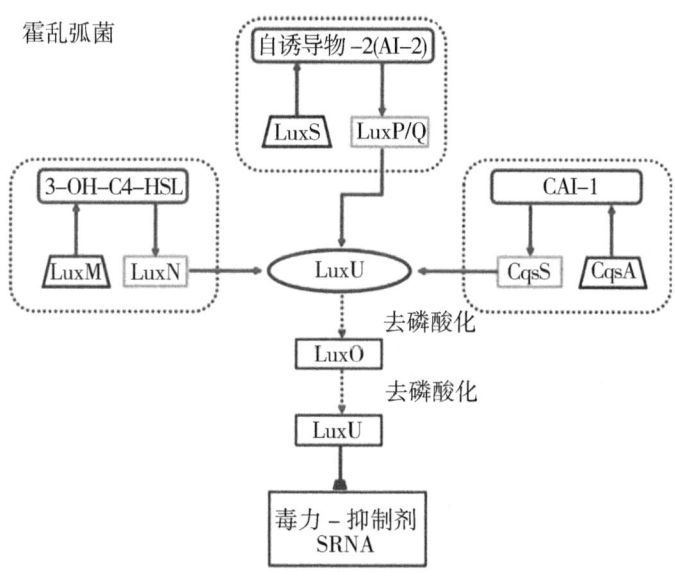

图 1-3 霍乱弧菌中的群体感应网络

注：霍乱弧菌存在 3 种不同的群体感应系统：AHL、AI-2 和 CAI-1 型群体感应系统。

2.3 使用其他自诱导物的 QS 系统

4-羟基-2-烷基喹啉（HAQs）是另一类 QS 信号，已在几种假单胞菌和伯克霍尔德菌（*Burkholderia*）中发现（Kim et al., 2010）。这类物质包括 4-羟基-2-庚基喹啉（HHQ）及其相应的二羟基化衍生物,如 2-庚基-3,4-二羟基喹啉(也称为 PQS,假单胞菌喹啉信号)。在铜绿假单胞菌中,HHQ 由 PqsABCD 前体邻氨基苯甲酸生物合成,然后,PQS 通过单氧酶 PqsH 对 HHQ 进行羟基化。HHQ 和 PQS 激活多毒力因子调节器 PqsR(也称为 MvfR),驱动 QS 分子产生以及毒素(铜绿假单胞菌素)和生物膜的形成(Allegretta et al., 2017)。

可扩散信号因子(DSF)家族包括不同链长和分支的顺式-2-不饱和脂肪酸。DSF 信号作为分布在革兰氏阴性细菌细胞间通信机制的一种广泛信使(Zhou et al., 2017)。这类分子的生物合成是由 DSF 合酶负责的,其具有双重酶活性,既作为 3-羟酰基-ACP 底物的脱硫酯酶也可作为脱水酶。DSF 信号家族的生物合成途径可能来自经典的脂肪酸合成途径。

在野油菜黄单胞菌(*Xanthomonas campestris*)中,当 DSF 在细胞环境中积累后,DSF 被识别并与传感器激酶 RpfC 结合,触发磷酸化级联机制来激活 RpfG。RpfG 作为反应调节因子,降解环二鸟苷酸(c-di-GMP),是整体转录因子 Clp 的抑制性配体。事实上,被抑制的 Clp 激活了几个基因的转录,包括编码毒力因子产生的基因。这种类型的 QS 系统已在黄单胞菌属(*Xanthomonas* sp.)、苛养木杆菌(*Xylella fastidiosa*)、溶酶菌(*Lysobacter enzymogenes*)和嗜麦芽寡养单胞菌(*Stenotrophomonas maltophilia*)中被鉴定(Zhou et al., 2017)。实际上,条件性致病菌洋葱伯克霍尔德菌(*B. cenocepacia*)和土耳其克罗诺杆菌(*Cronobacter turicensis*)具有 DSF 依赖的 QS 通路,二者的区别主要在于利用了新的感应器 RpfR,调节细胞内 c-di-GMP 的水平(Zhou et al., 2017)。

此外,最近成功报道了铜绿假单胞菌也使用 DSF 型的 QS 系统,并具有信号产生、感知和传导的特征基因簇(Zhou et al., 2017)。

自诱导物 3(AI-3)是由人类肠道菌群产生的(Parker et al., 2017)。AI-3 型信号通过肠出血性大肠杆菌组氨酸传感器激酶 QseC 感知。QseC 调节三种反应调节因子(RRs)的活性,即 QseB、QseF 和 KdpE,控制毒力决定因子的表达。实际上,QseC 对这些 RRs 的磷酸化触发了一个信号级联,导致靶基因的转录,如负责鞭毛运动、由肠细胞消失(LEE)位点编码的Ⅲ型分泌系统和志贺毒素的基因。有趣的是,通过细菌检测宿主释放到的肾上腺素和去甲肾上腺素,也会诱导同样的 QseC 依赖性机制。

一些弧菌种类,如霍乱弧菌和哈维弧菌,包含一个基于信号分子 CAI-1 的 QS 系统,也称为霍乱弧菌自诱导物-1(图 1-3)。CAI-1 的生物合成需要 CqsA 合成酶和底物(*S*)-2-氨基丁酸酯和癸酰辅酶 A。CqsA 产生氨基-CAI-1,从而在随后的 CqsA 独立步骤中将氨基-CAI-1 转化为 CAI-1。CAI-1 与传感器激酶 CqsS 相互作用,通过靶向对 LuxU 和 LuxO 的磷酸级联过程驱动毒力因子的表达。在弧菌种中,存在 AHL-、AI-2 和 CAI-1 依赖的 QS 通路共存整合,它们都共享一个共同的 LuxU 依赖的下游级联通路,用于毒力编码基因的转录调控。

3 自诱导物的化学性质

自诱导物是细菌响应种群密度变化而产生的可扩散化学信号,用于不同物种内部和之间的交流。

细菌在细胞外环境中释放各种小分子产物,被归类为 QS 信号分子,其必须满足 Winzer 等(2002)提出的四个标准。

(i) 自诱导物的产生是在特定的生长阶段，特定的生理条件下，或是因环境改变而产生的；

(ii) 自诱导物在细胞外环境中扩散和积累，并与特定的细菌受体结合；

(iii) 只有当自诱导物的浓度达到一个阈值（即群体感应）时，才会产生集体行为；

(iv) 细胞响应包括除了代谢或解毒分子之外的一系列事件。

在革兰氏阴性细菌中，QS 主要由 AHLs（也称为自诱导物-1）介导；到目前为止，不同类型的其他自诱导物也已经被发现。下面介绍主要类型的自诱导物。

3.1 N-酰基高丝氨酸内酯

AHL 信号分子（也称为自诱导物-1）是由一个高丝氨酸内酯环（HSL）与酰基侧链上的酰胺键连接组成。不同的 AHLs 主要是其长度（Britstein et al., 2016）、酰基链 C3 位上的取代以及不饱和度等方面存在差异（图 1-4A）（Saurav et al., 2016）。这些差异赋予了 LuxR 转录调控因子的信号特异性。酰基侧链可能是由脂肪酸生物合成产生的，由 4~18 个碳原子组成，通常每两个碳原子增加一次（C4、C6、C8 等）。大多数酰基侧链是直链、饱和或单不饱和的，并且是偶数，与微生物细胞中容易获取的脂肪酸相对应。一个特定长度的酰基侧链用 Cn 表示该链中碳的数量（如辛醇用 C8 表示）。取代类型和位置被指定为 3-O（3-氧）或 3-OH（3-羟基），HSL 指 D/L-高丝氨酸内酯（如 3-O-C6-HSL）（图 1-4B）。AHL 指的是 N-酰基-HSL，具有特定的链长或取代程度。

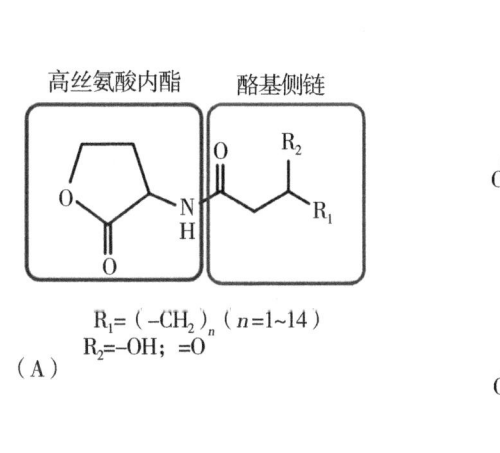

图 1-4 N-酰基高丝氨酸内酯类信号分子（AHL）

注：(A) 酰基高丝氨酸环内酯的普通结构；(B) 紫色色素杆菌、哈维弧菌和铜绿假单胞菌 AHLs 的化学结构。

在 α-变形杆菌纲如勒索甲基杆菌（*Methylobacterium extorquens*）、什叶派二硝酸杆菌（*Dinoroseobacter shibae*）中的 AHLs 含有双不饱和酰基链（Wagner-Döbler et al., 2005）。C7HSL 来自豆科根瘤菌（*Rhizobium leguminosarum*）（Schripsema et al., 1996），是首次报道的带有奇数

酰基的AHL；通常，与偶数相比，除在硫酸杆菌属（*Sulfitobacter* sp.）D13，奇数AHLs出现很少，其中，9-C17:1-HSL被大量合成（Ziesche et al.，2015）。来自微粒气单胞菌（*Aeromonas culicicola*）的 isoC9-HSL 和 3-OH-isoC9-HSL 中存在甲基分支（Thiel et al.，2009）。C3位的取代分为三种类型：（a）简单酰基，（b）3-羟基酰基，（c）3-氧酰基；这种对酰基链的取代除一个 *L*-型的高丝氨酸内酯外，还引入了一个新的立体中心。到目前为止，在这个中心的绝对构型已被确定为来自豆科根瘤菌的(3*R*,7*Z*)-*N*-(3-羟基-7-十四酰)-高丝氨酸内酯。短链 AHLs，如 C4-HSL，可通过细胞膜自由扩散，而 3-O-C12-HSL 通过载体和外排泵输出（Smith and Iglewski，2003）。

3.2 环呋喃酮化合物

与其他自诱导物不同，在许多革兰氏阳性和革兰氏阴性细菌（约70种）中发现了一些特殊细菌物种专一的自诱导物2（AI-2），其是一种跨种自诱导物，属于"通用自诱导物"（Guo et al.，2013）。AI-2是(2*R*,4*S*)-2-甲基-2,3,3,4-四羟基四氢呋喃的硼酯，由两个融合的五元环构成，通过多种极性相互作用稳定在 LuxP 结合位点内。AI-2由线性(*S*)-4,5-二羟基戊二酮（DPD）(*S*)自发环化成两种四氢氧化四氢呋喃的异构体（*R*-THMF 和 *S*-THMF），其存在于一个动态平衡中。在环境硼存在的情况下，形成了硼酸盐复合物等如 THMF-硼酸盐（Carrano et al.，2009）（图1-5）。AI-2信号分子的一个特点是：不同细菌有不同的AI-2受体，识别不同形式的AI-2。到目前为止，两个AI-2受体已经被鉴定出来：LuxP 结合 THMF-硼酸盐，而 LsrB 结合缺乏硼酸盐的 R-THMF（Rui et al.，2012）。

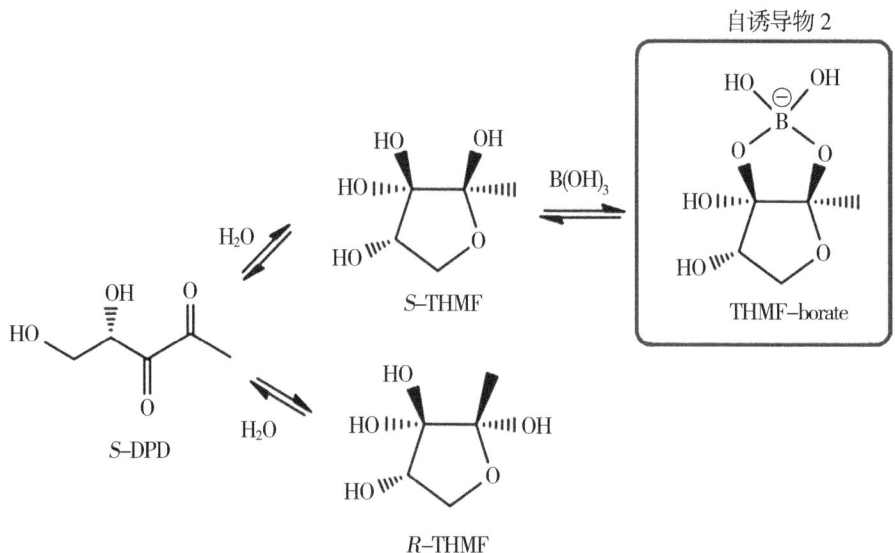

图1-5　AI-2的形成与结构

注：AI-2是由(*S*)-4,5-二羟基戊二酮（DPD）自发环化成2-甲基-2,3,3,4-四羟基四氢呋喃（*R*-THMF 和 *S*-THMF）的两种同分异构体而形成的。这些分子在动态平衡中共存，并且在环境中存在硼的情况下可以形成硼酸盐络合物。

3.3 其他

喹诺酮类是由杂二环芳香化合物喹啉衍生的分子结构。2-羟基喹啉和4-羟基喹啉分别以2(1H)-喹诺酮和4(1H)-喹诺酮的形式存在，并从植物中分离出许多生物碱的核心结构。几种不同的动物和细菌物种也产生喹诺酮类的化合物。这些差异不仅在碳环和杂芳香环的取代基上，而且还包括其他环融合到喹诺酮类核上（Heeb et al., 2011）。

2-庚基-3-羟基-4(1H)-喹诺酮（PQS）及其直接前体，2-庚基-4-羟基喹啉（HHQ），是主要参与QS的HAQs（4-羟基-2-烷基喹啉）。虽然有其他活性的AQ（烷基喹啉）类似物，如C9同类物，2-壬基-3-羟基-4(1H)-喹诺酮（C9-PQS）和2-壬基-4-羟基喹啉（NHQ），但在相似的浓度下均由铜绿假单胞菌产生（Ilangovan et al., 2013）。在C3和C6位置的取代会对其活性产生影响（图1-6）（Kamal et al., 2017）。

可扩散信号因子（DSF）家族信号包括不同链长、分支类型和双键构型的不饱和脂肪酸。脂肪酸的α,β-双键和顺式构型对QS信号活性至关重要，但其表现为一种物种特异性的方式。第一个家族成员是弯曲黄单胞菌（X. campestris）中顺式-11-甲基-十二烯酸，其他家族成员包括洋葱伯克霍尔德菌中顺式-2-十二烯酸、铜绿假单胞菌中顺式-2-十烯酸，以及苛养木杆菌中顺式-2-十六烯酸（图1-6）（Dow, 2017）。

AI-3是原核和真核细胞之间的跨界信号（Hughes and Sperandio, 2008）；其帮助细菌（如大肠杆菌和鼠伤寒沙门菌）在感染过程中与哺乳动物的肾上腺素/去甲肾上腺素"交叉通信（Sperandio et al., 2003）。AI-3是一种芳香胺化合物，但其完整结构尚未确定。霍乱弧菌的主要QS信号是CAI-1，其结构已被鉴定为(S)-3-羟基三聚糖-4-1，它代表了另一类QS信号分子，即AHK（α-羟基酮）（Kong et al., 2011）。CAI-1的生物活性对侧链长度敏感，13个碳分子的活性比12个碳分子高8倍，但羟基的构型对活性影响相对较小（差2倍）（图1-6）。

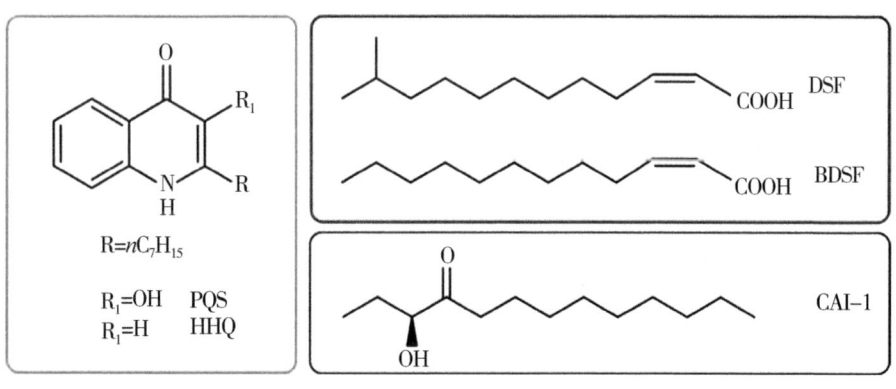

图1-6 其他QS信号分子

注：4-羟基-2-烷基喹啉类、扩散信号因子类和α-羟基酮类信号分子家族的代表。

3.4 海洋环境中的群体感应

在近40年前，QS首次在一种海洋细菌中被发现，但在过去的10年中，人们才开始对QS在海洋中所发挥的作用产生兴趣。QS参与海洋碳和有机磷循环，在珊瑚礁生态系统的健康中，多种真核生物及其细菌之间的营养相互作用，以及与藻华相一致的大规模生物发光事件中发挥作用（Hmelo，2017）。海洋中研究最充分的 QS 系统是由 AHLs 介导的表面附着的（生物膜）群落。

一般认为，AHL-QS 系统只用于生长在封闭的生态位（如海绵中）的细菌，其信号分子不会扩散出去，能达到阈值水平。此外，在海水高 pH 值下，独立生存的细菌产生 AHL 信号分子会发生内酯水解，即内酯环的打开，阻止 AHL 信号分子的有效相互作用。相反，宿主组织内的 pH 值允许 AHLs 作为信号分子发挥功能。因此，群体形成不仅受到 AHL 积累的影响，还受到 AHL 转化的影响，而 AHL 转化取决于环境条件（如 pH 值、温度）和 N-酰基链长度。此外，不同长度的 AHLs 对 pH 值具有不同的稳定性，较长的侧链 AHLs 和缺乏 3-氧取代基的 AHLs 在较高 pH 值下不易被水解失活。

几种海洋无脊椎动物和大型藻类样本的粗样本和提取物的生物分析测定表明，相关细菌在原位产生 AHLs（La Sarre and Federle，2013；Taylor et al.，2004），这些最初的观察结果随后被灵敏和先进的质谱技术所证实。

通过高效液相色谱-高分辨率串联质谱（LC-HRMS/MS），利用表面诱导解离（SID）和在 m/z 102.05 产生离子的特征性高丝氨酸内酯，很容易鉴定到与角质海绵（*Sarcotragus*）有关的副球菌属（*Paracoccus* sp.）Ss63 中的 AHL（Saurav et al.，2016）。

到目前为止，在海洋雪提取物（Guo et al.，2013）、蓝藻细菌（cyanobacterium）、束毛藻（*Trichodesmium*）菌落（Van Mooy et al.，2012）、叠层石（Decho et al.，2009）、海绵（Britstein et al.，2016）和海葵（Ransome et al.，2014）的提取物中鉴定出 AHLs 的结构。

除 AHLs 外，在海洋细菌培养物中观察到 aryl-HSLs（如 p-香豆酰-HSL），而 AI-2 仅在弧菌物种中作为信号发挥作用（Schaefer et al.，2008）。

4 群体淬灭分子

抑制 QS 系统可能成为设计新型抗毒性药物的一项具有挑战性的任务，其理念是通过设计能降低病原菌毒力的药物，而不是抑制其生长或杀死它们。有两种类型的分子可以抑制 QS 系统，根据它们的结构和作用机制可分为：

（A）在结构上模拟 AHLs 或 AIPs 的分子。这些分子由于结构相似性与受体结合，干扰 QS 通信，或加速 LuxR 周转，或共价修饰并使 LuxS 失活。

（B）酶抑制剂。能够抑制 AHL 生物合成中必需的酶小分子，如抑制烯酰-ACP 还原酶的三氯生。

此外，酶降解是抑制 QS 系统的另一种方法。从各种革兰氏阴性细菌菌株中分离出能催化 AHLs 水解的细菌 AHL 酶。

在此，报道了文献中最有趣的群体淬灭（QQ）分子的例子。

4.1 AHL 受体抑制剂的研究：LasR 拮抗剂

针对 LuxR 受体的 QS 抑制剂在设计关闭细菌毒力机制的药物引起了极大的关注。这种 QQ 方法的策略是开发类似于天然信号和已知 QQ 分子的类似物，维持信号-受体相互作用，但通过形成无效的配体-LuxR 复合物，破坏下游的生化事件。过去几年的研究表明，QS 抑制机制广泛分布于许多原核生物中，以便在竞争者中取得优势（Delago et al.，2016）。这些自然产生的 QQ 系统依赖于小分子，可以用作设计新一代抗毒性药物的先导化合物。在这种情况下，几项研究表明，将微生物发酵罐海洋海绵作为参与微生物-微生物相互作用和宿主-微生物相互作用的 QQ 过程的一个富有成效的环境。氯氟烷内酯（Costantino et al.，2017）是第一个从海绵全息体生物组中分离出来的第一个 QQ 分子；它在体外生物测定中作为 LuxR 拮抗剂起作用，因此，表明未来利用海绵作为 QS 抑制化合物的天然工厂是值得研究的。

尽管 QQ 是治疗抗病毒的潜在策略之一，但在治疗一些最重要的人类病原体（霍乱弧菌、铜绿假单胞菌）方面，通过使用体内和体外的动物模型已取得了积极的进展（Duan and March, 2010；Saeidi et al., 2011；Skindersoe et al., 2008）。

目前，迫切需要铜绿假单胞菌中抑制 AHL 介导的 QS，该菌目前通过医院获得、引起机会诱导性感染和囊性纤维化相关的慢性肺部感染且威胁着人类健康。许多研究都集中在以 LasR 受体为目标的 QQ 分子上，因其在铜绿假单胞菌中不同时发生的 QS 回路中起到中心调节因子作用（图 1-2）。

LasR 配体可结合结构多样性的分子，从而使预测激动剂和拮抗剂活性的化学需求变得困难。总的来说，SAR（结构-活性关系）和分子建模研究强调，激动剂和拮抗剂特性都需要充分的形成氢键和范德华相互作用能力，以便配体能够在 LasR 中适应，并平衡非天然配体取代基产生的任何体积效应。

总的来说，从过去的研究中获得了一些重要的见解，我们试图在本书中进行总结。

模拟 AHL 结构的分子需要适当的酰基链长度才能对 LasR 产生激动/拮抗生物活性（图 1-7A）（Passador et al., 1996）。AHL 模拟物表现出一个酰基链（超过 6 个碳），与内源性 LasR 配体 3-oxo-C12-HSL 相似，可以与 LasR 相互作用。例如，非 3-oxo-C10-HSL 是 AHLs 中最有效的 LasR 抑制剂，具有烷基链（Geske et al., 2008a）。天然 AHL 中酰基链第 3 位的碳基对活性很重要，但不是必需的，在大多数情况下，该位点的修饰会导致拮抗性的 AHL 激发配体（Passador et al., 1996）。

Kline 等（1999）表明，延长灵活链的几何形状是 LasR 激活非常必要的，因为受限的烯醇 AHL 类似物，被锁定成不可逆的共轭体，不能与结合袋相互作用。

Geske 等（2008a）引入芳香性功能团，设计非天然酰基团的 AHL 类似物发现了新的 LasR 抑制剂，如苯基乙酰高丝氨酸内酯（PHLs）和吲哚类 AHLs。有趣的是，PHLs 苯环的轻微修饰，以及一个取代基从对位转移到间位，都增加了对 LasR 的抑制活性（图 1-7A）（Geske et al., 2008a）。

一些研究阐明了对 AHL 类似物中内酯环修饰的影响（图 1-7B）。除少数情况外（例如，γ-内酰胺环产生任何一种活性的丧失），内酯环替换为生物等效体组基团通常都能被很好地接受（Passador et al., 1996）。据报道，引入 γ-硫内酯或环戊酮可以激活 LasR 受体，而环己酮和环戊烷，后者完全缺乏 H 键受体，是设计 LasR 拮抗剂的有用替代物，保持适当大小的酰基链（Ishida et al., 2007；Passador et al., 1996；Suga and Smith, 2003）。更有趣的是，内酯环被苯胺取代的合成类似物中（Smith et al., 2003），显示出对 LasR 受体的抑制效果。这些发现导致了一种假设，即苯环和一般的不饱和度，可以代替内酯部分产生 LasR 拮抗剂，如卤代呋喃酮。

此外，内酯环的立体化学值得研究。众所周知，AHLs 的 L-立体异构体是 LuxR 受体的天然配体。然而，并非所有的 D-AHLs 都是无活性的（Geske et al., 2008a），甚至已经发现

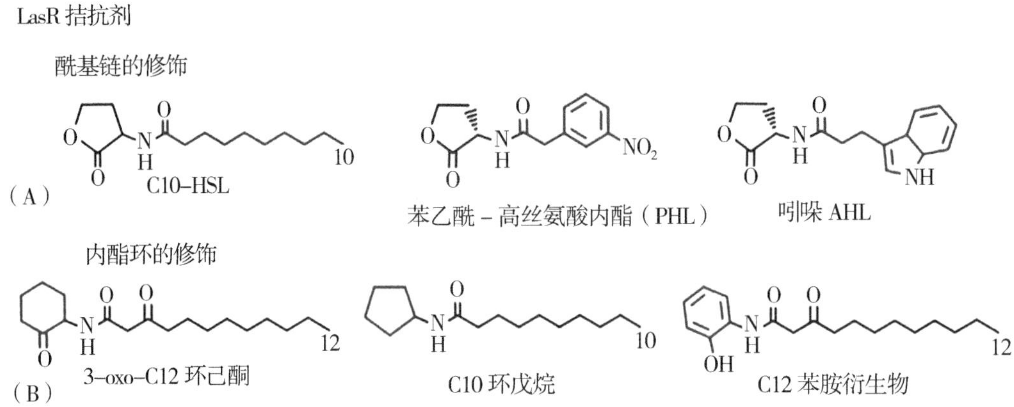

LasR 拮抗剂

酰基链的修饰

(A) C10-HSL　　苯乙酰-高丝氨酸内酯（PHL）　　吲哚 AHL

内酯环的修饰

(B) 3-oxo-C12 环己酮　　C10 环戊烷　　C12 苯胺衍生物

图 1-7　模拟内源性 AHLs 的 LasR 拮抗剂

注：LasR 拮抗剂的化学结构来源于对典型 AHL 型自诱导物
酰基链（A）和内酯环（B）的修饰。

具有 D 立体化学的非天然 AHL 衍生物可以作为 LasR 的激动剂（Geske et al., 2008b）。

多年来，经过努力旨在阐明配体与 LasR 结合域的关键相互作用，因此，突破了受体在配体识别后的构象变化。在这里，根据 Bottomley 等（2007）的模型，报道了原生配体和一些 QS 抑制剂与 LasR 相互作用的分子机制。该模型通过结合 X 射线数据和分子模型研究而提出。

LasR 配体结合域显示出对称的二聚体结构，每个单体都具有 α-β-α 折叠（6 个 α 螺旋和 5 个 β 折叠），大致类似于其他 LuxR 同源物，如 TraR［来自根瘤农杆菌（*Agrobacterium tumefaciens*）］和 SdiA（来自大肠杆菌）。LasR 的天然配体 3-oxo-C12-HSL 与 β 折叠平行放置，在 β 折叠和螺旋 α3、α4 和 α5 之间形成结合口袋。3-oxo-C12-HSL 可以在结合口袋中形成 6 个 H 键，包括 Tyr-56、Trp-60 和 Arg-61（水介导的 H 键）、Asp-73、Thr-75 和 Ser-129。值得注意的是，Tyr-56、Trp-60、Asp-73 和 Ser-129 在所有的 LuxR 蛋白中都高度保守，因为这些残基对于结合位点共享的高丝氨酸内酯核心与结合位点的相互作用至关重要。尽管有这些相似之处，LasR 结合域包括一些额外的残基（如 Leu-40、Tyr-47、Cys-79 和 Thr-80），这些残基在 TraR 和 SdiA 中缺失（对较短的 AHL 配体有选择性），并负责选择性地调节 3-oxo-C12-HSL 的长酰基链。

即使 LuxR 同源物在配体的结合袋中共享调节 HSL 的关键氨基酸残基，但这部分根据 LuxR 类型建立不同的 H 键相互作用。在 LasR 受体中，自诱导物通过其 1-oxo 基与 Tyr-56 和 Ser-129 同时形成 H 键，通过其 3-oxo 基与 Arg-61 侧链同时形成 H 键。另一方面，在 TraR 受体中，HSL 的 1-oxo 基只与 Tyr-53（对应 LasR 中的 Tyr-56）形成 H 键，而 3-oxo 基则与 Thr-129 侧链和 Ala-38 的主链羰基相互作用。由于这些不同的 H 键网络，相关自诱导物的酰基链从 AHL 的 3-oxo 基开始，在 LasR 和 TraR 的结合位点上向相反的方向延伸。虽然具有高度同源性，但 LasR 和 TraR 以不同的方式包含它们的配体。此外，LasR 和 TraR 的结合口袋完全根据其认知配体，即 3-oxo-C12-HSL 和 3-oxo-C8-HSL 的大小进行模型化；事实上，与 TraR（440Å^3）相比，LasR 具有更大的结合口袋（670Å^3）。

呋喃酮（Costantino et al., 2017；Wu et al., 2004）、青霉素和青霉酸（Rasmussen et al.,

2005)(图1-8)是众所周知的针对 LuxR 受体的 QS 抑制剂。当前的观点认为,这些抑制剂既不通过从结合位点取代天然的 AHL 配体,也不通过与 LuxR 同源物中假定的变构位点(实际上是不存在的)相互作用来发挥作用。AHL 被深深隐藏在结合袋中,因此,其不能被拮抗分子去除。有可能的是,QSIs 可成功地在与新生的 LuxR 蛋白结合时与天然配体竞争,干扰 AHL 介导的功能受体的折叠和包装。最近,Suneby 等(2017)通过体外试验证明,几种拮抗剂以一种不能结合 DNA 构象的方式,结合和限制 LasR 发挥作用。

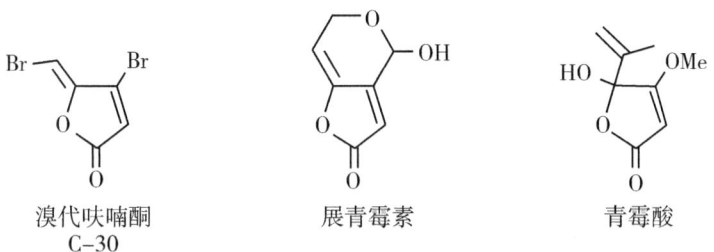

溴代呋喃酮 C-30　　展青霉素　　青霉酸

图 1-8　靶向 LuxR 受体的 QS 抑制剂的化学结构

据报道,溴化呋喃酮能抑制毒力因子产生(如铜绿假单胞菌素、蛋白酶)和生物膜形成,并增强铜绿假单胞菌对抗生素的敏感性。Bottomley 等(2007)的对接研究清楚地表明,卤化呋喃酮与 LasR 的相互作用类似于 3-oxo-C12-HSL,通过与 Trp-60 的 H 键结合以及与几个氨基酸残基的芳香侧链的疏水相互作用。然而,由于 AHLs 的化学特性不同,呋喃酮不能参与其他基本的氢键形成,并且缺乏一个长酰基链来形成正确的 LasR 疏水核心。因此,当呋喃酮被结合时,可能会形成一个异常的配体-受体复合物。尤其,如果溴原子具有亲脂性,允许呋喃酮在 LasR 结合域中调节,那么这种卤素的空间立体阻碍会破坏 LasR 蛋白中 Tyr-93 和 Leu-110 之间的水介导的内部氢键,极大地影响蛋白质的稳定性。因此,已经观察到,内酯环的 C4 位点的修饰导致 LasR 拮抗剂的产生,从而打破了适当折叠诱导的关键相互作用。

展青霉素是另一种模拟 AHL 结构的 QQ 分子。LasR-展青霉素复合物的硅模型(Bottomley et al.,2007)揭示了展青霉素与 Trp-60 建立一个规范的氢键,以及通过二氢吡喃环的氧和含有 Tyr-93 和 Asp-73(或 Thr-75)的羟基分别形成两个额外的 H 键。然而,展青霉素和呋喃酮共同的特点是二者没有与 3-oxo-C12-HSL 一样长的酰基链,因此,阻碍了亲脂袋的正确构象。

QSI 中发现的另一种新兴趋势是设计具有新型化学支架的 QQ 分子,而不是模仿 AHL 结构,如三苯基支架的化合物(O'Reilly and Blackwell,2016)。LasR 蛋白的突变研究表明,Trp60、Tyr56 和 Ser129 作为 H 键结合位点,决定受体的激活或抑制,在非内酯 LasR 调控剂的情况下是至关重要的(Gerdt et al.,2014)。

通常认为,基于三苯支架的化合物 TP-1 可作为 LasR 激动剂,以(a)类型与 Tyr-56、Trp-60 和 Ser-129 形成三个氢键,凭借 LasR 受体的三苯基部分,(b)占据整个结合袋(Bottomley et al.,2007)。虽然 TP-1 共享一个共同的核心,但 TP-5 作为 LasR 的拮抗剂发挥作用。事实上,与 TP-1 相比,TP-5 的三苯基系统中缺乏一个甲烷基会导致旋转自由度的丧失和空间干扰的增加(图1-9)。因此,TP-5 不能与结合域相互作用,建立所有的折叠促进。Zou 和 Nair(2009)推测 TP-5 作为抑制剂,其通过氯原子与 Leu-125 之间的空间碰撞破坏了 LasR 的天然构象,而(b)显示氯原子与 Trp60 侧链 NH 形成 H 键的排列不良。之后的研究证明了之前的假设,O'Reilly 和 Blackwell(2016)报道了 Trp60 突变到 Phe 并不影响 TP-5 对 LasR 的拮抗活性,因此,TP-5 不与 Trp60 形成氢键与结合位点相互作用。

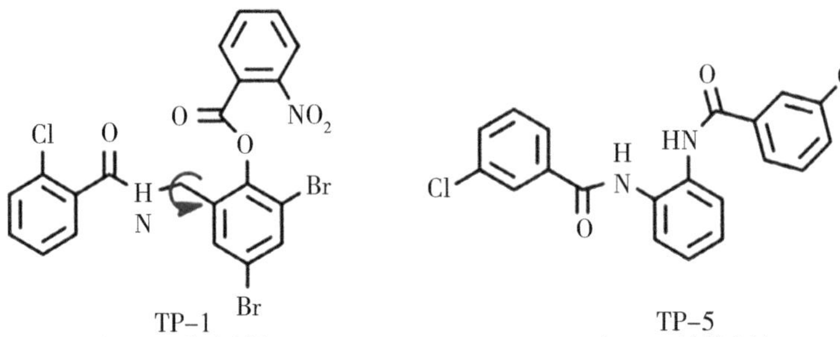

图 1-9　具有非内酯化学结构的 LasR 调节剂的化学结构

注:TP-1(LasR 激动剂)和 TP-5(LasR 拮抗剂)是基于三联苯骨架的化合物。

4.2　QS 酶的抑制作用

QS 酶抑制可以通过两种可能的策略实现:
- AHLs 或其他 QS 信号分子的酶降解和失活。
- 抑制 QS 信号分子的生物合成。

4.2.1　AHLs 的酶降解和失活

AHL-内酯酶、AHL-酰化酶和 AHL-氧化还原酶是发挥作用的关键酶(La Sarre and Federle,2013)。AHL-内酯酶基本上是负责水解高丝氨酸内酯环的酯键得到酰基-高丝氨酸化合物的金属蛋白。碱性 pH 值可以自然地促进这种水解,如果 pH 值变为酸性,内酯环会复原,这个反应是可逆的。由于所有的 AHL 分子中都存在高丝氨酸内酯,内酯酶表现出广泛的作用范围。第一个鉴定出的 AHL 内酯酶是来自芽孢杆菌属(*Bacillus sp.*)菌株 240B1 的 AiiA(自诱导失活)(Dong et al.,2000)。*aiiA* 同源基因广泛存在于芽孢杆菌物种中。在许多致病菌,包括铜绿假单胞菌、胡萝卜腐欧文菌(*Erwinia carotovora*)、泰国芽孢杆菌(*B. thailandensis*)中异源表达 AiiA,结果显示,AHL 信号的释放减少以及毒力表达降低。此外,表达 AiiA 的转基因植物对胡萝卜腐欧文菌的感染敏感性明显降低,表明使用 AiiA 作为一种抗病毒治疗策略具有广泛的应用潜力(Dong et al.,2000)。

AHL-酰化酶以不可逆的方式水解存在于酰基链和内酯环之间的酰胺键,产生一条脂肪酸链和相应的高丝氨酸内酯。与内酯酶不同的是,它们具有高度的底物特异性,这取决于酰基链的长度和最终出现在酰基链第三位置的取代程度。2023 年,从罗尔斯通菌属(*Ralstonia sp.*)XJI2B 植物中第一个鉴定出 AHL 酰化酶 AiiD(Lin et al.,2003)。AiiD 酰化酶与几种头孢菌素和青霉素酰化酶具有相似性。实际上,研究表明,AiiD 并不能降解青霉素 G 或氨苄西林,表明 AHLs 是独特的底物(Lin et al.,2003)。与 AHL 内酯酶一样,推测 AHL 酰化酶可以干扰细菌性病原体的 QS 系统。在铜绿假单胞菌 PAO1 中表达 AiiD 降低了 3OC12HSL 和 C4HSL 的积累,从而改变了毒力因子的产生和群体运动能力。此外,AiiD 在铜绿假单胞菌中表达降低了其产生弹性蛋白酶和藻蓝蛋白的能力,并使线虫瘫痪。

AHL-氧化还原酶通过氧化或还原修饰酰基链使 AHL 失活(图 1-10)。

这一类酶是针对 AHL 中数量最少，研究也较少的。AHL 失活最初是在红斑红球菌（*Rhodococcus erythropolis*）中被发现的，其中含有 3-oxo 取代基的 AHLs 通过 β 位的酮基还原而迅速降解，产生相应的 3-羟基衍生物 AHLs（Uroz et al.，2005）。

负责这种活性的基因尚未被鉴定。而且还发现 NADH 依赖的酶 BpiB09，通过宏基因组分析鉴定出其能够使 3OC12HSL 失活。在铜绿假单胞菌中表达 bpiB09 降低了其游动能力，藻蓝蛋白的产生和生物膜的形成，从而降低对秀丽隐杆线虫（*Caenorhabditis elegans*）的致病性（Bijtenhoorn et al.，2011），明确证实了氧化还原酶可以作为 QS 系统的抑制剂的证据。

图 1-10　酶法降解 AHLs 是革兰氏阴性细菌中采用的 QQ 策略
注：AHL 氧化还原酶通过氧化还原反应使修饰酰基链的 AHL 失活。

4.2.2　非 AHLs 群体感应信号的酶促降解和失活

到目前为止，许多研究集中在 AHLs 的失活，而只有几篇文章报道了其他 QS 自诱导物酶促失活的研究。最近的研究表明，干扰 AI-2 介导 QS 的可能性。特别是，研究了 LsrK 利用 AI-2 信号的能力。LsrK 是负责 AI-2 磷酸化的细胞质酶。

AI-2 被运输到细胞内，随后被 LsrK 磷酸化（Xavier et al.，2007）。Roy 等（2010）发现，通过向大肠杆菌或鼠伤寒杆菌培养物外源性提供 LsrK，其可通过阻止 AI-2 的输入来抑制 QS 的激活（磷酸化 AI-2 的负电荷阻止其通过 Lsr 转运体进行运输）。

在苛养木杆菌和几种黄单胞菌中，DSF 家族成员被用于 QS。需要强调的是，DSF 信号会影响这两种病原体的毒力（Deng et al.，2011）。Newman 等（2008）确定 *carA* 和 *carB* 是抑制 DSF 信号活性的基因。CarA 和 CarB 是负责合成氨基甲酰基磷酸盐异二聚体复合物的亚单位，其是精氨酸和嘧啶生物合成所必需的前体（Llamas et al.，2003）。CarA 和 CarB 在多种细菌物种中具有灭活 DSF 的作用（Newman et al.，2008）。

4.2.3　AHLs 分子生物合成抑制

AHLs 分子是由属于 LuxI 和 AinS 家族的 AHL 合成酶合成（Gilson et al.，1995；Parsek et al.，1999）。

S-腺苷甲硫氨酸（SAM）和酰基载体蛋白（ACP）是负责合成 AHLs 的底物。SAM 是生成高丝氨酸内酯环部分的氨基供体，而酰基-ACP 是 AHL 信号的酰基侧链前体。因此，AHLs 生物合成抑制取决于：

- SAM 生物合成的抑制
- 对酰基-ACP 产生的干扰
- SAM 的各种类似物，如 S-腺苷高半胱氨酸，S-腺苷半胱氨酸和西尼霉素，已被证明是铜绿假单胞菌 RhlI 蛋白催化 AHL 合成的有效抑制剂。
- FabI（NADH 依赖的烯酰-ACP 还原酶）是一种负责催化酰基-ACP 生物合成最后步骤的细菌酶。在铜绿假单胞菌中，FabI 参与了丁酰基-ACP 的生物合成，用于产生 C4-HSL。三氯森是一种能够抑制 C4HSL 产生的 FabI 抑制剂（Hoang and Schweizerm，1999）。

- 铜绿假单胞菌对三氯森具有抗性（Schweizer，2003），其可被认为是寻找其他衍生物的先导化合物。

4.2.4 对 AI-2 的生物合成有抑制作用

S-腺苷同型半胱氨酸（SAH）是负责 AI-2 生物合成的底物，该过程由两种酶催化：

- 5ʹ-甲基硫代腺苷/S-腺苷同型半胱氨酸核苷酶，MTAN（也称为 Pfs），该酶催化从 SAH 中去除腺嘌呤，生成 S-核糖基-L-同型半胱氨酸（SRH）。
- 金属酶 LuxS 将 SRH 转化为同半胱氨酸和 DPD（Schauder et al.，2001），其在水溶液中是不稳定的，并自发地重排成多个不同的环呋喃酮化合物异构体。这些环状呋喃酮化合物的异构体被称为 AI-2（Guo et al.，2013）。

LuxS 抑制作用发生在两个 SRH 类似物上：S-脱水核糖基-L-同型半胱氨酸和 S-同核糖基-L-半胱氨酸，二者都是竞争性抑制剂，阻止催化机制的第一步和最后一步（Alfaro et al.，2004）。

对新抑制剂的研究表明，氨基酸部分和连接到金属中心的部分在 LuxS 抑制中起至关重要的作用（Shen et al.，2006）。

MTAN 酶参与 AHL 和 AI-2 自诱导物的生物合成。在 AHL 生物合成中，MTAN 负责去除 AHL 合成的副产物甲基硫腺苷（MTA），而在 AI-2 生物合成中，MTAN 催化去除 SAH 中的腺嘌呤生成 SRH。MTA 类似物可以抑制 MTAN（Ferro et al.，1976）。特别地，一些研究证明，MTA 水解的过渡态类似物强烈抑制了包括肺炎链球菌（*Streptococcus pneumoniae*）、大肠杆菌和霍乱弧菌在内的多种细菌的 MTAN 活性（Gutierrez et al.，2009；Singh et al.，2006）。黏菌素-A（ImmA）类似物旨在模拟一种早期的解离过渡状态，在该状态中，核糖基和腺嘌呤键部分断裂，而 DADMe-ImmA 类似物模拟一种晚期过渡状态，其中腺嘌呤完全解离。MTA 也是人类 MTA 磷酸酶的底物，因此，一些 MTA 类似物可能会抑制人类酶引起的毒性。然而，细菌 MTA 核苷酶和人类 MTA 磷酸酶之间存在结构差异，这使得选择性的靶向细菌酶成为可能（Lee et al.，2004；Longshaw et al.，2010）。

5 结论

细菌 QS 是一种细胞间的通信过程，调控着包括毒素、蛋白酶、生物膜形成等多种毒力因素以及抗生素的耐受性。在 QS 系统中，分泌的信号分子（如 AHLs）协调整个细菌群体的行为，以同步方式表达毒力决定因素。抑制 QS 信号可能使病原体变得无害，因为它们失去了在宿主体内产生协调的毒力反应能力，这被视为一种有前景的疾病控制策略。过去几年的研究表明，QS 抑制机制，也称为群体淬灭（QQ）机制，在许多原核生物中广泛分布。这些自然发生的 QQ 机制通过在竞争中获得优势至关重要，并依赖于小分子，这些小分子可以作为设计新一代抗菌药物的先导化合物。QQ 化合物能够干扰病原体的毒性调控机制，而不是通过杀死它们来实现。因此，在不断暴发的抗生素耐药性感染的背景下，这种新兴的治疗方法显著降低了病原体中产生抗生素耐药性的选择压力。

术语表

芳香胺化合物　具有芳香环和胺官能团的化合物。

二氢吡喃环 含有五个碳原子和一个氧原子的单不饱和六元环。
烯醇化合物 双键配对中的一个碳原子与羟基连接的化合物。
呋喃酮 五元α,β-不饱和内酯。
杂芳环 含有至少一个杂原子（非碳原子）的芳环。
异构形式 共享相同分子式但原子在空间中的结构排列不同，因此性质不同的化学化合物。
内酯 羟基羧酸的环酯。
宏基因组分析 对一组微生物的DNA进行分析。
分子模型研究 通过计算化学和图形化可视化技术，对分子结构和性质进行在线模拟探索，目的是在给定的设置下理解出可能的3D表示。
手性中心 具有三个或更多不同配体的原子；交换这两个配体会导致另一个立体异构体。

缩写词

ACP	载酰蛋白
AHK	α-羟基酮
AHL	N-酰基高丝氨酸内酯
AI-2	自诱导物-2
AI-3	自诱导物-3
AiiA	自诱导器失活
CAI-1	霍乱自诱导物-1
DPD	4,5-二羟基-2,3-戊二酮
DSF	可扩散信号因子
HAQ	4-羟基-2-烷基喹啉
HHQ	4-羟基-2-庚基喹啉
HSL	高丝氨酸内酯
LC-HRMS/MS	液相色谱法-高分辨率串联质谱法
NIIQ	2-工基-4-羟基喹啉
PHL	苯乙酰-高丝氨酸内酯
PQS	喹诺酮假单胞菌信号
QQ	群体淬灭
QS	群体感应
SAM	S-腺苷甲硫氨酸
SID	表面诱导解离
SRH	S-核糖基同型半胱氨酸
THMF	四羟基甲基四氢呋喃

参考文献

ALFARO J F, ZHANG T, WYNN D P, et al., 2004. Synthesis of LuxS inhibitors targeting bacterial cell-cell communication. Org. Lett., 6 (18): 3043-3046.

ALLEGRETTA G, MAURER C K, EBERHARD J, et al., 2017. In-depth profiling of MvfR-regulated small molecules in *Pseudomonas aeruginosa* after quorum sensing inhibitor treatment. Front. Microbiol., 8: 1-12.

BIJTENHOORN P, MAYERHOFER H, MULLER-DIECKMANN J, et al., 2011. A novel metagenomic short-chain dehydrogenase/reductase attenuates *Pseudomonas Aeruginosa* biofilm formation and virulence on *Caenorhabditis elegans*. PLoS One, 6 (10): e26278.

BOTTOMLEY M J, MURAGLIA E, BAZZO R, et al., 2007. Molecular insights into quorum sensing in the human pathogen *Pseudomonas aeruginosa* from the structure of the virulence regulator LasR bound to its autoinducer. J. Biol. Chem., 282 (18): 13592-13600.

BRAMEYER S, KRESOVIC D, BODE H B, et al., 2014. LuxR solos in *Photorhabdus* species. Front. Cell. Infect. Microbiol., 4: 1-11.

BRITSTEIN M, DEVESCOVI G, HANDLEY K M, et al., 2016. A new N-acyl homoserine lactone synthase in an uncultured symbiont of the red sea sponge *Theonella swinhoei*. Appl. Environ. Microbiol., 82: 1274-1285.

BRITSTEIN M, SAURAV K, TETA R, et al., 2018. Identification and chemical characterization of N-acyl-homoserine lactone quorum sensing signals across sponge species and time. FEMS Microbiol. Ecol., 94 (2): 1-7.

CARRANO C J, SCHELLENBERG S, AMIN S A, et al., 2009. Boron and marine life: a new look at an enigmatic bioelement. Mar. Biotechnol., 11: 431-440.

COSTANTINO V, SALA G D, SAURAV K, et al., 2017. Plakofuranolactone as a quorum quenching agent from the Indonesian sponge *Plakortis* cf. lita. Mar. Drugs., 15: 1-12.

DECHO A W, VISSCHER P T, FERRY J, et al., 2009. Autoinducers extracted from microbial mats reveal a surprising diversity of N-acylhomoserine lactones (AHLs) and abundance changes that may relate to diel pH. Environ. Microbiol., 11: 409-420.

DELAGO A, MANDABI A, MEIJLER M M, 2016. Natural quorum sensing inhibitors-small molecules, big messages. Israel J. Chem., 56: 310-320.

DENG Y, WU J, TAO F, et al., 2011. Listening to a new language: DSF-based quorum sensing in gram-negative bacteria. Chem. Rev., 111: 160-179.

DONG Y H, XU J L, LI X Z, et al., 2000. AiiA, an enzyme that inactivates the acylhomoserine lactone quorum-sensing signal and attenuates the virulence of *Erwinia carotovora*. Proc. Natl. Acad. Sci. U. S. A., 97: 3526-3531.

DOW J M, 2017. Diffusible signal factor-dependent quorum sensing in pathogenic bacteria and its exploitation for disease control. J. Appl. Microbiol., 122: 2-11.

DUAN F, MARCH J C, 2010. Engineered bacterial communication prevents *Vibrio cholerae* virulence in an infant mouse model. Proc. Natl. Acad. Sci. U. S. A., 107 (25): 11260-11264.

FERRO A J, BARRETT A, SHAPIRO S K, 1976. Kinetic properties and the effect of substrate analogues on 50-methylthioadenosine nucleosidase from *Escherichia coli*. Biochim. Biophys., Acta 438: 487-494.

FUQUA C, GREENBERG E P, 2002. Listening in on bacteria: Acyl-homoserine lactone signaling. Nat. Rev. Mol. CellBiol., 3: 685-695.

FUQUA W C, WINANS S C, GREENBERG E P, 1994. Quorum sensing in bacteria: the LuxR-LuxI family of cell density-responsive transcriptional regulators. J. Bacteriol., 176: 269-275.

GERDT J P, MCINNIS C E, SCHELL T L, et al., 2014. Mutational analysis of the quorumsensing receptor LasR reveals interactions that govern activation and inhibition by non-lactone ligands. Chem. Biol., 21

(10): 1361-1369.

GESKE G D, O'NEILL J C, BLACKWELL H E, 2008a. Expanding dialogues: from natural autoinducers to nonnatural analogues that modulate quorum sensing in gram-negative bacteria. Chem. Soc. Rev., 37 (7): 1432-1447.

GESKE G D, O'NEILL J C, MILLER D M, et al., 2008b. Comparative analyses of N-Acylated Homoserine lactones reveal unique structural features that dictate their ability to activate or inhibit quorum sensing. Chembiochem, 9 (3): 389-400.

GILSON L, KUO A, DUNLAP P V, 1995. AinS and a new family of autoinducer synthesis proteins. J. Bacteriol., 177 (23): 6946-6951.

GUO M, GAMBY S, ZHENG Y, et al., 2013. Small molecule inhibitors of AI-2 signaling in bacteria: Stateof-the-art and future perspectives for anti-quorum sensing agents. Int. J. Mol. Sci., 14: 17694-17728.

GUTIERREZ J A, CROWDER T, RINALDO-MATTHIS A, et al., 2009. Transition state analogs of 50-methylthioadenosine nucleosidase disrupt quorum sensing. Nat. Chem. Biol., 5: 251-257.

HEEB S, FLETCHER M P, CHHABRA S R, et al., 2011. Quinolones: from antibiotics to autoinducers. FEMS Microbiol. Rev., 35: 247-274.

HENSE B A, SCHUSTER M, 2015. Core principles of bacterial autoinducer systems. Microbiol. Mol. Biol. Rev., 79: 153-169.

HMELO LR, 2017. Quorum sensing in marine microbial environments. Annu. Rev. Mar. Sci., 9: 257-281.

HOANG T T, SCHWEIZER H P, 1999. Characterization of *Pseudomonas aeruginosa* enoyl-acyl carrier protein reductase (FabI): A target for the antimicrobial triclosan and its role in acylated homoserine lactone synthesis. J. Bacteriol., 181: 5489-5497.

HUGHES D T, SPERANDIO V, 2008. Inter-kingdom signaling: communication between bacteria and their hosts. Nat. Rev. Microbiol., 6: 111-120.

ILANGOVAN A, FLETCHER M, RAMPIONI G, et al., 2013. Structural basis for native agonist and synthetic inhibitor recognition by the *Pseudomonas aeruginosa* quorum sensing regulator PqsR (MvfR). PLoS Pathog., 9 (7): e1003508.

ISHIDA T, IKEDA T, TAKIGUCHI N, et al., 2007. Inhibition of quorum sensing in *Pseudomonas aeruginosa* by N-acyl cyclopentylamides. Appl. Environ. Microbiol., 73 (10): 3183-3188.

KAMAL A A M, PETRERA L, EBERHARD J, et al., 2017. Structure-functionality relationship and pharmacological profiles of *Pseudomonas aeruginosa* alkylquinolone quorum sensing modulators. Org. Biomol. Chem., 15, 4620-4630.

KIM K, KIM Y U, KOH B H, et al., 2010. HHQ and PQS, two *Pseudomonas aeruginosa* quorum-sensing molecules, down-regulate the innate immune responses through the nuclear factor-κB pathway. Immunology, 129: 578-588.

KLINE T, BOWMAN J, IGLEWSKI B H, et al., 1999. Novel synthetic analogs of the *Pseudomonas* autoinducer. Bioorg. Med. Chem. Lett., 9 (24): 3447-3452.

KONG S, APPLICATION F, DATA P, 2011. (12) United States Patent, 2: 12-15.

LA SARRE B, FEDERLE M J, 2013. Exploiting quorum sensing to confuse bacterial pathogens. Microbiol. Mol. Biol. Rev., 77: 73-111.

LEE J E, SETTEMBRE E C, CORNELL K A, et al., 2004. Structural comparison of MTA Phosphorylase and MTA/AdoHcy Nucleosidase explains substrate preferences and identifies regions exploitable for inhibitor design. Biochemistry, 43 (18): 5159-5169.

LIN Y H, XU J L, HU J, et al., 2003. Acyl-homoserine lactone acylase from *Ralstonia* strain XJ12B represents a novel and potent class of quorum-quenching enzymes. Mol. Microbiol., 47: 849-860.

LLAMAS I, SUAREZ A, QUESADA E, et al., 2003. Identification and characterization of the carAB genes responsible for encoding carbamoylphosphate synthetase in *Halomonas eurihalina*. Extremophiles, 7: 205-211.

LONGSHAW A I, ADANITSCH F, GUTIERREZ J A, et al., 2010. Design and synthesis of potent "sulfur-free" transition state analogue inhibitors of 50-methylthioadenosine nucleosidase and 5'-methylthioadenosine phosphorylase. J. Med. Chem., 53: 6730-6746.

NEALSON K H, PLATT T, HASTINGS J W, 1970. Cellular control of the synthesis and activity of the bacterial luminescent system. J. Bacteriol., 104: 313-322.

NEWMAN K L, CHATTERJEE S, HO K A, et al., 2008. Virulence of plant pathogenic bacteria attenuated by degradation of fatty acid cell-to-cell signaling factors. Mol. Plant-Microbe Interact., 21: 326-334.

O'REILLY M C, BLACKWELL H E, 2016. Structure-based design and biological evaluation of Triphenyl scaffoldbased hybrid compounds as hydrolytically stable modulators of a LuxR-type quorum sensing receptor. ACS Infect. Dis., 2 (1): 32-38.

PARKER C T, RUSSELL R, NJOROGE J W, et al., 2017. Genetic and mechanistic analyses of the Periplasmic domain of the Enterohemorrhagic *Escherichia coli*. J. Bacteriol., 199: 1-15.

PARSEK M R, GREENBERG E P, 2000. Acyl-homoserine lactone quorum sensing in gram-negative bacteria: a signaling mechanism involved in associations with higher organisms. Proc. Natl. Acad. Sci. U. S. A., 97 (16): 8789-8793.

PARSEK M R, GREENBERG E P, 2005. Sociomicrobiology: the connections between quorum sensing and biofilms. Trends Microbiol., 13 (1): 27-33.

PARSEK M R, VAL D L, HANZELKA B L, et al., 1999. Acyl homoserine-lactone quorum sensing signal generation. Proc. Natl. Acad. Sci., 96: 4360-4365.

PASSADOR L, TUCKER K D, GUERTIN K R, et al., 1996. Functional analysis of the *Pseudomonas aeruginosa* autoinducer PAI. J. Bacteriol., 178 (20): 5995-6000.

RANSOME E, MUNN C B, HALLIDAY N, et al., 2014. Diverse profiles of N-acyl-homoserine lactone molecules found in cnidarians. FEMS Microbiol. Ecol., 87: 315-329.

RASMUSSEN T B, SKINDERSOE M E, BJARNSHOLT T, et al., 2005. Identity and effects of quorum-sensing inhibitors produced by *Penicillium* species. Microbiology, 151 (5): 1325-1340.

ROY V, FERNANDES R, TSAO C Y, et al., 2010. Cross species quorum quenching using a native AI-2 processing enzyme. ACS Chem. Biol., 5: 223-232.

RUBY E G, LEE K H, 1998. The *Vibrio fischeri-Euprymna scolopes* light organ association: current ecological paradigms. Appl. Environ. Microbiol., 64: 805-812.

RUI F, MARQUES J C, MILLER S T, et al., 2012. Stereochemical diversity of AI-2 analogs modulates quorum sensing in *Vibrio harveyi* and *Escherichia coli*. Bioorg. Med. Chem., 20: 249-256.

SAEIDI N, WONG C K, LO T-M, et al., 2011. Engineering microbes to sense and eradicate *Pseudomonas aeruginosa*, a human pathogen. Mol. Syst. Biol., 7: 521.

SAURAV K, BURGSDORF I, TETA R, et al., 2016. Isolation of marine *Paracoccus* sp. Ss63 from the sponge *Sarcotragus* sp. and characterization of its quorum-sensing chemical-signaling molecules by LC-MS/MS analysis. Israel J. Chem., 56: 330-340.

SCHAEFER A L, GREENBERG E P, OLIVER C M, et al., 2008. A new class of homoserine lactone quorum-sensing signals. Nature, 454: 595-599.

SCHAUDER S, SHOKAT K, SURETTE M G, et al., 2001. The LuxS family of bacterial autoinducers: Biosynthesis of a novel quorum-sensing signal molecule. Mol. Microbiol., 41 (2): 463-476.

SCHRIPSEMA J, DE RUDDER, K E E, et al., 1996. Bacteriocin small of *Rhizobium leguminosarum* belongs to the class of *N*-acyl-*L*-homoserine lactone molecules, known as autoinducers and as quorum sensing co-transcription factors. J. Bacteriol., 178: 366-371.

SCHWEIZER H P, 2003. Efflux as a mechanism of resistance to antimicrobials in *Pseudomonas aeruginosa* and bacteria: unanswered questions. Genet. Mol. Res., 2 (1): 48-62.

SHEN G, RAJAN R, ZHU J, et al., 2006. Design and synthesis of substrate and intermediate analogue inhibitors of *S*-ribosylhomocysteinase. J. Med. Chem., 49 (10): 3003-3011.

SINGH V, SHI W, ALMO S C, et al., 2006. Structure and inhibition of a quorum sensing target from *Streptococcus pneumoniae*. Biochemistry, 45 (43): 12929-12941.

SKINDERSOE M E, ALHEDE M, PHIPPS R, et al., 2008. Effects of antibiotics on quorum sensing in *Pseudomonas aeruginosa*. Antimicrob. Agents Chemother., 52 (10): 3648-3663.

SMITH R S, IGLEWSKI B H, 2003. *Pseudomonas aeruginosa* quorum sensing as a potential antimicrobial target. J. Clin. Investig., 112: 1460-1465.

SMITH K M, BU Y, SUGA H, 2003. Induction and inhibition of *Pseudomonas aeruginosa* quorum sensing by synthetic autoinducer analogs. Chem. Biol., 10 (1): 81-89.

SPERANDIO V, TORRES A G, JARVIS B, et al., 2003. Bacteria-host communication: the language of hormones. Proc. Natl. Acad. Sci., 100: 8951-8956.

SUGA H, SMITH K M, 2003. Molecular mechanisms of bacterial quorum sensing as a new drug target. Curr. Opin. Chem. Biol., 7 (5): 586-591.

SUNEBY E G, HERNDON L R, SCHNEIDER T L, 2017. *Pseudomonas aeruginosa* LasR DNA binding is directly inhibited by quorum sensing antagonists. ACS Infect. Dis., 3 (3): 183-189.

TAYLOR M W, SCHUPP P J, BAILLIE H J, et al., 2004. Evidence for acyl Homoserine lactone signal production in Bacteria associated with marine sponges. Appl. Environ. Microbiol., 70: 438-439.

THIEL V, KUNZE B, VERMA P, et al., 2009. New structural variants of homoserine lactones in bacteria. Chembiochem, 10: 1861-1868.

TOMASZ A, 1965. Control of the competent state in Pneumococcus by a hormone-like cell product: an example for a new type of regulatory mechanism in bacteria. Nature, 208: 155-159.

UROZ S, CHHABRA S R, CÁMARA M, et al., 2005. *N*-acylhomoserine lactone quorum-sensing molecules are modified and degraded by *Rhodococcus erythropolis* W2 by both amidolytic and novel oxidoreductase activities. Microbiology, 151: 3313-3322.

VAN MOOY B A S, HMELO L R, SOFEN L E, et al., 2012. Quorum sensing control of phosphorus acquisition in *Trichodesmium* consortia. ISME J., 6: 422-429.

WAGNER-DOBLER I, THIEL V, EBERL L, et al, 2005. Discovery of complex mixtures of novel long-chain quorum sensing signals in free-living and host-associated marine alphaproteobacteria. Chembiochem, 6: 2195-2206.

WINZER K, HARDIE K R, WILLIAMS P, 2002. Bacterial cell-to-cell communication: Sorry, can't talk now-gone to lunch! Curr. Opin. Microbiol., 5: 216-222.

WU H, SONG Z, HENTZER M, et al., 2004. Synthetic furanones inhibit quorum-sensing and enhance bacterial clearance in *Pseudomonas aeruginosa* lung infection in mice. J. Antimicrob. Chemother., 53 (6): 1054-1061.

XAVIER K B, MILLER S T, LU W, et al., 2007. Phosphorylation and processing of the quorum-sensing molecule autoinducer-2 in enteric bacteria. ACS Chem. Biol., 2: 128-136.

ZHOU L, ZHANG L, CÁMARA M, et al., 2017. The DSF family of quorum sensing signals: diversity, biosynthesis, and turnover. Trends Microbiol., 25: 293-303.

ZIESCHE L, BRUNS H, DOGS M, et al., 2015. Homoserine lactones, methyl Oligohydroxybutyrates, and other extracellular metabolites of macroalgae-associated Bacteria of the *Roseobacter* clade: Identification and functions. Chembiochem., 16: 2094-2107.

ZOU Y, NAIR S K, 2009. Molecular basis for the recognition of structurally distinct autoinducer mimics by the *Pseudomonas aeruginosa* LasR quorum sensing signaling receptor. Chem. Biol., 16 (9): 961-970.

第2章 群体感应(QS)分子的鉴定分析方法

Adele Cutignano

National Research Council of Italy—Institute of Biomolecular Chemistry, Pozzuoli, Italy

1 前言

群体感应（QS）是细菌界中广泛存在且保守的细胞间通信策略（Camilli, 2006; Hmelo, 2017; Taga and Bassler, 2003; Waters and Bassler, 2005）。这一过程涉及自诱导物（AIs）的小分子化合物在局部环境中产生和释放，这些化合物通过细胞膜扩散到相邻的细胞中，并以细胞密度/浓度依赖性的方式调节基因表达。QS 参与多种生物活动，如生物膜形成、生物发光、孢子产生、运动性、遗传交换以及抗生素和毒力因子的产生等。值得注意的是，QS 在病原体中起到一个关键作用，与细菌致病性的发生密切相关，因此，QS 越来越多地被视为是抗生素治疗的替代靶点。

革兰氏阳性细菌和革兰氏阴性细菌都依赖于化学信号来协调种群行为，尽管它们产生的 QS 分子属于不同的化合物类别（Hawver et al., 2016）（图 2-1）。通常，在大多数革兰氏阴性细菌中信号分子（AI-1）是 N-酰基-L-高丝氨酸内酯（AHLs）（Fuqua et al., 1994; Fuqua and Greenberg, 2002; Papenfort and Bassler, 2016），而革兰氏阳性细菌的信号分子是短自诱导肽（AIPs）（Bofinger et al., 2017; Kleerebezem et al., 1997; Sturme et al., 2002; Verbeke et al., 2017）。

每种细菌都会产生和感知特定模式的自诱导物（AI）。不同种属的革兰氏阴性细菌产生的 AHLs 可能因 N-酰基侧链的长度和不饱和度而异。通常 N-酰基侧链含有 4~18 个碳原子，仅有 1 篇文献报道 N-酰基侧链含有 19 个碳原子（Doberva et al., 2017）；其中 3 号碳原子可以携带一个羟基或氧基（图 2-1）。然而，来自周围物种的特定信号分子可以相互识别，由（S）-4,5-二羟基-2,3-戊二酮（DPD）和平衡连接的呋喃糖基硼酸二酯（AI-2）衍生的信号分子是由革兰氏阳性细菌和革兰氏阴性细菌产生的独特的、非物种特异性的化学物质，这些分子可以介导种间通信，从而构成一种普遍的 QS 信号（Bassler et al., 1997; Schauder et al., 2001; Vendeville et al., 2005）（图 2-1）。除这 3 种自诱导物（AI-1、AIPs、AI-2）外，还从细菌种属中发现了一些次要的 QS 信号分子，如图 2-1 所示（Bofinger et al., 2017; Deng et al., 2011; He and Zhang, 2008; Pesci et al., 1999; Winans, 2011）。随着对微生物与宿主间相互作用的深入了解，这些信号分子的临床意义也越来越明显。复杂的通信网络使共生细菌、病原体及其宿主紧密联系在一起，从而在抵御入侵微生物攻击组织的过程中发挥重要作用（Curtis and Sperandio, 2011; Goswami, 2017; Sperandio et al., 2003; Williams, 2007）。其中一个例子是肾上腺素/去甲肾上腺素/自诱导物-3（AI-3）信号系统，该系统调节大肠杆菌 EHEC 在其哺乳动物宿主中的致病性（Curtis and Sperandio, 2011; Reading et al., 2007; Sifri, 2008; Sperandio et al., 2003）。对于这种自诱导物-3，推测其具有芳香性（Reading and Sperandio，2006），但迄今为止，尚未有文献对其结构进行证实。因此，该结论是不确定的。

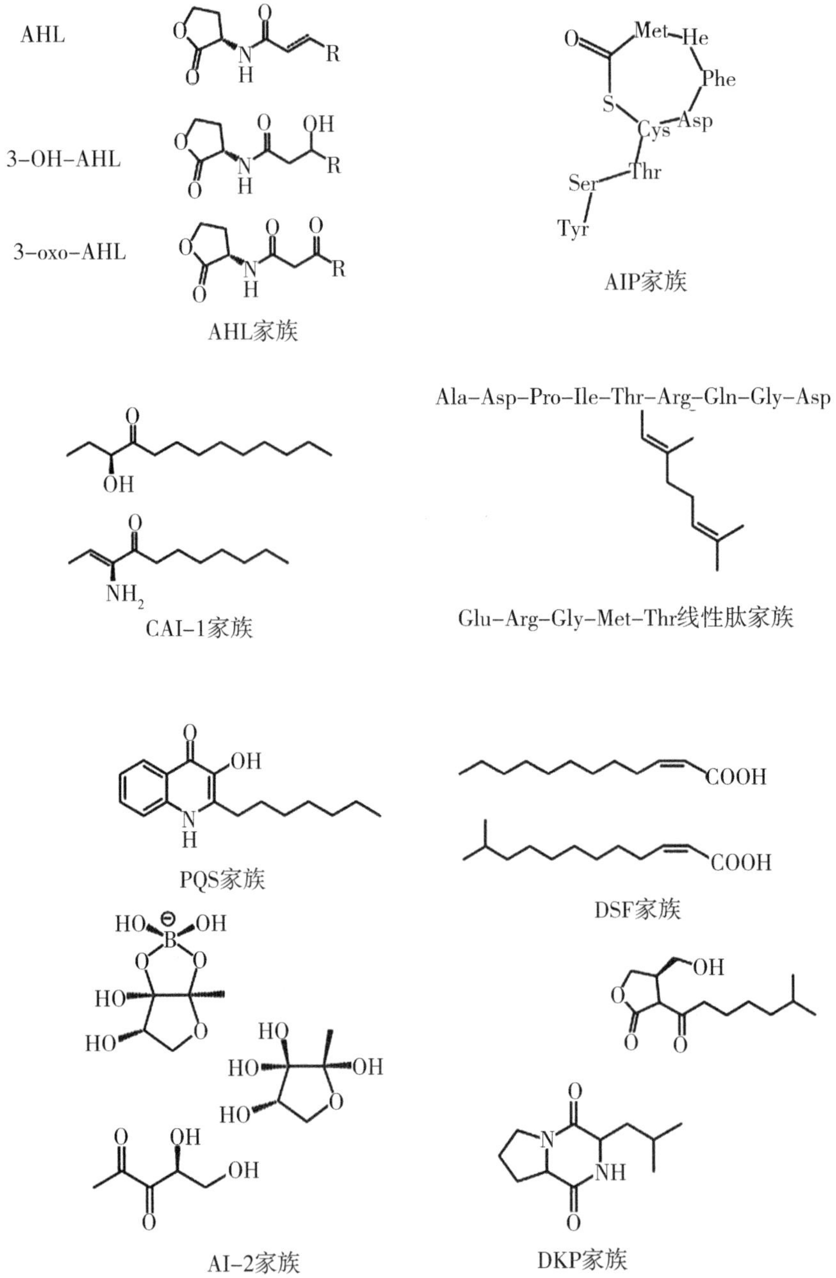

图 2-1　自诱导物的代表性结构和化学类别

所有这些情况的共同特点是信号分子在非常低的浓度（通常是 nmol/L 范围）下产生的，其检测需要高灵敏的检测技术才能将其检出。生物传感器的快速发展使得筛选微生物产生的自诱导物成为可能。然而，对于这些细菌代谢物的特异性鉴定和定量，质谱法结合气相色谱（GC-MS）或液相色谱（LC-MS/MS）是最佳选择。样品提取、净化和预浓缩可以极大地提高从复杂基质中对自诱导物的鉴定,同时也允许可靠的定量。本章将重点介绍研究这些化合物的共同策略,特别是 AHLs,以及基于最先进技术的未来研究趋势。

2 样品的制备

处理极低丰度的天然代谢物时,遇到的瓶颈之一是在分析前选择正确的方案对生物样品进行清理和富集。这一步骤要求避免由于化合物稀释的问题或细胞培养上清液中存在其他成分(生长介质的胞外产物或成分)或其他不同环境基质中存在其他成分,而造成的化合物被忽略,这些成分可能掩盖或损害目标代谢物的正确鉴定(Cutignano et al.,2015)。另外,由于盐和大分子(例如:蛋白质和多糖)的存在,无法直接分析含有 QS 信号分子的培养基,这些物质可能会堵塞分析柱。可以采用不同的策略,包括液液萃取(LLE)和固相萃取(SPE)(Sitnikov et al.,2016)。

2.1 液液萃取(Liquid-Liquid Extraction,LLE)

LLE 是获得小 QS 信号分子富集部分的最直接、最传统和相对简单的方法。通常使用亲油性有机溶剂(如乙酸乙酯或二氯甲烷)提取培养基上清液或环境水样中的水相(Feng et al.,2014;Kumari et al.,2008;Ortori et al.,2007;Shaw et al.,1997;Wang et al.,2017)。在此阶段,考虑到定量分析的需要,应在进一步操作前加入内标以校正回收率和仪器偏差。一旦提取物被蒸发至干燥,可用少量溶剂(乙腈 1~2 mL)重新溶解,并直接进行分析,例如薄层色谱(TLC)。然而,通常还会进一步经过色谱进行预浓缩和清理,这些步骤是通过固相萃取(SPE)色谱进行的。

2.2 固相萃取(SPE)

在过去 20 年里,特别是在定量分析之前,SPE 是一种广泛采用的技术,用于增加痕量分析物的浓度。其广泛应用的两个巨大优势:一是,具有各种化学性质不同的吸附剂的预装柱,从而将其使用范围扩展到化学多样性丰富的物质;二是,通过自动化过程进行此步骤,避免因人工操作造成的差异性变化。此外,SPE 具有很高的回收率,这在处理生物基质中痕量出现的化学成分时尤为重要。事实上,与 LLE 相比,SPE 可以提高至少 2~10 倍的检测灵敏度(Schupp et al.,2005)。

通过 SPE 色谱法利用吸附剂提取 QS 的 AHLs,包括硅胶、辛烷基(RP-18)、氧化铝、二醇、苯基、离子交换、聚合物苯乙烯-二乙烯苯(PSDVB)、一级-二级胺(PAS)和聚酰胺(PA)基树脂(Li et al.,2006;Schupp et al.,2005;Wang et al.,2017)。反相柱获得的结果最好,而正相材料和具有离子交换活性的吸附剂一般是不可取的,因为其总保留较高比例的加载量(Li et al.,2006)。一般来说,没有一种单一的吸附剂可以回收和鉴定所有的化合物。Schupp 等(2005)报道了聚酰胺对短链和部分中链 AHLs 具有优良的回收性能(DPA)树脂,而对于链长为 C6~C12 的 AHLs,RP-18 和改性 PSDVB 树脂(HLB)的回收率均高于 90%,对 C4-HSL 和 C14-HSL 的回收率较差(Li et al.,2006;Wang et al.,2017)。

现有 SPE 的不同洗脱方法的报道(Fekete et al.,2007;Gould et al.,2006;Li et al.,2006;Schupp et al.,2005;Wang et al.,2017)。通常,二氧化硅基柱用非极性溶剂(即正

己烷随乙酸乙酯量的增加而增加）洗脱，并在正己烷-乙酸乙酯（65∶35）洗脱液中检测到 AHLs（Schupp et al.，2005）；另外，端盖十八烷基柱上的优化方案依赖于 AHL 的初始甲醇/水洗涤（15∶85，v/v）和正己烷/异丙醇（25∶75，v/v）洗脱（Fekete et al.，2007；Li et al.，2006）。2017 年，Wang 等对亲脂性和亲水性平衡相互作用的 HLB 吸附剂上进行分馏。用甲醇/水（5∶95，v/v）洗涤后，目标化合物用含 2% 乙酸的甲醇洗脱，回收率良好。

3 QS 分子的检测、鉴定和定量

3.1 AHL 检测的分析方法

3.1.1 生物传感器

早在 20 多年前，在费氏弧菌中发现了细胞对 AHL 自诱导物的反应机制，是由 LuxI-LuxR 家族的两个蛋白（及其他同源蛋白）介导的，即 LuxI-AHL 合成酶和 LuxR-AHL 转录激活因子（Fuqua et al.，1994）（图 2-2）。

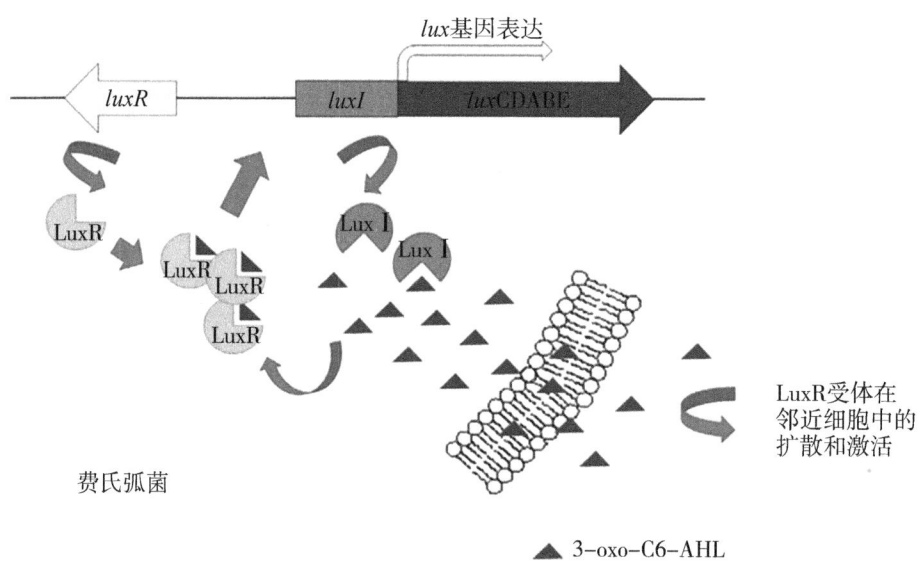

图 2-2 费氏弧菌 LuxI/LuxR 群体感应系统的示意

注：根据 Galloway 等（2011）进行修改。

LuxR 型蛋白是细胞质受体，能检测并结合由配对的 LuxI 型合成酶产生的 AHLs，并与 lux-boxes 的短 DNA 序列结合，触发数百个（多达 600 个）基因的表达。AHL 介导的细胞间信号传导的例子包括能生物发光的海洋细菌［费氏发光杆（弧）菌 *Photobacterium (Vibrio) fischeri*］和哈维弧菌中的生物发光诱导（Fuqua et al.，1994），以及细菌紫色色素杆菌中色素紫醋酸盐产生的调控（McClean et al.，1997）。细菌生物传感器不产生 AHLs，但能在 AHLs 存在的情况下进行识别和响应，促进了革兰氏阴性细菌中大量基于 AHLs 的 QS 系统被发现。这些生物传感器可通过克隆一个有功能的 LuxR 型蛋白和一个合适的启动子来获得，例如，同源的 luxI 合成酶的启动子，其正调控报告基因的转录，如紫色杆菌素的产生、生物发光或半乳糖苷酶（Steindler and Venturi，2007）。换言之，通过使用基因工程细菌或利用组成型表达的报告基因，将识别活动耦合到一个传感器元件上，将其转化为可读且最终可测量的输出。LuxR 蛋白优先结合同源 LuxI 合成酶产生的 AHLs，但在一定程度上对表现出不同类型侧链的密切相关的 AHLs 类似物作出反应；因此，每个生物传感器都是量身定做的，可用于筛选产生小范围内的 AHL 家族自诱导物。另外，通过使用一组生物传感器，可以感知更广泛的 AHLs 光谱。基于细胞的生物传感器能以不同的方式被使用。一种是细菌生产者和生物传感器菌株的平板交叉画线：两个菌株之间距离最近的点将表现出最强烈的反应。另一种是薄层色谱（TLC）叠加法。从生物产生菌中获得的提取物和/或色谱馏分与化学标准品一起加载到 TLC 板上，连同化学标准，允许在色谱室中分解成分子组分。AHLs 的检测可以通过使用生物传感器菌株的琼脂悬浮液进行 TLC 叠加来实现。与 AHL 标准品相比，所检测斑点的色谱阻滞因子（Rf）有助于暂时确定自诱导物的链长和取代基团（McClean et al.，1997；Shaw et al.，1997）。此外，报告基因显示的强度可在定量分析中进行解释。尽管这些基于细胞的生物传感系统简单且稳定，但也存在一些缺点，包括反应速度慢、细菌细胞成分的化学干扰和批次间的变异性。因此，在过去 20 年中，已经提出了基于含有编码绿色荧光蛋白 *gfp* 基因的质粒来检测 AHL 的替代工具（Andersen et al.，2001）。这些传感器利用细胞内的生物机制，具有巨大的优势，即可以在单细胞水平上用于检测 AHL 的产生。最近，开发出来一种"无细胞"形式的生物传感器，其利用在溶液中自由漂浮的工程化 DNA 电路和细胞机制，用于检测铜绿假单胞菌的临床样本（Wen et al.，2017）。在目前研究中，在纳摩尔水平上对囊性纤维化患者痰液中的 QS 信号分子进行了定量测定，其结果与基于串联质谱的其他敏感技术的结果相当，但可能成本更低。

3.1.2 放射性标记分析

酰基高丝氨酸内酯骨架的生物合成分为两类内酯合成酶：利用合成酶的酰基-CoA 和利用合成酶的酰基-ACP。然而，它们都利用了 *S*-腺苷-*L*-甲硫氨酸（SAM），因此，是一种保守的底物（图 2-3）。

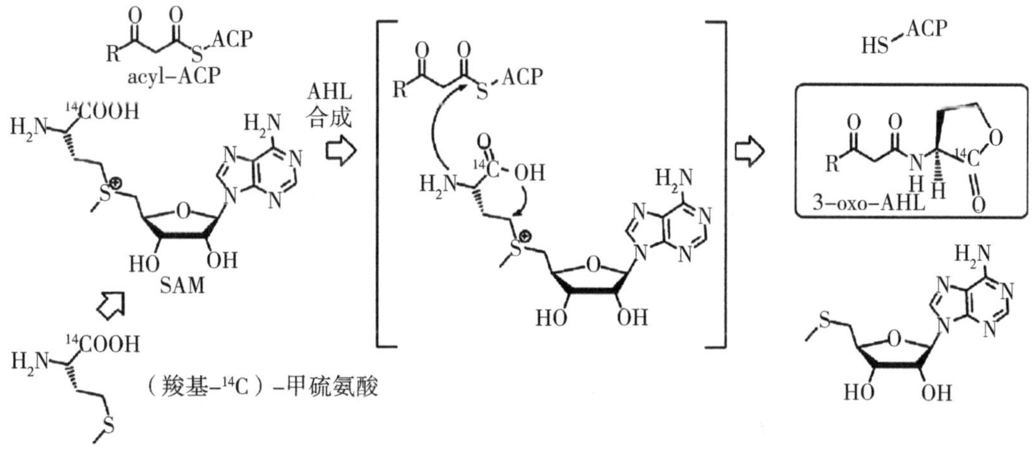

图 2-3 （羧基-^{14}C）-甲硫氨酸用于放射性标记分析的 AHLs 生物合成方案

^{14}C-AHL 放射性标记分析是基于活细胞摄取［1-^{14}C］-*L*-甲硫氨酸并转化为放射性标记的 *S*-腺苷甲硫氨酸。后者的代谢中间产物又通过 AHL 合成酶被整合到 AHL 信号代谢产

物中。一旦被有机溶剂提取，与 AHLs 相关的放射性可以很容易地通过闪烁体探测器来揭示，而不会对特定的链长或氧化状态产生任何偏差。放射性标记分析过去被应用于生物膜细菌中 AHLs 的检测；事实上，可用的生物材料数量非常低，检测不同类型 AHLs 所需的测试数量非常少，在此背景下阻碍了生物检测的使用（Greenberg and Parsek，2001；Jones et al.，1993；Schaefer et al.，2001，2002）。然而，尽管这种方法简单和灵敏，但这种方法目前正在被更快和更安全的质谱方法所取代。

3.1.3 比色法测定

为了开发一种廉价且易于使用的方法，Yang 等（2006）提出了一种对酯类化合物分析的先前程序进行改进的比色分析方法。该方法最初是为了内酯化合物和内酯酶活性检测而提出的，并缩小到 96 孔板中，需要极少量的样品（20~50 μL），对内酯分子的检测限为 1 nmol。比色法是基于羧酸酯/内酯转化为相应的羟肟酸；后者在铁离子存在下，形成红色到紫色的络合物，可通过分光光度法测量。尽管比色法非常敏感，但上述对 AHL 检测的特异性不足，可能导致对 QS 信号分子的高估或误识别。

3.2 AHL 鉴定和定量分析方法

3.2.1 气相色谱-质谱法（GC-MS）

气相色谱-质谱联用（GC-MS）是一种将气相色谱分离的高分辨能力与质谱仪特有的高灵敏检测相结合的联用分析技术。它适用于挥发性和热稳定分子的分离，可通过解释根据其质荷比（m/z）比值检测到的每个离子种类的质谱来识别。在处理已知分子时，标准品和/或光谱数据库的可用性允许通过比较保留时间和裂解模式来快速识别未知分子。事实上，GC-MS 是基于一种硬电离技术，利用电子轰击（EI）或化学电离（CI）产生分子离子，并与源自化学键断裂的碎片离子一起被检测和记录。因此，GC-MS 是鉴定和定量细菌提取物中 N-酰基-L-高丝氨酸内酯的一种可靠和非常灵敏的技术（Osorno et al.，2012）。由于内酯环的裂解，在 m/z 143 和 102 处产生的诊断离子使得 Cn-AHL 系列的所有化合物均表现出典型的裂解规律（Thiel et al.，2009a）（图 2-4）。而 3-羟基-AHLs 的质谱图则主要由 m/z 172 处 α 的裂解产生的羟基（Wagner-Döbler et al.，2005）（图 2-4）。因此，当质谱分析仪为四极杆型时，如在 GC-MS 平台中经常遇到的那样，最合适的试验设计将考虑在全扫描和选定离子监测（SIM）模式下使用 m/z 143 或 172 处的突出碎片来获取光谱数据。

后一种采集模式利用了基于四极杆的质量分析器作为质量过滤器的独特电子特性。在源中形成的离子在分析仪中积累，只有那些具有确定的 m/z 比的离子会被选择并允许到达检测器。只有当预设的 m/z 离子高于灵敏度阈值时，才会出现峰值，但由于其他离子不被监测，所有的采集时间都用于列出离子，信噪比显著提高；这一事实，加上快速的扫描速率，使 SIM 的灵敏度比全扫描采集模式提高了 100 倍。

已经报道了一些方法用于将不稳定的 3-oxo-AHLs 转化为相应更稳定的五氟苯甲酰肼（PFBO）衍生物进行特定的 GC-MS 分析（Charlton et al.，2000）。然而，在一些方案中，所有 AHL 的分析都没有在使用任何化学衍化法，也没有在 SIM 模式下监测最丰富的诊断性片段（Cataldi et al.，2004，2007；Celio et al.，2006；Chi et al.，2017；Pomini and Marsaioli，2008；Ran et al.，2016；Rani et al.，2011）。尽管分支的 AHLs 表现出比非分支化的对应物更短的洗脱时间，但它们显示出与相应的线性高丝氨酸内酯相似的质谱。然而，它们的鉴定

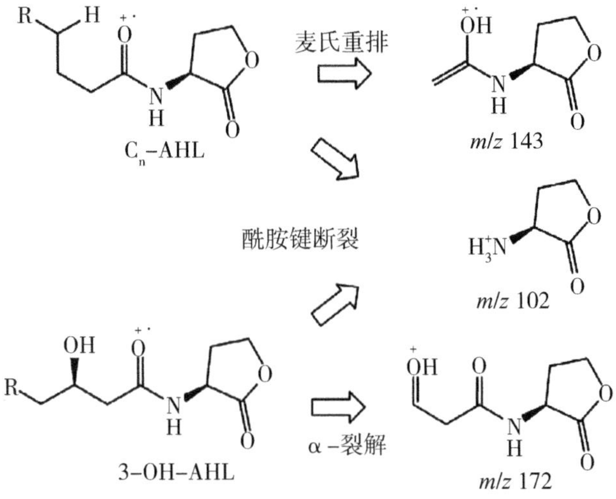

图 2-4　饱和的未取代的 Cn-AHL 分子在 m/z 143 和 102 处产生诊断性离子，
3-OH-AHLs 在 m/z 172 和 102 处产生诊断性离子的电子撞击断裂示意

可以通过 GC-MS 分析来完成，该经验模型利用保留指数值来计算不同脂质中长烷基链上甲基分支的位置（Schulz，2001）。根据该模型，首次在微粒气单胞菌中鉴定出了 *iso*C9-HSL 和 3-OH-*iso*C9-HSL。此外，对较高质量区域的仔细检查，揭示了在 ω-1 位置有甲基分支的烷基链上存在一个 [M-43]$^+$ 离子（Thiel et al.，2009a）。色谱分离通常在 20~30 cm 长的熔融硅胶毛细管柱上进行，使用氦气作为载气和温度梯度从 100~300℃。在某些情况下，使用 C7-HSL 作为内标进行定量（Cataldi et al.，2007），但在少数细菌物种中存在奇数碳链 AHLs 的存在（Dickschat，2010），因此，使用氘代类似物作为内标参考更加合适（Gould et al.，2006）。应用领域涵盖了与微藻相关的海洋细菌庞蒂球菌属（*Ponticoccus* sp.）杀藻剂活性的生态学研究（Chi et al.，2017），用于在痰样本中鉴定铜绿假单胞菌 AHL 标记物的生物医学筛查（Rani et al.，2011），以及对高丝氨酸内酯的新结构变体进行化学表征（Thiel et al.，2009a）。

GC-MS 分析也被用来确定 AHLs 中手性碳的绝对构型。该基团的所有分子都具有以内酯环的 C-3 为代表的立体中心，而 3-OH-AHL 衍生物在烷基侧链的羟基化位置显示出额外的手性中心。手性分析采用火焰离子化检测器（FID）的气相色谱，与合成标准品进行比较，对含有非极性 β-环糊精的手性柱进行分析（Pomini and Marsaioli，2008；Malik et al.，2009）。有趣的是，利用新型的单滴微萃取技术，在洋葱伯克霍尔德菌 LA3 菌株中发现了 *D*-高丝氨酸酰基内酯和 *L*-衍生物（Malik et al.，2009），表明需要手性分析技术来区分可能具有不同生物活性的对映体。另外，通过 3-OH-AHL 的酸性醇解和相应脂肪酸甲酯的释放来确定 C30 的绝对构型；对同一手性 GC 相进行色谱分析，与外消旋体和对映体纯合成标准品进行比较，确定天然化合物的碳立体属性（Thiel et al.，2009a）。通过对其二甲基硫醚衍生物（DMDS）的 GC-MS 分析，实现了单不饱和和不常见的双不饱和类似物中双键的定位（Wagner-Döbler et al.，2005；Thiel et al.，2009a）；甲基化基团之间的优先 α-裂解可以直接分配双键位置（图 2-5）。通过与选择性合成制备的 E/Z 标准品进行比较，确定了远端不饱和的烯烃几何构型，假设 C2 位的反式构型在生物来源的类似酸中是常见的。

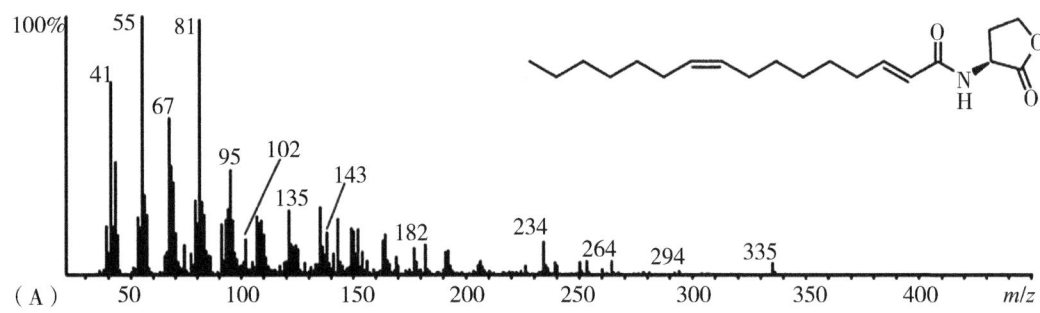

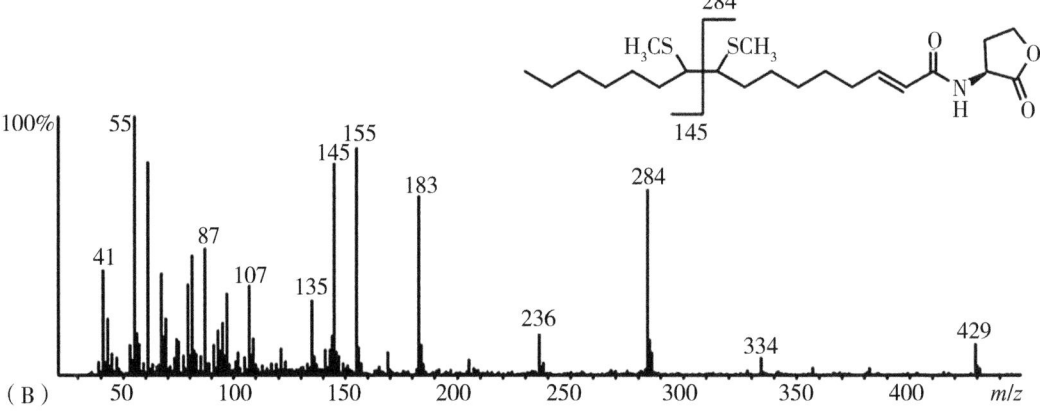

图 2-5 EI-MS 图谱

注：(A) C16:2-AHL 和 (B) DMDS 加合物的 C16:2-AHL 的内部双键定位。改编自 Thiel et al., 2009a。

3.2.2 液相色谱-质谱法（LC-MS）

与其他分析技术相比，液相色谱-质谱联用（LC-MS）已被证明是最可靠和敏感的 QS 信号分子检测、鉴定和定量测量方法。据 LC-MS 的一般定义下，包括不同和独特的技术解决方案，它们提供不同程度的灵敏度、质量分辨率和准确性。因此，可根据领域的适用性来满足不同的科学需求。两个连接的分析模块（即色谱系统和质谱仪）可根据预期获得的输出进行配置、定制和接口化，用于目标分析。直到 2004 年，液相色谱与质谱仪的连接只通过高效液相色谱（HPLC）系统实现。HPLC 是一种众所周知且广泛应用的技术，用于将复杂混合物分离成其单独的组分，利用它们在移动液相和固相之间的不同亲和性，并以洗脱时间来衡量。固相通常填充在不同长度和直径的钢柱中，自从第一次应用以来，现在可以分离出许多化学类别的令人印象深刻的一系列分析物。尽管在许多技术上引入了改进，但所有固相都可以归因于这三种相互作用类型中的一种：尺寸排斥、离子交换和吸附色谱。后者可以在正相（NP）或反相（RP）中进行。在 NP 色谱中，固定床是极性较高的二氧化硅或以硅基组分，洗脱相主要由非极性溶剂（如正己烷）组成。适用于亲脂性化合物的分离。与此相反，RP 色谱依赖于一个非极性固定相，由极性溶剂（包括甲醇、乙腈和水）的混合物洗脱。通过使用甲醇/水或乙腈/水流动相，在等温或梯度洗脱程序中，这种适合于分离极性到低极性化合物的相通常被选择用于 AHL 分析。与 LC 组件一样，MS 部分也可以在几种不同的工艺方案中使用。对 LC 与 MS 配对影响最大的组分是界面，在这个界面上是分子转化成离子。通常，在 LC-MS 中，界面被加热并允许在大气压（API）电场中发生电离。这可以通过两种方式获得——化学电离（APCI）和电喷雾电离（ESI）。ESI 是目前最广泛的模式适用于多种化学物质，包括酰基高丝氨酸内酯。它是一种软电离，这意味着只有分子离子在源中形成；因此，一方面，分子质量和公式（在精确的质量测量中）很容易计算出来，另一方面，要获得结构信息需要进一步质量碎裂（串联 MS 或 MS/MS），在质谱分析过程中，尽管它有可能在内源进行碎裂（CID），但在一个不同的地方和不同的时间通常也会发生。

质谱仪的核心由分析器构成。最常见的分析仪是基于四极杆、飞行时间和离子阱的分析仪。在不进行更深入的技术描述的情况下，对大多数目标和定量方法，在所有情况下，当灵敏度是关键时，如 QS 检测，基于四极杆的分析器是首选，特别是当三个四极杆单元串联耦合（QqQ，三重四极杆）时。此外，QqQ 由于其极快的扫描速度而提供了很高的分析速度，尽管质量分辨率通常很低（1 个质量单元）。2004 年，采用超高效液相色谱（UPLC）获得最佳的 LC-MS 耦合方法。基于直径小于 2 μm 颗粒的 UPLC 技术为色谱分离开辟了新的时代，可缩短分析运行时间，但仍保留了最佳的峰分离度。Fekete 等（2007）的工作为 UPLC 在 QS 分析中的应用提供了范例。在 AHL 分析采用的策略中，首先使用 RP-18 SPE 固相萃取小柱对细菌上清液进行预纯化，然后用 BEH C18 色谱柱进行 UPLC 分析，洗脱液为水和乙腈的混合液。使用的梯度非常快，在 1 min 中 ACN 从 10% 到 100%，在等温洗脱中保持 100%；然而，饱和和未取代的 AHLs 的同源物成员之间的分辨率是可观的。此外，因 AHLs 的理论 log$P$ 值，对大多数 AHLs 的保留时间进行了预测。然而，为了确认细菌提取物中的 AHL 类型，采用了一种集成的离线方法，对质子化的 $[M+H]^+$ 分子离子进行了额外的 FTICR-MS（傅里叶变换离子回旋共振-质谱联用仪）测量。FTICR-MS 是一种超高分辨率质谱技术，其分辨率可以达到 100 万 FWHM（半峰全宽），质量分辨率低于 0.5 mg/kg。这意味着已知和未知的代谢物可以在复杂的混合物中直接进行简单的化学表征；因此，FTICR-MS 被用来鉴定细菌食酸菌属（*Acidovorax* sp.）N35 的主要 AHL 产物（如 3-OH-C10-AHL）。然而，由于扫描速率不同，与 FTICR-MS 的耦合在技术上是不可行的，因此，开发出一种基于纳米电喷雾芯片在线耦合器进行快速筛选（Li et al.，2007）。然而，同样的 UPLC 方法通过使用更易于管理的在线平台 UPLC-QqQ（Abbamondi et al.，2016），利用 MRM 采集扫描模式成功应用于 AIIL 检测。多反应监测（MRM）是基于串联质谱鉴定目标代谢物的最强有力的方法之一，具有较高的灵敏度和特异性。在 MRM 分析中，只有当选择的碎片发生时才会触发数据采集。因此，第一，三重四极杆作为质量过滤器，分别选择分子离子和预期产物离子；第二，中间四极杆作为碰撞室，是分子、母离子发生可控碎裂的空间。只有当母离子如前所述产生产物离子时，检测器才会采集并登记信号；这种行为，连同四极杆分析器的快速扫描特性，反映在对监测离子对跃迁的选择性和灵敏度的显著提高。在 Abbamondi 等（2016）的研究中，AHL 分析的标志性跃迁是产生 m/z 102 高丝氨酸内酯片段的跃迁，相应地，在 Fekete et al（2007）报道的 UPLC 条件下处理了同源系列的合成标准衍生物，包括未取代、羟基和氧代 AHLs（图 2-6，Grauso and Cutignano，未发表的数据）。

基于生物传感器和质谱联合方法，通过 TLC 覆盖层和 UPLC-MS/MS（MRM）分析，检测到了嗜盐细菌嗜盐单胞菌（*Halomonas smyrnensis*）AAD6 培养基中与 AHL 相关的 QS 活性，并明确鉴定为一种不寻常的长链 C16-HSL（Abbamondi et al.，2016）。

根据相同的基本原理，其他几项研究采用了 MRM 采集模式的三重四极杆质谱分析，使

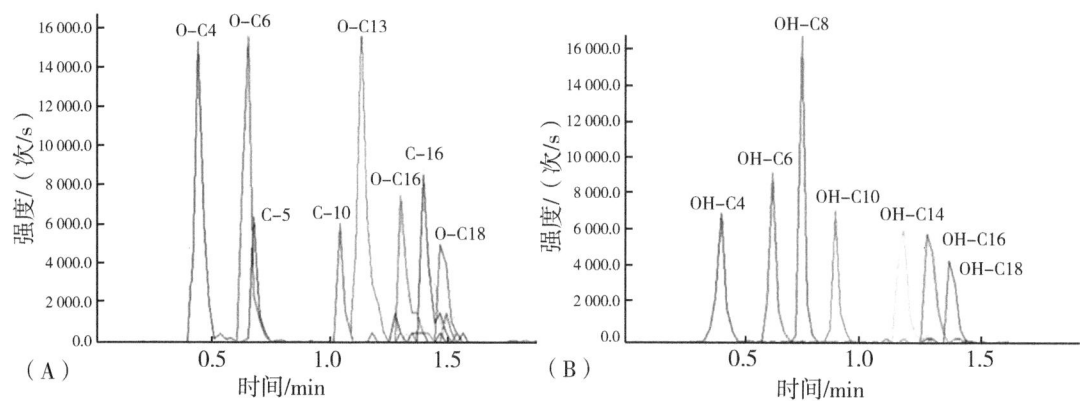

图2-6　UPLC-ESI+-MS/MS（MRM）特征图谱

注：采用BEH-C18色谱柱，以水/乙腈为流动相进行梯度洗脱，建立了AHLs标准品的UPLC-ESI+-MS/MS（MRM）特征图谱。（A）5氧代AHLs和3C-AHLs的混合池；（B）7种OH-AHLs的混合池。

用了来自不同制造商的类似UPLC色谱支架，但柱尺寸和色谱运行长度可变（Chan et al.，2014，2016），或者使用传统的HPLC在RP-18（Kumari et al.，2008；Morin et al.，2003；Tan et al.，2014）或更多的极性相上，如Atlantis T3（Sun et al.，2018）。

三重四极杆的替代品是以混合构型的形式出现，如四极杆-线性离子阱质谱仪（QqQLIT）。这些仪器有利于将传统的三重四极杆和线性离子阱（LIT）分析仪的优点结合在一个LC运行中。因此，除了通常的三重四极杆扫描模式，包括MRM扫描模式外，还包括产物离子（PI）、前体离子（PC）和中性丢失（NL）扫描模式，利用线性离子阱的高离子存储能力、扫描速率和分辨率，从而获得未知化合物的结构信息。事实上，当三重四极杆作为LIT工作时，可以获得用于结构鉴定的高质量产物离子谱。该HPLC-QqQLIT技术被应用于测定假结核耶尔森菌（*Y. pseudotuberculosis*）中24种AHLs的组成和相对丰度，从而得到已知由单一物种产生的AHLs最广谱的细菌（Ortori et al.，2007）。

3.2.3　毛细管电泳质谱（Capillary electrophoresis-mass spectrometry，CE-MS）

毛细管电泳（CE）在无涂层的熔融石英毛细管上与质谱联用是一种分析技术，它提供了几个有吸引力的优点：分离效率、检测灵敏度和化学鉴定。然而，一个主要的缺点是电泳分离所需的缓冲溶液与ESI质谱电离过程的兼容性。事实上，电泳分离本质上是基于带电物质在电场中的差异迁移，因此，用于色谱分离的缓冲介质对pH值稳定性和迁移率的差异有很大影响。另外，CE中使用的常用缓冲液（例如，磷酸盐和TRIS缓冲液）与仅使用挥发性缓冲液的ESI不兼容。此外，高物质的量浓度通常可以提高系统的缓冲能力，但ESI在缓冲液浓度约为10 mmol/L时效果更好。更复杂的是，AHLs不具有可带电的基团，是由酰胺键稳定的仲胺。因此，提出了一种通过胶束电动色谱（MEKC）来克服这一问题的方法（Frommberger et al.，2003）。使用十二烷基硫酸钠（SDS）10 mmol/L形成胶束，用胶束溶液（PF-MEKC）部分填充毛细管形成胶束，得到了最佳的结果。在相对较短的时间（20 min）分析中，在洋葱伯克霍尔德菌中检测到了两种AHLs，但仍无法使用所提出的分析工具获得定量数据。通过对内酯环进行碱性水解后，在荧光假单胞菌（*P. fluorescens*）培养的上清液中通过CE-MS对AHLs检测进行技术改进。以保护质谱界面，在20 mmol/L碳酸铵缓冲液中进行分离,通过C7-HSL内标和阴离子交换-SPE进行样品预浓缩来获得定量测量（Frommberger et al.,2005）。

3.2.4 基质辅助激光解吸电离质谱法（Matrix-assisted laser desorption ionization-mass spectrometry, MALDI-MS）

基于MALDI-MS的方法传统上被应用于生物分子（肽/蛋白、寡糖/多糖、核酸）的鉴定,这些分子在电离过程中不会发生碎裂。从技术角度来看,样品与一定量的基质混合[例如,α-氰基-4-羟基肉桂酸（α-CHCA）和2,5-二羟基苯甲酸（DHB）],并在平板上共结晶。通过激光照射发生电离,迅速加热基质,然后基质与样品一起蒸发。普遍认为,由于基质峰的干扰,MALDI最初不是为小分子（低于500Da）的电离而设计的。为了解决这个问题,已经报道了几种方法（Zhang et al.,2010）。然而,通过MALDI-TOF-MS在海洋细菌成团泛菌（Pantoea agglomerans）的提取物中鉴定出3-oxo-AHL分子（Jiang et al.,2015）,或者用铜绿假单胞菌中Girard试剂衍生后再进行定量。

3.2.5 核磁共振光谱学

核磁共振（NMR）是用于阐明有机结构的主要技术。然而,其检测限度在mmol/L级别上,因此,其浓度范围远高于质谱技术。因此,在获得分离纯化的化合物NMR光谱之前,需要大量的细菌培养物进行AHL提取和纯化（Pearson et al.,1994）。尽管在开始时,结构表征QS信号和识别新的分子骨架至关重要（Doberva et al.,2017；Shen et al.,2016）,但在分析方法中旨在识别和量化已知的自诱导物,其作用处于次要水平。

3.3 其他QS分子鉴定和定量分析方法

来自强毒力革兰氏阳性金黄色葡萄球菌（Staphylococcus aureus）的自诱导肽（AIPs）控制属于超抗原家族的肠毒素的分泌：这些蛋白质毒素触发T细胞的异常活化,导致从特应性皮炎到中毒性休克综合征的一系列临床症状。值得注意的是,来自一组的AIPs交叉抑制来自其他组菌株的基因表达,因此,一种检测和区分AIPs的方法在感染中具有临床相关性,有望成为实用性的诊断工具。因此,在Kalkum等（2003）的研究中,利用基质辅助激光解吸电离（MALDI）-四极杆离子阱仪器开发了一种多级质谱方法,实现了依靠MALDI离子源研究肽分子的传统方法（Chaurand et al.,1999）。在第一种情况下,它是按照假说驱动方法使用的。事实上,AIPs是作为大的前肽合成的,并在排泄过程中依次水解为成熟的QS-肽（Sturme et al.,2002）；因此,对DNA序列的了解可以预测来自生物合成前体裂解的较短肽的所有可能的氨基酸序列；这些假定多肽的m/z值可通过在细菌上清中的目标筛选进行搜索,并列出MS^2（MS/MS）和MS^3（MS/MS/MS）确认实验。在第二种情况下,将中间葡萄球菌（S. intermedius）AIP的未知结构作为非肽内酯进行应用。在这两种情况下,MALDI-离子陷阱在短时间内利用少量样品分离特定离子,并将其转化为背景干扰较低的碎片的能力得到了充分发挥。然而,传统MALDI质谱的局限性在于其对化合物鉴定的定性信息,不如其他定量质谱平台可靠。为了克服这一问题,最近研究提出将四极杆或线性离子阱和轨道阱质量分析器在线与UPLC组合构成的混合配置应用于肽分析,以确保色谱运行时间短（10min以内或更少）（Junio et al.,2013；Olson et al.,2014；Todd et al.,2016）。因此,基于假定的较长前体肽（agrD基因）的基因序列,预测表皮葡萄球菌（S. epidermidis）和耐甲氧西林金黄色葡萄球菌MRSA菌株的一系列假定成熟的AIPs,然后在生物测定报告菌株选择的活性组分中寻找与其精确质量相匹配的 m/z 离子。利用该策略，通过串联 MS 试验和合成方法检测到不同的 AIPs，并建立了其氨基酸序列（Olson et al., 2014）。该方法用于监测金黄色葡萄球菌中 AIP-I 的时间依赖性，并比较其在不同菌株中的水平（Junio et al., 2013），以及通过定量测定在不同抗生素处理下细菌滤液中的多肽对 AIP 产生的抑制作用（Todd et al., 2016）。

小环二肽（二酮哌嗪，DKPs）具有激活基于 LuxR 的 AHL 生物传感器的能力，已从革兰氏阴性细菌，特别是海洋环境中被报道（Abbamondi et al., 2014；Holden et al., 1999；Tommonaro et al., 2012）。Gu 等（2013）报道，它们的分子鉴定可以通过对未活化分子进行 GC-MS 分析并通过化学合成进行确认。

在三重四极杆平台上，利用 GC-MS 和 HPLC-MS/MS 技术对 AI-2 信号进行了直接或后衍生化定量分析。考虑到这类自诱导物的化学性质，由互相转化的硼化和非硼化分子混合而成，池中活性成分的鉴定一直很困难。除了它们的丰度较低外，主要的困难之一是这些化合物的亲水性质，这阻碍了采用有机溶剂萃取法测定样品浓度和在亲脂性介质中将其转化为极性较小衍生物的化学反应。因此，发展了替代策略，通过与 1,2-苯二胺在水溶液缓冲液中反应来衍生化二羟基戊二酮（DPD），得到其（3-甲基喹喔啉-2-基）乙烷-1,2-二醇衍生物（De Keersmaeckert et al., 2005）。该产物最初通过 HPLC-MS/MS 直接鉴定（Hauck et al., 2003），或者在二醇基团进一步与 N-甲基-N-（三甲基硅烷基）三氟乙酸酯（MSTFA）衍生后，通过 GC-MS 鉴定（Thiel et al., 2009b）。后续步骤是为了使喹喔啉二醇衍生物具有挥发性并适合于 GC-MS 分析。通过应用这一方法并使用同位素标记的标准模拟物，开发了一种定量方法，并在哈维弧菌 BB152 和变形链球菌（S. mutans）UA159 中进行了验证，显示出与生物传感器系统相当的敏感性。最近，提出通过 HPLC 与串联质谱仪上的 QqQ 结合的改进方法进行后衍生化 AI-2 的定量测定（Campagna et al., 2009；Xu et al., 2017）。

通过 GC-电子捕获质谱和 LC-MS 对喹诺酮类家族（alkyl quinolone，AQ）的 QS 信号分子，包括 4-羟基-2-烷基喹啉及其 N-氧化物进行鉴定（Lépine et al., 2004；Taylor et al., 1995；Vial et al., 2008）。由于试验条件的热不稳定性和波动性较差，GC-MS 方法要求将天然化合物转化为双三氟甲基苯甲酰衍生物，这些化合物长期以来被用于鉴定和定量一系列羟基化和含氮化合物，对于 N-氧化物，进一步还原为母体羟基喹啉。发现烷基链长度与保留时间之间呈线性关系，这对寻找低丰度化合物提供了方便。与往常一样，已报道的 LC-MS 分析方法不需要任何衍生化处理。这两种方法均已用于氘代内标的 AQs 定量。有趣的是，通过靶向 HPLC-ESI-MS/MS（MRM）程序，同时鉴定铜绿假单胞菌中 AQ 和 AHLs 家族成分（Ortori et al., 2011）。毫无疑问，当兴趣在于广泛的 QS 信号时，这个优势在于提供了一个细菌信号反应的清晰轮廓。最近，一种基于分散液-液微萃取和离子液体基质（ILM）的微型 AQs 鉴定和定量平台被提出（Leipert et al., 2018）。后者构成了 MALDI 样品制备的另一种基质，使人们能够克服传统固体样品中分析物和基质的不均匀分布，从而产生更可靠的定量测量（Zabet-Moghaddam et al., 2004）。尽管它不能检测 AHLs，但它是一个快速、可重复的、直接的方法，同时识别临床样本中不同的信号分子、毒素和毒力因子（如 AQs 和紫青素），并在囊性纤维化患者的生物样本上进行验证，作为基础研究或诊断目的的工具。

4 结论

细菌中的群体感应（QS）通信是过去 20 年中最有趣的发现之一。尽管在微生物之间以及微生物与寄主之间的分子信息是由简单的化学物质驱动的，但检测、鉴定和测量所有这些不同的低分子量代谢物的水平仍需要进行大量分析工作。近年来，技术平台，特别是在质谱领域的巨大进展，为化学和生化研究的新时代打开了大门，以便更深入地理解生物体相互作用及其调控机制。事实上，QS 引起了一种全局的代谢重调，诱导了通常由沉默基因簇编码的次生代谢物的合成，这可能最终导致新型天然产物的发现。在不同 QS 信号分子的影响下，基于 MS 代谢组学监测包括种内和种间通信变化的分析方法尚处于起步阶段。截至目前，相关研究报道较少，但预计将有更多的研究会涉及这一主题（Davenport et al., 2015；Tang et al., 2018）。QS 介导的代谢网络调控的重要性在于其可以跨越微生物领域的界限，涉及非致病菌和致病菌与宿主之间的跨界互作。深入了解这一课题将有助于揭示宿主细胞如何对 QS 分子等外部线索做出反应，从而为设计新型基于细胞的治疗方法铺平道路（McNerney and Styczynski, 2018）。总体而言，不仅在细菌中，而且在其他微生物（如真菌和微藻）中，对 QS 信号传导的化学基础进行更深入和广泛的研究，有望对生态学、农业和医学产生巨大影响。

术语表

毛细管电泳（CE） 一种依赖于窄孔（20~200 μm ID）毛细管来高效分离小分子和大分子的分析技术。分离是通过应用高压来实现的，促进缓冲溶液和离子物质的电渗透和电泳流动。

电喷雾电离（ESI） 一种"软"电离技术，用于质谱分析，通过在高压下雾化液体，没有或很少破碎，从而产生分子离子。

傅里叶变换离子回旋共振质谱（FTICR-MS） 一种基于离子放置在固定磁场中的回旋频率的质谱分析仪。其特点是具有高分辨率，质量精度为 1~2 mg/kg。

气相色谱-质谱（GC-MS） 一种结合气相色谱和灵敏度检测的结合分析技术。根据其质荷比（m/z）值，适用于挥发性和热稳定分子的分离。

高效液相色谱（HPLC） 一种用于分离、识别和量化复杂混合物中的单个组分的分析技术。它需要一个泵送装置来允许包含样品的溶剂通过一个填充了吸附剂材料的柱。不同的溶剂组成（流动相）和吸附剂材料（固定相）允许分离不同的组分，根据其化学性质和与固定相的特定相互作用，在不同的时间从柱中洗脱。

液相色谱-质谱（LC-MS） 一种分析技术，其中液相色谱是在线的质谱分析。它是分子检测、鉴定和定量测量中最可靠和最灵敏的方法之一。在 LC-MS 方法的一般定义下，包含了不同的技术解决方案，提供了不同程度的灵敏度、质量分辨率和准确性。当液相色谱法与多级质谱分析仪（LC-MS/MS）耦合时，其可以检测由碰撞诱导的碎片产生的产物离子。

基质辅助激光解吸/电离（MALDI） 一种"软"电离技术，用于质谱分析，从碎片率最低的大分子中获得离子。使用脉冲激光照射与合适的基质混合的样品。

质谱（MS） 一种强大的分析技术，用于测量离子的质量与电荷（m/z）比。其基本仪器配置包括离子源、分析仪和探测器系统。它被用于识别一个样品中的未知化合物，以阐

多反应监测（MRM） 一种多阶段实验设计，与三重四极杆（QqQ）质谱分析仪一起使用，用于专门检测和同时量化数百种高灵敏度的混合物目标分析物。需要在碎片反应中预先的前体和碎片离子来检测和登记信号。

核磁共振波谱（NMR） 一种光谱技术，利用某些原子核（如 ^1H 和 ^{13}C）的磁性能，当一个样品被放置在一个强磁场中时，通过测量电磁辐射的吸收，来收集有关有机分子的结构信息。

选择离子监测（SIM） 一个单级质量实验，其中一个或多个具有特定质量与电荷比（m/z）比的离子进行监测。在 SIM 采集模式下，源中形成的离子会在分析仪中积累，分析仪作为一个滤波器，只允许将那些具有选定 m/z 比的离子转移到检测器中。

固相萃取（SPE） 一种样品制备过程，将混合物中的化合物溶解或悬浮在液相中，并使用吸附剂材料根据其物理和化学性质相互分离的方法。它被用于从各种基质中提取或浓缩和净化样本，用于分析，包括生物液体（尿液、血液）、细菌培养基、植物和动物组织以及土壤。

薄层色谱法（TLC） 一种传统的分离混合物中非挥发性化合物的色谱技术。在玻璃板或铝箔上进行，上面覆盖着固定相（通常是二氧化硅），装载在样品的基点上，并在含有洗脱混合物（流动相）的色谱室中开发。不同组分的不同延迟因子（Rf），即相对于溶剂移动的距离，取决于它们的化学性质和对流动相的亲和性。

三重四极子（QqQ） 一种由三个四极子串联组成的质谱分析仪。通过应用包括 MRM 在内的不同实验设计，它适用于混合物中代谢物的高灵敏度和有针对性的定量分析。

超高性能液相色谱（UPLC/UHPLC） 一种现代技术，可以在速度、分辨率和灵敏度方面提高 HPLC 的性能。它依赖于由直径小于 2 μm 的颗粒填充的柱，并且需要比 HPLC 更高的压力（15~18 kpsi，1 kpsi≈6.895 MPa）。特别适合与现代质谱仪的耦合。

缩写词

AHL	N-酰基-L-同丝氨酸内酯
AIP	自诱导肽
AQ	烷基喹诺酮
CE	毛细管电泳
ESI	电喷射离子化（作用）
FTICR-MS	傅里叶变换离子回旋共振质谱法
FWHM	半峰全宽
GC-FID	气相色谱法-火焰电离检测器
GC-MS	气相色谱-质谱法
HPLC	高效液相色谱法
HSL	高丝氨酸内酯
LC-MS	液相色谱-质谱法
LC-MS/MS	液相色谱法-串联质谱法
LLE	液液萃取

MALDI	基质辅助激光解吸/电离技术
MRM	多重反应监测
MS	质谱分析
NMR	核磁共振
QqQ	三重四极
QqQLIT	三重四极-线性离子阱
QS	群体感应
Rf	持水系数
SIM	选择性离子监测
SPE	固体萃取
TOF	飞行时间
TLC	薄层分析法
UPLC	超性能液相色谱法

参考文献

ABBAMONDI G R, DE ROSA S, IODICE C, et al., 2014. Cyclic dipeptides produced by marine sponge-associated Bacteria as quorum sensing signals. Nat. Prod. Commun., 9 (2): 229-232.

ABBAMONDI G R, SUNER S, CUTIGNANO A, et al., 2016. Identification of N-Hexadecanoyl-L-homoserine lactone (C16-AHL) as signal molecule in halophilic bacterium *Halomonas smyrnensis* AAD6. Ann. Microbiol., 66 (3).

ANDERSEN J B O, HEYDORN A, HENTZER M, et al., 2001. Gfp-based N-acyl Homoserine-lactone sensor systems for detection of bacterial communication. Appl. Environ. Microbiol., 67 (2): 575-585.

ANDRADE-EIROA A, CANLE M, LEROY-CANCELLIERI V, et al., 2016. Solid-phase extraction of organic compounds: a critical review. Part ii. TrAC: Trends Anal. Chem., 80: 655-667.

BASSLER B L, GREENBERG E P, STEVENS A M, 1997. Cross-species induction of luminescence in the quorum-sensing bacterium *Vibrio harveyi*. J. Bacteriol., 179 (12): 4043-4045.

BOFINGER M R, DE SOUSA L S, FONTES J E N, et al., 2017. Diketopiperazines as cross-communication quorum-sensing signals between *Cronobacter sakazakii* and *Bacillus cereus*. ACS Omega, 2 (3): 1003-1008.

CAMILLI A, 2006. Bacterial small-molecule signaling pathways. Science, 311 (5764): 1113-1116.

CAMPAGNA S R, GOODING J R, MAY A L, 2009. Direct quantitation of the quorum sensing signal, Autoinducer-2, in clinically relevant samples by liquid chromatography-tandem mass spectrometry. Anal. Chem., 81 (15): 6374-6381.

CATALDI T R I, BIANCO G, FROMMBERGER M, et al., 2004. Direct analysis of selected N-acyl-L-homoserine lactones by gas chromatography/mass spectrometry. Rapid Commun. Mass Spectrom., 18 (12): 1341-1344.

CATALDI T R I, BIANCO G, PALAZZO L, et al., 2007. Occurrence of N-acyl-L-homoserine lactones in extracts of some gram-negative bacteria evaluated by gas chromatography-mass spectrometry. Anal. Biochem., 361 (2): 226-235.

CELIO S, TROXLER H, DURKA S S, et al., 2006. Free 3-nitrotyrosine in exhaled breath condensates of children fails as a marker for oxidative stress in stable cystic fibrosis and asthma. Nitric Oxide Biol. Chem.,

15 (3): 226-232.

CHAN K G, CHENG H J, CHEN J W, et al., 2014. Tandem mass spectrometry detection of quorum sensing activity in multidrug resistant clinical isolate *Acinetobacter baumannii*. Sci. World J., 891041.

CHAN X Y, HOW K Y, YIN W F, et al., 2016. N-acyl Homoserine lactone-mediated quorum sensing in *Aeromonas veronii* biovar *sobria* strain 159: Identification of LuxRI homologs. Front. Cell. Infect. Microbiol., 6: 1-6.

CHARLTON T S, DE NYS R, NETTING A, et al., 2000. A novel and sensitive method for the quantification of N-3-oxoacyl homoserine lactones using gas chromatography-mass spectrometry: application to a model bacterial biofilm. Environ. Microbiol., 2 (5): 530-541.

CHAURAND P, LUETZENKIRCHEN F, SPENGLER B, 1999. Peptide and protein identification by matrix-assisted laser desorption ionization (MALDI) and MALDI-post-source decay time-of-flight mass spectrometry. J. Am. Soc. Mass Spectrom., 10 (2): 91-103.

CHI W, ZHENG L, HE C, et al., 2017. Quorum sensing of microalgae associated marine *Ponticoccus* sp. PD-2 and its algicidal function regulation. AMB Express Springer Berlin Heidelberg, 7 (1).

CURTIS M M, SPERANDIO V, 2011. A complex relationship: the interaction among symbiotic microbes, invading pathogens, and their mammalian host. Mucosal Immunol., 4 (2), 133-138.

CUTIGNANO A, NUZZO G, IANORA A, et al., 2015. Development and application of a novel SPE-method for bioassay-guided fractionation of marine extracts. Mar. Drugs, 13 (9): 5736-5749.

DAVENPORT P W, GRIFFIN J L, WELCH M, 2015. Quorum sensing is accompanied by global metabolic changes in the opportunistic human pathogen *Pseudomonas aeruginosa*. J. Bacteriol., 197 (12): 2072-2082.

DE KEERSMAECKERT S C J, VARSZEGI C, VAN BOXEL N, et al., 2005. Chemical synthesis of (S)-4, 5-dihydroxy-2,3-pentanedione, a bacterial signal molecule precursor, and validation of its activity in *Salmonella typhimurium*. J. Biol. Chem., 280 (20): 19563-19568.

DENG Y, WU J, TAO F, et al., 2011. Listening to a new language: DSF-based quorum sensing in gram-negative bacteria. Chem. Rev., 111 (1): 160-179.

DICKSCHAT J S, 2010. Quorum sensing and bacterial biofilms. Nat. Prod. Rep., 27 (3): 343-369.

DOBERVA M, STIEN D, SORRES J, et al., 2017. Large diversity and original structures of acyl-Homoserine lactones in strain MOLA 401, a marine *Rhodobacteraceae bacterium*. Front. Microbiol., 8: 1-10.

FEKETE A, FROMMBERGER M, ROTHBALLER M, et al., 2007. Identification of bacterial N-acylhomoserine lactones (AHLs) with a combination of ultra-performance liquid chromatography (UPLC), ultra-high-resolution mass spectrometry, and in-situbiosensors. Anal. Bioanal. Chem., 387 (2): 455-467.

FENG H, DING Y, WANG M, et al., 2014. Where are signal molecules likely to be located in anaerobic granular sludge? WaterRes., 50: 1-9.

FROMMBERGER M, SCHMITT-KOPPLIN P, MENZINGER F, et al., 2003. Analysis of N-acyl-L-homoserine lactones produced by *Burkholderia cepacia* with partial filling micellar electrokinetic chromatography-electrospray ionization-ion trap mass spectrometry. Electrophoresis, 24 (17): 3067-3074.

FROMMBERGER M, HERTKORN N, ENGLMANN M, et al., 2005. Analysis of N-acylhomoserine lactones after alkaline hydrolysis and anion-exchange solid-phase extraction by capillary zone electrophoresis-mass spectrometry. Electrophoresis, 26 (7-8): 1523-1532.

FUQUA C, GREENBERG E P, 2002. Listening in on bacteria: acyl-homoserine lactone signalling. Nat. Rev. Mol. CellBiol., 3 (9): 685-695.

FUQUA W C, WINANS S C, GREENBERG E P, 1994. Quorum sensing in Bacteria: the LuxR-LuxI family of cell density-responsive transcriptional regulators. J. Bacteriol., 176 (2): 269-275.

GALLOWAY W R J D, HODGKINSON J T, BOWDEN S D, et al., 2011. Quorum sensing in gram-negative bacteria: small-molecule modulation of AHL and AI-2 quorum sensing pathways. Chem. Rev., 111 (1): 28-67.

GOSWAMI J, 2017. Quorum sensing by super bugs and their resistance to antibiotics, a short review. Global J. Pharm. Pharmaceut. Sci., 3 (3): 1-7.

GOULD T A, HERMAN J, KRANK J, et al., 2006. Specificity of acyl-homoserine lactone synthases examined by mass spectrometry. J. Bacteriol., 188 (2): 773-783.

GU Q, FU L, WANG Y, et al., 2013. Identification and characterization of extracellular cyclic dipeptides as quorum-sensing signal molecules from *Shewanella baltica*, the specific spoilage organism of *Pseudosciaena crocea* during 4℃ storage. J. Agric. Food Chem., 61 (47): 11645-11652.

HAWVER L A, JUNG S A, NG W L, 2016. Specificity and complexity in bacterial quorum-sensing systems. FEMS Microbiol. Rev., 40 (5): 738-752.

HE Y W, ZHANG L H, 2008. Quorum sensing and virulence regulation in *Xanthomonas campestris*. FEMS Microbiol. Rev., 32 (5): 842-857.

HMELO L R, 2017. Quorum sensing in marine microbial environments. Annu. Rev. Mar. Sci., 9 (1): 257-281.

HOLDEN M T G, RAM CHHABRA S, DE NYS R, et al., 1999. Quorum-sensing cross talk: Isolation and chemical characterization of cyclic dipeptides from *Pseudomonas aeruginosa* and other gram-negative bacteria. Mol. Microbiol., 33 (6): 1254-1266.

JIANG J, WU S, WANG J, et al., 2015. AHL-type quorum sensing and its regulation on symplasmata formation in *Pantoea agglomerans* YS19. J. Basic Microbiol., 55 (5): 607-616.

JONES S, YU B, BAINTON N J, et al., 1993. The lux autoinducer regulates the production of exoenzyme virulence determinants in *Erwinia carotovora* and *Pseudomonas aeruginosa*. EMBO J., 12 (6), 2477-2482.

JUNIO H A, TODD D A, ETTEFAGH K A, et al., 2013. Quantitative analysis of autoinducing peptide I (AIP-I) from *Staphylococcus aureus* cultures using ultrahigh performance liquid chromatography-high resolving power mass spectrometry. J. Chromatogr., B 930: 7-12.

KALKUM M, LYON G J, CHAIT B T, 2003. Detection of secreted peptides by using hypothesis-driven multistage mass spectrometry. Proc. Natl. Acad. Sci., 100 (5): 2795-2800.

KIM Y W, SUNG C, LEE S, et al., 2015. MALDI-MS-based quantitative analysis for ketone containing Homoserine lactones in *Pseudomonas aeruginosa*. Anal. Chem., 87 (2), 858-863.

KLEEREBEZEM M, QUADRI L E, KUIPERS O P, et al., 1997. Quorum sensing by peptide pheromones and two-component signal-transduction systems in gram-positive bacteria. Mol. Microbiol., 24 (5): 895-904.

KUMARI A, PASINI P, DAUNERT S, 2008. Detection of bacterial quorum sensing N-acyl homoserine lactones in clinical samples. Anal. Bioanal. Chem., 391 (5): 1619-1627.

LEIPERT J, BOBIS I, SCHUBERT S, et al., 2018. Miniaturized dispersive liquid-liquid microextraction and MALDI MS using ionic liquid matrices for the detection of bacterial communication molecules and virulence factors. Anal. Bioanal. Chem., 410: 4337-4748.

LEPINE F, MILOT S, DEZIEL E, et al., 2004. Electrospray/mass spectrometric identification and analysis of 4-hydroxy-2-alkylquinolines (HAQs) produced by *Pseudomonas aeruginosa*. J. Am. Soc. Mass Spectrom., 15 (6): 862-869.

LI X, FEKETE A, ENGLMANN M, et al., 2006. Development and application of a method for the analysis of

N-acylhomoserine lactones by solid-phase extraction and ultrahigh pressure liquid chromatography. J. Chromatogr. A, 1134 (1-2): 186-193.

LI X, FEKETE A, ENGLMANN M, et al., 2007. At-line coupling of UPLC to chip-electrospray-FTICR-MS. Anal. Bioanal. Chem., 389 (5): 1439-1446.

MALIK A K, FEKETE A, GEBEFUEGI I, et al., 2009. Single drop microextraction of homoserine lactones based quorum sensing signal molecules, and the separation of their enantiomers using gas chromatography mass spectrometry in the presence of biological matrices. Microchim. Acta, 166 (1-2): 101-107.

MCCLEAN K H, WINSON M K, FISH L, et al., 1997. Quorum sensing and *Chrornobacteriurn violaceum*: exploitation of violacein production and inhibition for the detection of N-acyl homoserine lactones. Microbiology, 143: 3703-3711.

MCNERNEY M P, STYCZYNSKI M P, 2018. Small molecule signaling, regulation, and potential applications in cellular therapeutics. Wiley interdisciplinary reviews: systems. Biol. Med., 10 (2): e1405.

MORIN D, GRASLAND B, VALLEE-REHEL K, et al., 2003. On-line high-performance liquid chromatography-mass spectrometric detection and quantification of N-acylhomoserine lactones, quorum sensing signal molecules, in the presence of biological matrices. J. Chromatogr., A 1002 (102): 79-92.

OLSON M E, TODD D A, SCHAEFFER C R, et al., 2014. *Staphylococcus epidermidis* agr quorum-sensing system: signal identification, cross talk, and importance in colonization. J. Bacteriol., 196 (19): 3482-3493.

ORTORI C A, ATKINSON S, CHHABRA S R, et al., 2007. Comprehensive profiling of N-acylhomoserine lactones produced by *Yersinia pseudotuberculosis* using liquid chromatography coupled to hybrid quadrupole-linear ion trap mass spectrometry. Anal. Bioanal. Chem., 387 (2): 497-511.

ORTORI C A, DUBERN J F, CHHABRA S R, et al., 2011. Simultaneous quantitative profiling of N-acyl-L-homoserine lactone and 2-alkyl-4 (1H) -quinolone families of quorumsensing signaling molecules using LC-MS/MS. Anal. Bioanal. Chem., 399 (2): 839-850.

OSORNO O, CASTELLANOS L A F, ARVALO FERRO C, et al., 2012. Gas Chromathography as a tool in quorum sensing studies, in Salih, B. (ed.) Gas Chromatography—Biochemicals, Narcotics and Essential Oils. InTech, 67-96.

PAPENFORT K, BASSLER B L, 2016. Quorum sensing signal-response systems in gram-negative bacteria. Nat. Rev. Microbiol., 14 (9): 576-588.

PEARSON J P, GRAY K M, PASSADOR L, et al., 1994. Structure of the autoinducer required for expression of *Pseudomonas aeruginosa* virulence genes. Proc. Natl. Acad. Sci., 91 (1): 197-201.

PESCI E C MILBANK J B J, PEARSON, J P, et al., 1999. Quinolone signaling in the cell-to-cell communication system of *Pseudomonas aeruginosa*. Proc. Natl. Acad. Sci., 96 (20): 11229-11234.

POMINI A M, MARSAIOLI A J, 2008. Absolute configuration and antimicrobial activity of acylhomoserine lactones. J. Nat. Prod., 71 (6): 1032-1036.

RAN T, ZHOU C, XU L, et al., 2016. Initial detection of the quorum sensing autoinducer activity in the rumen of goats in vivo and in vitro. J. Integr. Agric., 15 (10): 2343-2352.

RANI S, KUMAR A, MALIK A K, et al., 2011. Occurrence of N-acyl Homoserine lactones in extracts of bacterial strain of *Pseudomonas aeruginos* a and in sputum sample evaluated by gas chromatography-mass spectrometry. Am. J. Anal. Chem., 2 (2): 294-302.

READING N C, SPERANDIO V, 2006. Quorum sensing: the many languages of bacteria. FEMS Microbiol. Lett., 254 (1): 1-11.

READING N C, TORRES A G, KENDALL M M, et al., 2007. A novel twocomponent signaling system that activates transcription of an enterohemorrhagic *Escherichia coli* effector involved in remodeling of host actin. J. Bacteriol., 189 (6): 2468-2476.

SCHAEFER A L, GREENBERG E P, PARSEK M R, 2001. Acylated homoserine lactone detection in *Pseudomonas aeruginosa* biofilms by radiolabel assay. In: Methods in Enzymology. Academic Press, pp: 41-47.

SCHAEFER A L, TAYLOR T A, BEATTY J T, et al., 2002. Long-chain acyl-Homoserine lactone quorum-sensing regulation of *Rhodobacter capsulatus* gene transfer agent production. J. Bacteriol., 184 (23): 6515-6521.

SCHAUDER S, SHOKAT K, SURETTE M G, et al., 2001. The LuxS family of bacterial autoinducers: Biosynthesis of a novel quorum-sensing signal molecule. Mol. Microbiol., 41 (2): 463-476.

SCHULZ S, 2001. Composition of the silk lipids of the spider Nephila clavipes. Lipids, 36 (6): 637-647.

SCHUPP P J, CHARLTON T S, TAYLOR M W, et al., 2005. Use of solid-phase extraction to enable enhanced detection of acyl homoserine lactones (AHLs) in environmental samples. Anal. Bioanal. Chem., 383 (1): 132-137.

SHAW P D, PING G, DALY S L, et al., 1997. Detecting and characterizing N-acyl-homoserine lactone signal molecules by thin-layer chromatography. Proc. Natl. Acad. Sci., 94 (12): 6036-6041.

SHEN Q, GAO J, LIU J, et al., 2016. A new acyl-homoserine lactone molecule generated by *Nitrobacter winogradskyi*. Sci. Rep., 6 (1): 22903.

SIFRI C D, 2008. Healthcare epidemiology: Quorum sensing: Bacteria talk sense. Clin. Infect. Dis., 47 (8): 1070-1076.

SITNIKOV D G, MONNIN C S, VUCKOVIC D, 2016. Systematic assessment of seven solvent and solid-phase extraction methods for metabolomics analysis of human plasma by LC-MS. Sci. Rep., 6: 1-11.

SPERANDIO V, TORRES A G, JARVIS B, et al., 2003. Bacteria-host communication: The language of hormones. Proc. Natl. Acad. Sci., 100 (15): 8951-8956.

STEINDLER L, VENTURI V, 2007. Detection of quorum-sensing N-acyl homoserine lactone signal molecules by bacterial biosensors. FEMS Microbiol. Lett., 266 (1): 1-9.

STURME M H J, KLEEREBEZEM M, NAKAYAMA J, et al., 2002. Cell to cell communication by autoinducing peptides in gram-positive bacteria. Antonie Van Leeuwenhoek, 81 (1-4): 233-243.

SUN Y, HE K, YIN Q, et al., 2018. Determination of quorum-sensing signal substances in water and solid phases of activated sludge systems using liquid chromatography-mass spectrometry. J. Environ. Sci., 69: 85-94.

TAGA M E, BASSLER B L, 2003. Chemical communication among bacteria. Proc. Natl. Acad. Sci., 100 (Suppl. 2): 14549-14554.

TAN C H, KOH K S, XIE C, et al., 2014. The role of quorum sensing signalling in EPS production and the assembly of a sludge community into aerobic granules. ISMEJ., 8 (6): 1186-1197.

TANG X, GUO Y, WU S, et al., 2018. Metabolomics uncovers the regulatory pathway of acylhomoserine lactones-based quorum sensing in anammox consortia. Environ. Sci. Technol., 52 (4): 2206-2216.

TAYLOR G W, MACHAN Z A, MEHMET S, et al., 1995. Rapid identification of 4-hydroxy-2-alkylquinolines produced by *Pseudomonas aeruginosa* using gas chromatography-electroncapture mass spectrometry. J. Chromatogr. B Biomed. Sci. Appl., 664 (2): 458-462.

THIEL V, KUNZE B, VERMA P, et al., 2009a. New structural variants of homoserine lactones in bacteria. Chembiochem, 10 (11): 1861-1868.

THIEL V, VILCHEZ R, SZTAJER H, et al., 2009b. Identification, quantification, and determination of the absolute configuration of the bacterial quorum-sensing signal autoinducer-2 by gas chromatography-mass spectrometry. Chembiochem, 10 (3): 479-485.

TODD D A, ZICH D B, ETTEFAGH K A, et al., 2016. Hybrid quadrupole-Orbitrap mass spectrometry for quantitative measurement of quorum sensing inhibition. J. Microbiol. Methods, 127: 89-94 (d).

TOMMONARO G, ABBAMONDI G R, IODICE C, et al., 2012. Diketopiperazines produced by the halophilic archaeon, *Haloterrigena hispanica*, activate AHL bioreporters. Microb. Ecol., 63 (3): 490-495.

VENDEVILLE A, WINZER K, HEURLIER K, et al., 2005. Making "sense" of metabolism: Autoinducer-2, LuxS and pathogenic bacteria. Nat. Rev. Microbiol., 3 (5): 383-396.

VERBEKE F, DE CRAEMER S, DEBUNNE N, et al., 2017. Peptides as quorum sensing molecules: measurement techniques and obtained levels in vitro and in vivo. Front. Neurosci., 11: 1-18.

VIAL L, LEPINE F, MILOT S, et al., 2008. *Burkholderia pseudomallei*, *B. thailandensis*, and *B. ambifaria* produce 4-hydroxy-2-alkylquinoline analogues with a methyl group at the 3 position that is required for quorum-sensing regulation. J. Bacteriol., 190 (15): 5339-5352.

WAGNER-DOBLER I, THIEL V, EBERL L, et al., 2005. Discovery of complex mixtures of novel long-chain quorum sensing signals in free-living and host-associated marine Alphaproteobacteria. Chembiochem, 6 (12): 2195-2206.

WANG J, DING L, LI K, et al., 2017. Development of an extraction method and LC-MS analysis for N-acylated-L-homoserine lactones (AHLs) in wastewater treatment biofilms. J. Chromatogr. B Anal. Technol. Biomed. Life Sci., 1041-1042: 37-44.

WATERS C M, BASSLER B L, 2005. Quorum sensing: cell-to-cell communication in Bacteria. Annu. Rev. Cell Dev. Biol., 21 (1): 319-346.

WEN K Y, CAMERON L, CHAPPELL J, et al., 2017. A cell-free biosensor for detecting quorum sensing molecules in *P. aeruginosa*-infected respiratory samples. ACS Synth. Biol., 6 (12): 2293-2301.

WILLIAMS P, 2007. Quorum sensing, communication and cross-kingdom signalling in the bacterial world. Microbiology, 153 (12): 3923-3938.

WINANS S C, 2011. A new family of quorum sensing pheromones synthesized using S-adenosylmethionine and acyl-CoAs. Mol. Microbiol., 79 (6): 1403-1406.

XU F, SONG X, CAI P, et al., 2017. Quantitative determination of AI-2 quorum-sensing signal of bacteria using high performance liquid chromatography-tandem mass spectrometry. J. Environ. Sci., 52: 204-209.

YANG Y H, LEE T H, KIM J H, et al., 2006. High-throughput detection method of quorum-sensing molecules by colorimetry and its applications. Anal. Biochem., 356 (2): 297-299.

ZABET-MOGHADDAM M, HEINZLE E, THOLEY A, 2004. Qualitative and quantitative analysis of low molecular weight compounds by ultraviolet matrix-assisted laser desorption/ionization mass spectrometry using ionic liquid matrices. Rapid Commun. Mass Spectrom., 18 (2): 141-148.

ZHANG S, LIU J, CHEN Y, et al., 2010. A novel strategy for MALDI-TOF MS analysis of small molecules. J. Am. Soc. Mass Spectrom., 21 (1): 154-160.

第3章　海洋生物膜和环境中的群体感应

Raphaël Lami

Sorbonne Université, CNRS, Laboratoire de Biodiversité et Biotechnologies Microbiennes, LBBM, Banyuls-sur-Mer, France

1 前言

在海洋水域中群体感应（QS）被猜测能引起壮观现象。例如，在 15 400 km^2 的阿拉伯海中令人印象深刻的"奶海"被归因于发光的哈维弧菌。猜测这些细胞通过表达 QS 发光和繁殖从而对浮游植物水华做出响应（Miller et al., 2005）。然而，自 20 世纪 70 年代发现此现象以来，QS 不仅在这种壮观的生物发光中得到证明，而且在许多不同类型的细菌活动和系统发育多样化的海洋细胞中均得到证实。本章对海洋生物膜和环境中群体感应机制进行了综述。对海洋水域保持特别的关注，对涉及海洋群体感应化合物的广泛多样性，群体感应的广泛生物功能以及在这些过程中发生的大量生态位进行综述。本章的最后一部分将介绍应用，以海洋环境作为群体感应抑制化合物和酶的来源或专注于利用群体淬灭策略来操纵海洋细菌生理。

2 海洋环境中群体感应的发现和日益增长的研究兴趣

2.1 20世纪70年代：在海洋环境中发现群体感应——弧菌-乌贼模型

20 世纪 90 年代，群体感应的概念被提出，是指基于种群密度的细菌细胞的生理反应（Fuqua et al., 1994）。然而，这一概念得到详细阐述的大部分观察结果都是由 20 世纪 70 年代的海洋科学家在试验中获得的。在这十年中，人们收集了大量关于费氏弧菌菌株的数据，这些菌株定殖在夏威夷短尾乌贼的发光器官上并产生生物发光（Greenberg et al., 1979; Nealson et al., 1970）。特别地，在这个共生细菌群落中最初发现了密度依赖的表型。在周围海水中，这些细胞自由生活且稀缺时并不发光。然而，当它们达到高浓度时，类似于实验室中培养或当定殖于乌贼的光器官上时，它们才能够发光。

有趣的是，乌贼中费氏弧菌细胞丰度遵循昼夜节律模式。在晚上，费氏弧菌以高浓度（$10^{10} \sim 10^{11}$ 个细胞/mL）存在并释放出一种可扩散因子，被称为自诱导物（AI）与发光有关。在夜晚，大多数细菌细胞从光器官中排出，导致细菌浓度和可扩散因子非常低。在白天，未被排出的费氏弧菌的浓度非常低，不能产生扩散因子，乌贼不发光。然而，乌贼体内剩余的弧菌种群在有利条件下持续稳定增长，并在夜晚再次达到足以产生生物发光的细胞丰度。这种细菌-乌贼共生关系构成了一种细菌-动物共生关系。乌贼依赖弧菌的光来逃避捕食者或捕食猎物。作为回报，乌贼为弧菌提供宿主和营养物质（Graf and Ruby, 1998）。

自 20 世纪 70 年代第一次观察以来，这种原始的生物发光调节系统已在化学和遗传学上得到了完整的描述。在 1981 年，该扩散信号被鉴定为一种酰基高丝氨酸内酯（AHL），并被描述为 3-氧代-己酰基高丝氨酸内酯（3-oxo-C6-HSL）（Eberhard et al., 1981）。涉及这种现象的基因簇被描述为一个具有 8 个基因的双向转录操纵子，分别命名为 *luxA-E*、*luxG*、*luxI* 和 *luxR*。LuxA 和 LuxB 蛋白是荧光素酶的两个亚基，荧光素酶负责光的产生。Lux C-D-E 蛋白参与荧光素酶底物的合成，而 LuxG 是一种黄素还原酶。然而，在群体感应研究中，大多数人的兴趣集中在 LuxI 和 LuxR 蛋白上。LuxI 是负责 AIs 产生的 AI 合成酶，而 LuxR 是这种扩散信号的受体。当 AIs 在细菌细胞附近环境中达到阈值浓度时（反映了细胞丰度的增加），它们与 LuxR 受体结合，该受体作为转录因子激活所有 *lux* 基因的表达。这种可扩散信号被指定为 AI，因为其通过 *luxI* 的自诱导来促进自身的产生（图 3-1）（Eberhard et al., 1981；Engebrecht et al., 1983）。

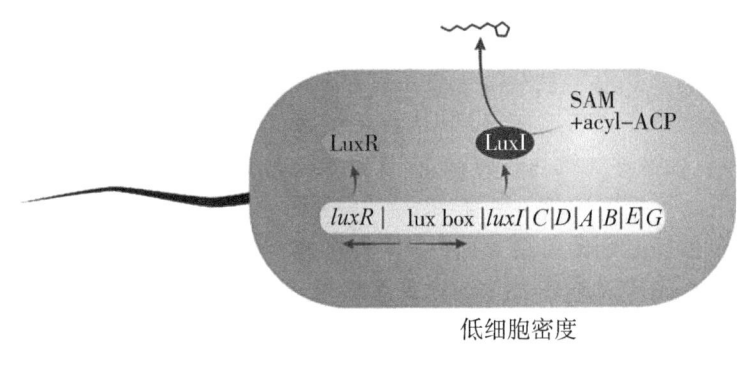

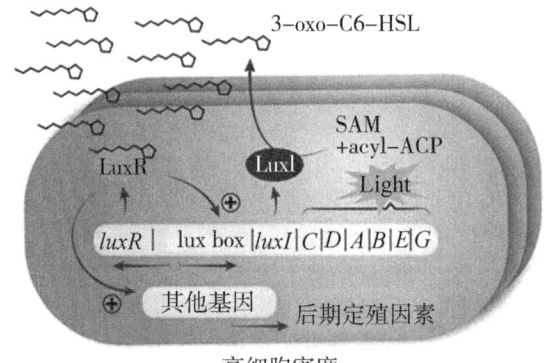

图 3-1 在模式物种费氏弧菌中首次发现基于 **luxI/luxR** 的群体
感应系统的示意图，可产生 **3-oxo-C6-HSL**

2.2 20 世纪 90 年代至 21 世纪 10 年代：对海洋环境中群体感应研究兴趣的日益增长

在这些最初发现和随后对群体感应遗传系统的全面阐明之后，10 多年来，对这一机制的研究并未引起科学界的兴趣。可能是由于群体感应在当时似乎是专门用于调控生物发光。20 世纪 90 年代，随着 DNA 测序方法的发展，在许多不同类型的细菌中发现了大量的 *luxI*

和 luxR 同源物，这重新引起了研究兴趣。渐渐地，似乎对费氏弧菌中 luxI-luxR 模型的开发与细菌菌株的多样性有关。这些观察结果促成了 1994 年群体感应概念的建立（Fuqua et al., 1994）。

然而，尽管对环境中弧菌-乌贼模型进行了完整的描述，但在群体感应领域的大部分科研工作都是在 20 世纪 90 年代提出，集中在医学或农学领域的菌株上。在 2005 年之前，研究人员很少在环境科学领域关注这些机制。在医学领域产生这种兴趣的一个重要原因是，在 20 世纪 90 年代，越来越多的关于葡萄球菌株和铜绿假单胞菌等致病菌中毒力和群体感应之间建立了越来越多的联系。在接下来的 10 年里，从海洋中分离出来的细菌成为环境科学领域关注的细菌。1998 年，首次报道了在自然环境水生生物中存在 AIs 的研究，标题为"群体感应自诱导物：它们在自然环境中起到什么作用？"这揭示了对自然水生生物膜的早期兴趣（McLean et al., 1997）。2001 年，提出群体感应可能在海洋颗粒附着的细菌中起作用的假设（Kiørboe, 2001）。2002 年，Gram 等首次报道了从海洋雪中分离出的玫瑰杆菌属（*Roseobacter*）和海杆菌属（*Marinobacter*）菌株中 AHLs 的产生。从那时起，越来越多的报告集中在海洋细菌群体感应的性质和作用，以及大量依赖培养（Rasmussen et al., 2014；Wagner-Döbler et al., 2005）和独立培养（Doberva et al., 2015；Muras et al., 2018a）的研究，重点强调了群体感应机制在海洋生物膜和环境中的重要性。

3 概述海洋环境和生物膜中参与群体感应的原核生物多样性和化合物

3.1 海洋环境中 AIs 特征分析的试验方法

从海洋环境中描述 AIs 的一个重要困难是其浓度很低，仅达到"pmol"水平。只有在少数情况下，研究人员能在原位采集的样品上直接鉴定出 AIs，如海洋雪（Jatt et al., 2015）、浮游植物细胞的藻类区（Bachofen and Schenk, 1998；Van Mooy et al., 2012）、附着有刺胞动物的黏液层（Ransome et al., 2014）或微生物垫（Decho et al., 2009）。然而，直接测量 AIs 浓度的方法仍然很少，通常采用间接检测方法，例如，Gram 等（2002）在一篇开创性论文中所描述的方法。在这个工作流程中，第一步是从所研究的环境中分离细菌菌株，如大量繁殖的浮游植物（Bachofen and Schenk, 1998）、藻类培养物、微生物垫或海洋雪（Gram et al., 2002；Schaefer et al., 2008；Wagner-Döbler et al., 2005）。然后，将分离细胞的上清液加入全细胞生物传感器的培养基中，例如，大肠杆菌 JB523、紫色色素杆菌 CV026 或哈维弧菌 JMH612。这些生物传感器非常多样化，在检测各种类型的 AHLs 时具有很大的敏感度差异。它们是转基因生物，能在 AIs 存在时产生信号，并最终在筛选菌株的上清液中释放出来。例如，在 AHLs 存在时，紫色色素杆菌产生紫色色素和大肠杆菌 JB523 发出绿色荧光信号（Steindler and Venturi, 2006）。

群体感应化合物特征下一步依赖于天然物质化学领域中所使用的工具。开创性的研究采用薄层液相色谱法（TLC）（Gram et al., 2002）。然而，最近的方法通常是基于液相色谱与质谱联用（LC-MS）（Schaefer et al., 2008）、气相色谱与质谱联用（GC-MS）（Wagner-Döbler et al., 2005）和 MS/MS 方法（Van Mooy et al., 2012）。某些情况下，在分析之前先进行微分馏步骤，这使得提取的化合物得到更好的分离和浓缩（Doberva et al., 2017）。当关注 AHLs 特征时，可以通过使用二甲基二硫醚的额外衍生化步骤来确定 AHLs 侧链上双键的位置（Neumann et al., 2013）。有时可以通过 1D 和 2D 核磁共振分析 AHLs 的确定性特征来实现，这取决于目标化合物的纯度和浓度是否足以允许进行此类分析。

3.2 AHLs 或自诱导物-1（AI-1）

在海洋环境中，AHLs 的产生主要来自革兰氏阴性细菌，并且存在于许多不同类型的海洋 α、β 和 γ-变形菌中。AI-1 是由 LuxI 家族酶产生的，并由 LuxR 家族受体检测到（Engebrecht et al., 1983）。此外，一些 AHLs 是由 *ainS* 类基因产生（Gilson et al., 1995），在某些情况下也可通过 *hdtS* 类基因产生（Laue et al., 2000；Rivas et al., 2007）。AHLs 由 *S*-腺苷甲硫氨酸和脂肪酸残基形成，作为 AIs 合成酶的底物。此外，已有研究证实一些细菌携带 *luxR* 孤儿基因，这意味着这些生物能够在海洋环境中捕获 AHLs 而不需要自身合成它们。因此，它们可以节省生产 AHL 的能量成本，同时"感知"附近环境中的化学对话，利用这种"窃听"系统来调节其生理机能。此外，在一些海洋细菌基因组中也发现一些"*luxI* 孤儿基因"（Cude and Buchan, 2013）。然而，在海洋水域中，LuxI 和 LuxR 孤体的生态作用和重要性仍需进一步研究。

AHLs 是由一个内酯环连接一个脂肪酸残基（酰基侧链），并与酰胺结合组成（图 3-2）。这些 AHLs 表现出许多不同长度的结构变体，在酰基侧链上含有 4~19 个碳（Doberva et al., 2017），可以是饱和的或不饱和的。这些化合物在 C3 位置（氢、羟基或酮基）上的取代也不同，有时是在酰基侧链上的其他碳原子上。此外，一些具有支链酰基侧链的 AHLs 已被描述，但据我们所知，目前在海洋菌株中尚未发现。其他研究者也报道了带有芳香酸残基的侧链（对香豆酸或肉桂酸）的 AHLs。例如，对香豆酰-HSL 的情况，该化合物在沼泽红假单胞菌（*Rhodopseudomonas palustris*）中被发现，同时在海洋细菌波美罗氏硅酸杆菌（*Silicibacter pomeroyi*）DSS-3 中也被发现（Schaefer et al., 2008）。

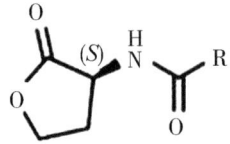

图 3-2 AHL 的一般化学结构

注：内酯环部分通过酰胺结合与脂肪酸残基（R）连接。
酰基侧链在长度、氧化状态和氢、羟基或羰基的存在方面是高
度可变的；这种可变性为 AHL 信号提供了其特异性。

大多数红杆菌科（Rhodobacteraceae）物种（一种涉及群体感应的主要海洋细菌群）能产生额外修饰的长链 AHLs（图 3-3）（Cude and Buchan, 2013）。在系统发育上，球形红杆菌（*Rhodobacter sphaeroides*）与各种海洋菌株非常接近，能合成 C14:1-HSL。海洋什叶派二硝酸杆菌释放 C18:2-HSL、C18:1-HSL 以及微量的 C16-HSL、C15-HSL 和 C14-HSL（Neumann et al., 2013；Patzelt et al., 2013；Wagner-Döbler et al., 2005）。海绵共生体鲁杰氏菌属（*Ruegeria* sp.）释放 3-OH-C14-HSL、3-OH-C14:1-HSL 和 3-OH-C12-HSL（Zan et al., 2012）。最近的研究表明，利用 UHPLC-HRMS/MS 方法揭示了红杆菌科菌株中 AHL 化合物的广泛多样性。MOLA401 菌株（归类为 *Palleronia rufa*；

与 Barnier C 个人通信交流手稿）可释放出大量多样的长链 AHLs，且包括 20 种不同假定类型的化合物，其中包括一种具有 19 个碳原子的酰基侧链（Doberva et al.，2017）。来自角质海绵的副球菌属 Ss63 释放出长链 AHLs，包括 12 种饱和的推测物和 4 种不饱和的推测物（Saurav et al.，2016b）。显然，在海洋红杆菌科菌株中 AHL 多样性的真实程度仍然有待解决，仍需进一步研究。

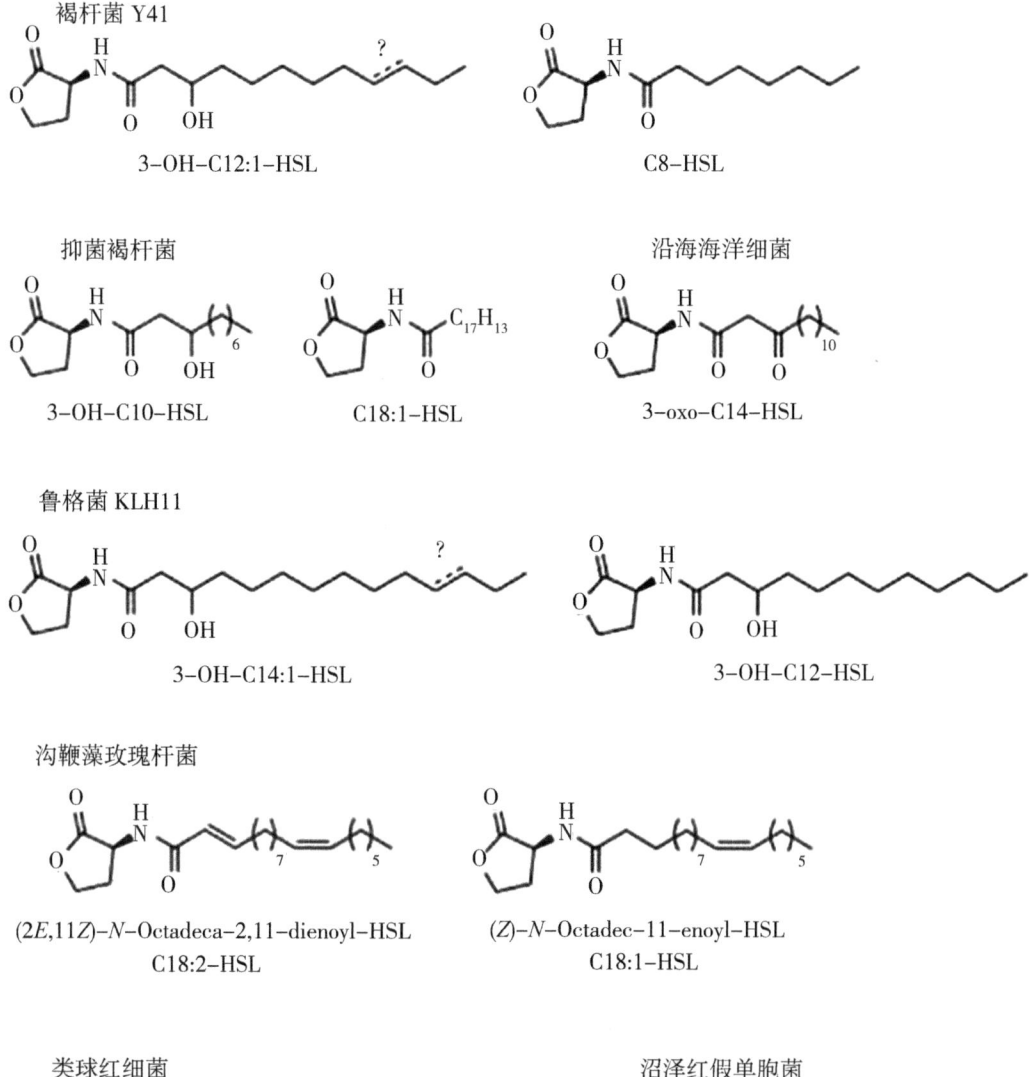

图 3-3　海洋红细菌科 AHL 多样性研究汇总

弧菌科菌株存在于许多类型的海洋环境和寄主中，常作为鱼类、软体动物和珊瑚的病原体，例如，鳗弧菌、创伤弧菌（*V. vulnificus*）、哈维弧菌。同时，它们也可作为乌贼的共生体，如费氏另类弧菌、波美罗氏弧菌（*V. aesturianus*）。第一个群体感应信号是在海洋乌贼相关的费氏弧菌中发现的（请参见本章前面的内容）。自从这一发现以来，一些弧菌已被充分描述作为群体感应研究的原核模型。然而，不同弧菌物种间 AHL 信号多样性的程度还远未被充分探索。对弧菌菌株进行收集筛选用于检测 AHL 的产生，发现有 9%~85% 的菌株在这些生物测试中呈阳性（Garcia-Aljaro et al., 2012；Girard et al., 2017；Purohit et al., 2013），这使得确定 AHL 类型在该属中的真实分布程度变得困难。然而，一些研究试图在不同弧菌菌株中描绘 AHL 的化学多样性。一般来说，在不同的弧菌菌株中 AHL 的酰基侧链长度比在红杆菌科菌株中检测到的要短：为 C4-HSL 至 C12-HSL（图 3-4）（Purohit et al., 2013；Rasmussen et al., 2014），在塔氏弧菌（*V. tasmaniensis*）LGP32 中为 C10-HSL 至 C14-HSL（Girard et al., 2017）。

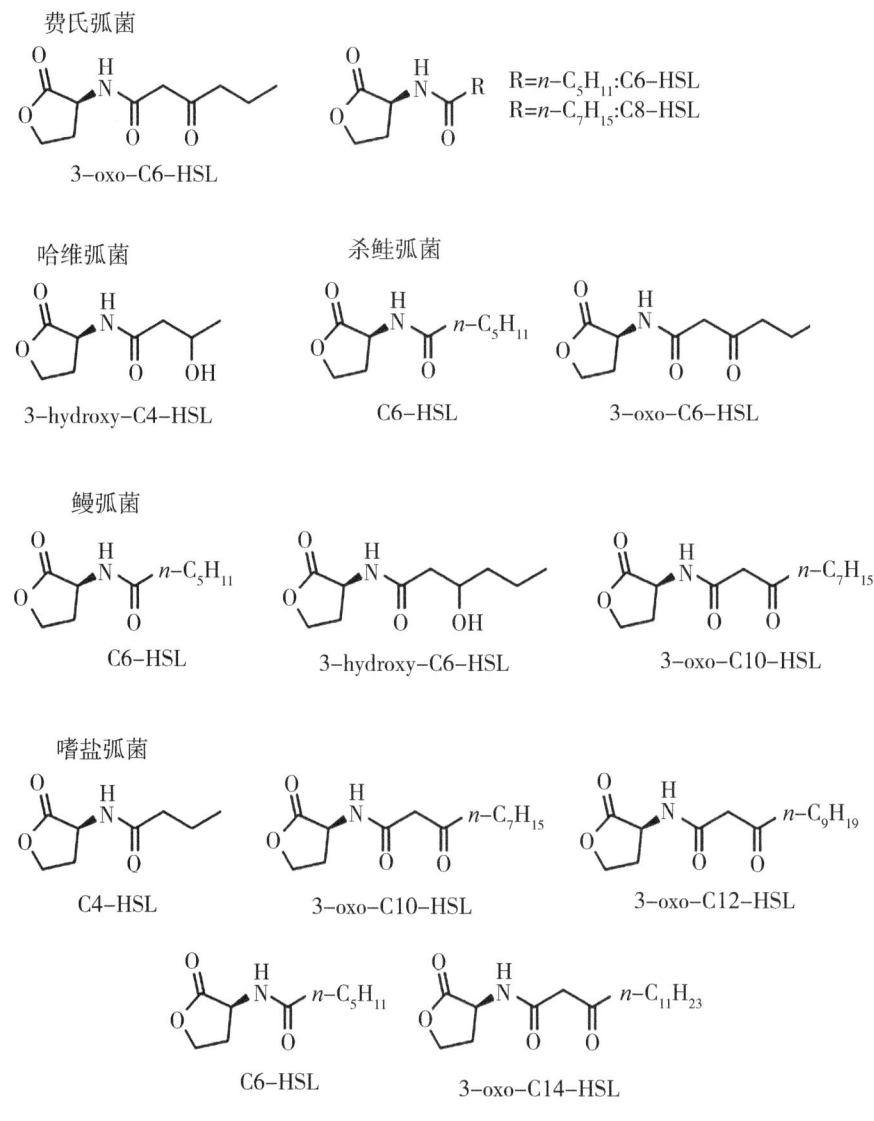

图 3-4　海洋红杆菌科 AHL 多样性研究汇总

在一些海洋中慢生根瘤菌（*Mesorhizobium*）（Krick et al., 2007）、拟杆菌门（Bacteroidetes）（Huang et al., 2008）和蓝细菌（*Cyanobacteria*）（Sharif et al., 2008）中发现了一种基于 AHL 的群体感应系统。此外，在水生古菌（Archaea）（Biswa and Doble, 2013）以及海洋革兰氏阳性细菌微杆菌（*Exiguobacterium*）（Paggi et al., 2003）中也发现了一些产生 AHL 的证据。相比之下，一些重要的海洋细菌类群似乎不产生 AHLs，如主导许多类型的海洋原核生物群落的 SAR11 成员［R. Lami，基于生物传感器恶臭假单胞菌（*P. putida*）F117 和大肠杆菌 MT102 未发表的数据］。

3.3 自诱导物-2（AI-2）

4,5-二羟基-2,3-戊二酮（DPD）是 AI-2 的前体，由 LuxS 酶合成。LuxS 酶将 S-核糖基同型半胱氨酸转化为同型半胱氨酸和 DPD，催化 AI-2 生物合成的关键步骤。AI-2 存在于许多不同类型的革兰氏阴性和革兰氏阳性细菌中；因此，AI 可能作为一种种间信号（Surette et al., 1999）。在硼酸的存在下，DPD 会生成(2S,4S)-2 甲基-2,3,3,4-四羟基四氢呋喃硼酸酯（S-THMF-硼酸酯）。在海洋环境中，这种复合物在弧菌中被用作 AI-2（Chen et al., 2002）。在没有硼的情况下，DPD 会产生(2R,4S)-2-甲基-2,3,3,4-四羟基四氢呋喃（R-THMF），这是肠道细菌中的 AI-2 信号化合物。局部环境中硼的存在与否可能会导致两种 AI-2 形式之间的自发非酶催化重排（Miller et al., 2004）。然而，一些研究人员对 AI-2 的信号功能提出了质疑，因为 DPD 也是甲基循环激活的副产物（Rezzonico and Duffy, 2008）。因此，AI-2 同时是一种信号、一个提示或一种代谢信物，这取决于产生或接收它的细菌物种。然而，产生的代谢物如何演变成一个信号仍然是化学通信领域应该解决的一个重要且开放的问题（Whiteley et al., 2017）。

在海洋环境中，通过对自然样本的直接测量已证实 DPD 的存在（Van Mooy et al., 2012）。基于培养的研究证实 AI-2 信号经常被弧菌所利用（Yang et al., 2011）。通过对全球海洋数据库中 *luxS* 翻译序列的系统发育多样性的研究表明，在海洋环境中希瓦氏菌属（*Shewanella*）相关物种构成了使用 AI-2 信号的一个主要类群（Doberva et al., 2015），在培养的菌株中也能观察到（Bodor et al., 2008）。嗜无机物硫卵菌（*Sulfurovum lithotrophicum*）和中嗜热硫氧化菌（*Caminibacter mediatlanticus*）是 ε-变形菌门，且 AI-2 是在深海热液环境中定殖的生物膜上产生的。这些菌株被发现表达基于 AI-2 的群体感应，并且直接从 RNA 提取物和随后对深海生物膜进行 RT-qPCR 中也发现了 *luxS* 转录本（Perez-Rodriguez et al., 2015）。尽管研究较少，但在海洋环境中涉及 AI-2 信号的细菌多样性的真实程度，以及 AI 的生态作用，仍需要进一步的研究。

3.4 其他 AIs

在海洋环境中，哈维弧菌合成 AI(Z)-3-氨基-2-en4-1。事实上，在霍乱弧菌中首次发现了一个在化学上非常相似的化合物，其合成(S)-3-羟基三聚糖 4-1，因此，给这个 AIs 家族命名为 CAI-1 或 Cholera AI-1（图 3-5）。合成 CAI-1 的酶属于 Cqs A 家族，存在于所有弧菌中。CAI-1 在天然海洋生物群落中是否发挥生态作用有待在未来的试验中进行评估（Higgins et al., 2007; Kelly et al., 2009; Ng et al., 2011; Wei et al., 2011）。

在海洋菌株中还发现了许多其他的 AIs，但数据尚未发表。3,5-二甲基吡嗪-2-ol（DPO）已被证明是霍乱弧菌中的一种 AI（图 3-5），且激活编码小 RNAs 的 *vqmR* 的表达

(Papenfort et al., 2017)。此外,对二硫代酸(tropodithietic acid,TDA)已被描述为红杆菌科中的一种 AI(Geng and Belas, 2010)。总的来说,这些观察结果清楚地表明,在海洋环境中,群体感应 AIs 不仅限于 AI-1 和 AI-2 家族。群体感应化学物质可能有更广泛多样性的信息仍有待发现,在与海洋微生物群落动态中可能发挥重要的生态作用。

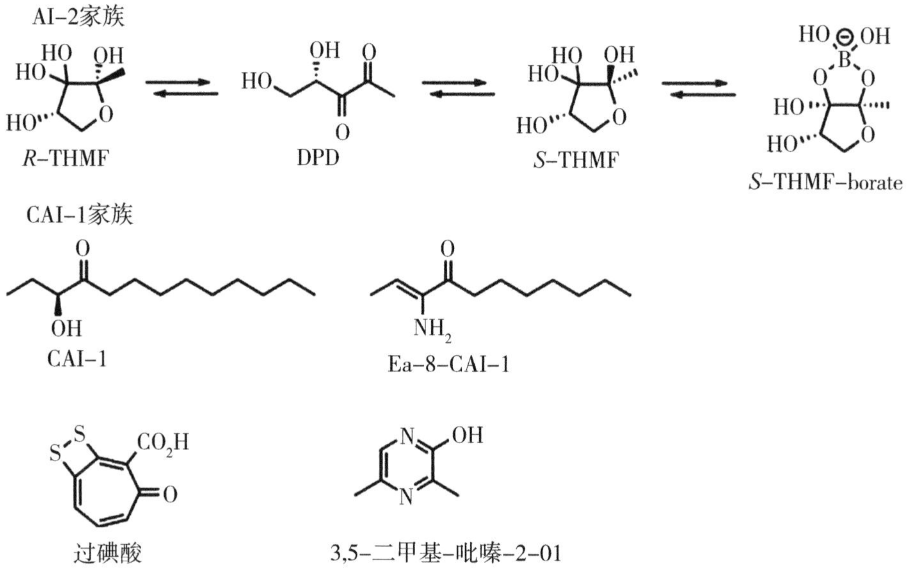

图 3-5　AI-2 和 CAI-1 自诱导物家族及自诱导物 TDA 和 DPO 的结构

3.5　AIs 在海洋环境中的扩散

海洋细菌一旦产生 AIs,就会被动地通过海洋环境扩散到达目标。根据热力学和化学特性,如溶解性、扩散性、分散势或化学稳定性(Harder et al., 2014),这些信号似乎不能扩散超过 10~100 μm 的"召唤距离"(Gantner et al., 2006)。根据菲克定律,短链 AHLs 比长链 AHLs 扩散更快。AHLs 的分子量和溶解性与其迁移率呈正相关,但其他特性可改变一般规律。一方面,酰基侧链上羟基或氧取代基的存在能增加其溶解性,从而增加其被动扩散的能力(Doberva et al., 2017)。另一方面,预测具有高分子量的长链 AHLs 也可被吸附在有机表面上(Decho et al., 2011;Harder et al., 2014)。

要区分 AHL 从一种细菌到另一种细菌的扩散过程,必须考虑许多其他因素。非生物过程可导致内酯环的快速碱基水解,特别是在碱性条件下,并将 AHLs 转化为无活性的 γ-羟基羧酸盐。然而,这一过程在酸性条件下是可逆的(Tait et al., 2005;Yates et al., 2002)。此外,值得注意的是,在其 3-碳上存在氧取代的 AHLs 可能会自发地发生克莱森缩合反应,形成四胺酸,从而抑制它们传递信息的能力(Kaufmann et al., 2005)。3-oxo-C6-HSL 的半衰期以天为单位根据以下公式测定:$1/(1\times10^7\times[OH^-])$(Schaefer et al., 2000)。然而,在人工海水中,3-oxo-C6-HSL 的降解速率要慢得多(根据公式为 0.094 vs 0.26 h^{-1}),表明 AHLs 比最初假设可能扩散更长的距离(Hmelo and Van Mooy, 2009)。在同一份报告中,提出了比以前发表的未取代 AHLs 降解速率更慢的情况。此外,一些研究指出,短链 AHLs 的半衰期比长链 AHLs 短,这表明长链 AHLs 可能比短链 AHLs 扩散的距离更长(Decho et al.,

2011)。

AHL 的扩散不仅受非生物因素的影响，也受生物因素的影响。例如，众所周知，许多细菌或真核生物会释放群体感应淬灭酶或化合物，从而影响 AIs 的扩散和接收。Hmelo 和 Van Mooy（2009）研究了 C6-HSL、3-oxo-C6-HSL 和 3-oxo-C8-HSL 在天然海水中的降解速率，发现其分别比人工海水高出 54%、23% 和 57%，这些作者将这种增加归因于海水中群体淬灭过程的发生，对于此，更多的细节将在本章的最后一节中提供。

总的来说，AHLs 在海洋环境中的扩散似乎是一个依赖于许多不同的非生物和生物变量的复杂现象。考虑的一个重要因素是，这些信息化学物质通常由复杂的外聚合物组成的三维生物膜中释放，这些生物膜赋予了这些生物结构的密度和厚度（Harder et al., 2014）。因此，一些作者倾向于认为色谱运动而不是被动扩散运动作为 AHLs 通过海洋微生境中传播的最佳概念模型（Decho et al., 2011）。生物膜的结构可以激发 AHLs 在微沟中隔离，导致局部重要的 AIs 浓度可以激活群体感应过程，即使在局部环境中只有很少的细胞（Charlton et al., 2000）。

3.6 哈维弧菌（和 *Harveyi* 支系）是研究海洋细菌群体感应的模型

哈维弧菌是海洋环境中的主要病原体，可引起养殖牡蛎、软体动物和虾类的许多疾病，造成大量的经济损失。哈维弧菌具有一个复杂的群体感应系统，包括 3 个相互依赖的通道：基于 AI-1、AI-2 和 CAI-1 的群体感应（图 3-6）。在所有情况下，AIs 与其同源的膜受体

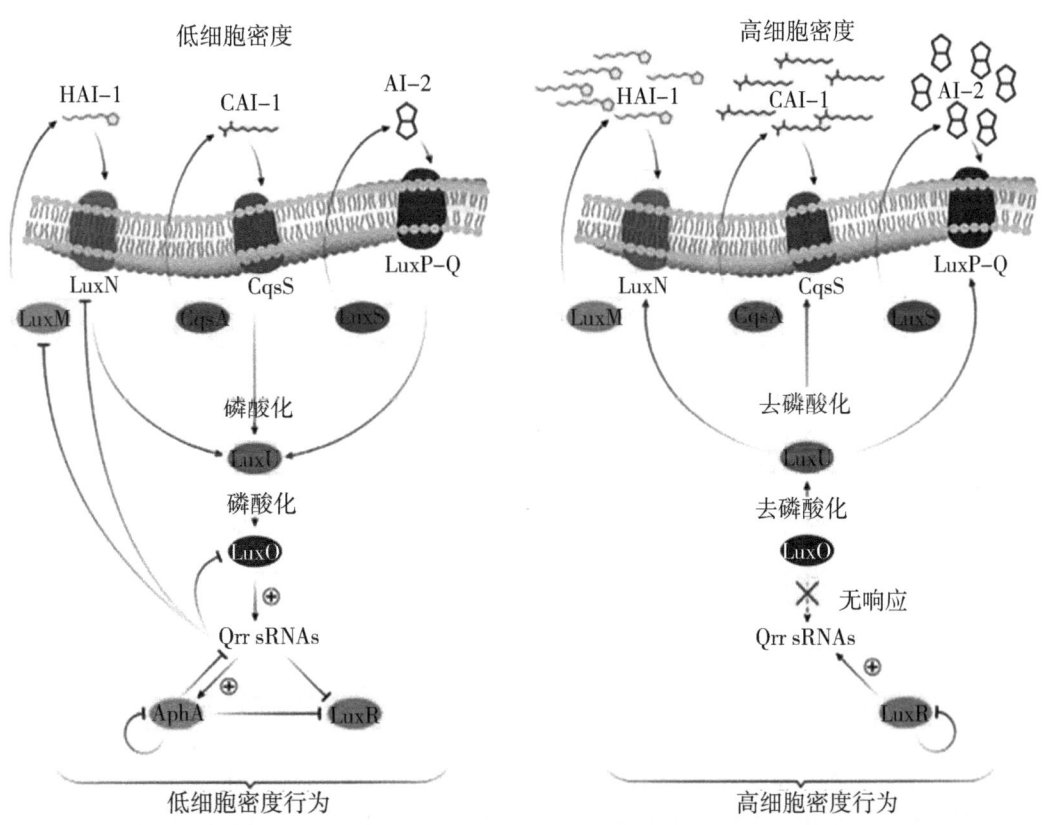

图 3-6 海洋模式菌株哈维弧菌群体感应信号通路及其在低细胞密度和高细胞密度下的作用模式

结合，从而激活细胞内磷酸化/去磷酸化信号转导级联反应。这个级联反应中的关键蛋白质是LuxO，作为反应调节器，其可整合来自3个通道中呈现出不同水平磷酸化的信号（Lilley and Bassler，2000）。在低细胞密度下磷酸化时，LuxO通过五个群体感应调节的sRNA（Qrr RNA），激活蛋白质AphA，从而调控许多基因，尤其是那些涉及大多数已鉴定的毒力因子的表达以及生物膜的形成（详细讨论见本章后文）。

在低细胞密度下，Qrr RNAs抑制主调节因子LuxR的转录（Lenz et al.，2004）。然而，当存在高浓度的群体感应信号时，LuxO去磷酸化，从而诱导 *lux* 基因的表达。更准确地说，细胞质中LuxR的浓度取决于这五种不同类型的小调控RNAs的浓度，其与LuxO的磷酸化水平相关。在这种高细胞密度下，AphA的产生被抑制，而LuxR依赖的基因被激活，就像那些参与生物发光产生相关的基因一样。

在本章第5节，我们将回顾由群体感应依赖表型的多样性及其功能。要记住的一个重要事实是AphA是控制低细胞密度行为的群体感应主调节因子，而LuxR是控制高细胞密度行为的群体感应主调节因子。从这个意义上说，哈维弧菌群体感应依赖表型的调控是复杂的。例如，在毒力因子中，群体感应在高细胞密度下正向调节金属蛋白酶和细胞外毒素。相反，在这种高细胞密度下，群体感应负向调节几丁质酶、磷脂酶、铁载体和三型分泌系统。这种对立可能反映了在不同的感染阶段需要不同类型的毒力因子（Natrah et al.，2011b）。

4 表面生物膜和共生功能体：微生物栖息地的多样性使海洋环境中的群体感应成为可能

群体感应发生在高浓度的细菌细胞中。因此，自由生活的海洋细菌细胞进行群体感应的可能性似乎较小。然而，在海洋环境中存在许多（微）生态位和栖息地，细菌细胞在高度结构化的生物膜上能够达到足够的丰度，从而诱导基于群体感应的信号传递。生物膜被组织在一个凝胶状的外大分子聚合物基质中，这种三维结构包括微米级的空间组织，并有效地浓缩了化合物（Decho et al.，2011；McLean et al.，1997）。此外，原核生物会定殖在海洋植物、脊椎动物和无脊椎动物上，对其他生物而言，细菌微生物群落重要性的认识促成了共生功能体概念的提出。同样，共生功能体为细菌浓度和原核生物相互作用的发生提供了可能，包括群体感应（Teplitski et al.，2016）。本节将描述一些示例。

4.1 （微）藻际和海洋植物的叶际

藻际指的是与周围藻类直接对应的区域。描述了一个深受藻类影响的微生物栖息地。在这个微环境中，细菌的丰度可以达到$10^8 \sim 10^{11}$个/mL（Rolland et al.，2016），从而使群体感应得以发生。Bachofen和Schenk（1998）开创性的研究揭示了蓝藻浮游植物水华中产生AHLs且浓度可达到10 mg/L（Bachofen and Schenk，1998）。此后，关于（微）藻际中群体感应过程的研究不断增多（Rolland et al.，2016）。最近的一项研究采用转录组学方法，观察到了与亚历山大微藻（*Alexandrium tamarense*）共培养的玫瑰杆菌（*Ruegeria pomeroyi*）中 *luxI* 和 *luxR* 基因表达增加。这项研究揭示了微藻生长与一些微藻相关的红杆菌科菌株中群体感应依赖的生理调节之间的潜在重要关系（Landa et al.，2017）。

从微藻和大型海藻藻际微环境中分离出许多能进行群体感应的菌株，阐明了它们所产生的化合物及其参与的细胞间信号传导。例如，从不同藻际中分离出不同物种的菌株可产生

AHL，如什叶派二甲杆菌、光养赫夫勒氏菌（*Hoeflea phototrophica*）和黏液玫瑰变色菌（*Roseovarius mucosus*）分别产生 C18:1-HSL 和 C14:1-HSL（Wagner-Döbler et al.，2005）。一些研究报道了蓝藻产生 AHL 的能力，尤其是产毒素的铜绿微囊藻（*Microcystis aeruginosa*），以及释放 C8-HSL 的黏杆藻属（*Gloeothece*）PCC6909（Sharif et al.，2008）。

不仅在藻际环境中存在参与群体感应信号的 AHLs，而且还存在大量种类繁多的化合物，而这些化合物的研究相对较少。例如，从浮游植物的藻际环境中分离到的 AI-2 产生菌很少。然而，最近报道从束毛藻分离的一些弧菌能产生 AI（Van Mooy et al.，2012）。已假设 AI-2 在控制对链状裸甲藻（*Gymnodinium catenatum*）的杀藻活性中具有潜在作用（Skerratt et al.，2002）。TDA 在许多红杆菌属（*Rhodobacterales*）物种中被作为 AI（Geng and Belas，2010），经常与微藻相关，类似于隶属于 *Phaeobacter*、硅酸杆菌属（*Silicibacter*）和鲁杰氏菌属的细菌菌株（Bruhn et al.，2005；Geng et al.，2008；Porsby et al.，2008）。

最近的研究调查了一种在沿海生态系统中起关键作用的海洋被子植物，地中海海草[欧海神草（*Posidonia oceanica*）]相关微生物群落中群体感应的普遍性。为了解决这个问题，从欧海神草的叶片和根茎中分离出 60 个菌株，包括附生菌株和内生菌株。在这些菌株中，检测到 6 株能产生 8 种不同类型的 AHLs，以及 19 株能激活 AI-2 的生物传感器。这些结果揭示了在海洋被子植物微生物群落中群体感应关系的重要性，同时也有助于更好地理解这些微生物与宿主之间可能存在的复杂关系（Blanchet et al.，2017）。

4.2 下沉的海洋雪颗粒周围的生物膜

海洋雪指直径包括大于 0.5 mm 且在深海中下沉的颗粒。这些聚集体由有机和无机碳组成，被认为在海洋和全球碳循环中发挥重要作用。重要的海洋细菌菌群定殖在海洋雪中，它们的代谢活动在很大程度上促进了这些聚集体的形成。例如，这些群落表达大量参与颗粒有机碳降解的水解酶。这些细菌群落形成生物膜，使得海洋雪成为海洋中高浓度细菌的热点地区，其细菌浓度可达 $10^8 \sim 10^9$ 个/mL，比周围海水高出 2~4 个数量级（Simon et al.，2002）。

Gram 等（2002）发表的一份报告中显示，从海洋雪中分离出的 43 株细菌中有 4 株产生 AHLs。他们强调这些菌株产生 C6-HSL 和 C8-HSL，并将其鉴定为属于玫瑰杆菌属和海杆菌属。这项开创性研究揭示了在颗粒附着的海洋微生物群落中存在群体感应调控。该文随后假设在这些海洋生境中群体感应可能调节细菌水解酶活性、生物膜形成或抗生素产生。2011 年，Hmelo 等发表了其他有趣的研究。在这项研究中，研究者直接从温哥华岛附近的太平洋收集的聚集体上检测到 AHLs。此外，还证明了在含有下沉海洋雪聚集体的烧瓶中添加外源 AHLs，增加细菌水解酶的活性。在类似的方法中，Jatt 等（2015）原位检测到 AHLs 产生，揭示了 3-oxo-C6-HSL 和 C8-HSL 存在于中国边缘海洋收集的海洋雪聚集体中。在各种分离菌株中检测到 AHLs，作者报道了一株隶属于菠萝泛菌（*P. ananatis*）的海洋菌株能够产生六种不同类型的 AHLs。在试验中，他们发现向培养基中添加外源 AHLs 可增加细胞外水解酶活性。特别是观察到碱性磷酸酶活性的上调。总的来说，这些数据表明，群体感应可能在调控定殖在海洋雪中细菌的活性起至关重要作用。从这个意义上说，群体感应可能是调控海洋碳循环的关键过程，亟须更多结合原位测量的研究来支持这个假设。

4.3 海洋微生物垫和其他类型的亚潮带生物膜

对于海洋微生物垫和各种类型的潮带下生物膜的群体感应研究仍然较少。这些生物膜被认为是地球上已知的最早的生命形式之一，存在于 3.4 亿~3.5 亿年前化石的记录中，如叠层石中。这些微生物垫在微生物群落组成和功能方面表现出高度的多样性，通常以良好的结构化和分层组合形式存在。因此，它们呈现出与群体感应调控完全兼容的重要细胞丰度。

McLean 等（1997）发表的开创性研究，首次报道了自然环境中 AHLs，重点关注水下生物膜。虽然这项研究并未专门针对海洋生物膜，但为水生环境提供了许多有趣的数据。研究揭示了未被天然生物膜覆盖的岩石上没有 AHLs 的存在，而被生物膜覆盖的岩石上有 AHLs。几年后，Decho 等（2009）发表的一份报告揭示了海洋垫中 AHLs 非常有趣的调控模式。他们能直接从海洋垫样品中原位收集和表征 AHLs，并检测到长链和短链 AHLs（C4-HSL、C6-HSL、oxo-C6-HSL、C7-HSL、C8-HSL、oxo-C8-HSL、C10-HSL、C12-HSL 和 C14-HSL）。更有趣的是，作者发现短链 AHLs 在白天浓度显著减少。一个可能的原因是，由于光合作用，当白天 pH 值高于 8.2 时，短链 AHLs 的结构可能会发生变化。亚潮生物膜的发展是动态的，研究表明，随着群落组成的变化，释放的 AHLs 模式也会发生改变。在亚潮生物膜中分离出许多产生 AHL 的细菌（Huang et al., 2008），尤其是产生 AHL 的弧菌属，似乎是这些生物膜中的先驱物种（Huang et al., 2009）。

许多研究表明，细菌生物膜的存在有利于无脊椎动物幼虫在表面上定殖。尤其是，一系列有趣的研究强调了海洋生物膜中产生的 AHLs 可以被不同类型的大型生物检测到，并引导其定殖（Hadfield, 2011; Hadfield and Paul, 2001; Wieczorek and Todd, 1998）。在化学引诱试验中，发现 AHLs 对巨藻 *Ulva* 的游动孢子起着重要作用。通过这些孢子检测 AHLs，导致它们泳动行为的改变和化学运动的激发，从而有利于它们在生物膜上的定殖。这些数据揭示了细菌生物膜中 AHLs 的产生可被真核生物感知，同时，这些研究也指出 AHLs 可作为跨界的化学信号（Joint et al., 2002; Tait et al., 2005）。类似地，研究表明，AHLs 有利于致密藤壶（*Balanus improvisus*）幼虫在细菌生物膜中定殖（Tait and Havenhand, 2013）。

4.4 珊瑚和其他刺胞动物相关的群落

许多刺胞动物物种在其体内寄居着与细菌相关的群落，这些成员使用群体感应信号进行交流。例如，沟迎风海葵（*Anemonia viridis*）和疣皮珊瑚（*Gorgonacea Eunicella verrucosa*）拥有非常多样化的细菌群落，包括群体感应信号 AIs 的生产者（Ransome et al., 2014）。然而，在群体感应领域，对刺胞动物物种的研究大多集中在珊瑚上，尤其是在致病性研究方面（如黑带病、白带病等）。

珊瑚物种拥有重要且密集的相关微生物群落，其浓度与周围海水比高出 10~1 000 倍（Rosenberg et al., 2007）。2011 年进行的一项研究报告指出，这些菌群中 30%的细菌能够进行群体感应。这些菌株中包括一株弧菌，通过 TLC 检测其可释放 3-OH-C10-HSL（Golberg et al., 2011）。类似地，在 2010 年发表的另一项研究中，从不同健康或患病的珊瑚（从黏液、组织和周围水样中）总共收集了 29 株弧菌，并测试了它们产生群体感应化合物的能力（Tait et al., 2010）。研究发现，所有分离的弧菌包括从健康和患病动物中的菌株均能释放 AI-2，其中 17 株能合成 AHL。有趣的是，在同一研究中，报道了温度可以抑制致病菌哈维弧菌 AHL 的产生。综上所述，这些数据表明，群体感应可能在珊瑚疾病和弧菌感染中起到一定作用，但这个假设背后的潜在机制仍有待阐明（Hmelo et al., 2011）。

最近，来自研究白带病和黑带病的研究人员提供了更多信息，这两种疾病是威胁珊瑚的多种微生物疾病中的两种。从受感染珊瑚的细菌群落中分离出一些具有群体感应能力的菌株

（Zimmer et al.，2014）。有趣的是，随后的研究表明，鹿角珊瑚（Acropora cervicornis）暴露 C6-HSL 后，其健康的微生物组会转化为一种导致白带病的微生物组（Meyer et al.，2016）。此外，通过 16S rRNA 基因测序发现，群体感应抑制化合物的添加，显著改变了受感染珊瑚的微生物群落组成（Certner and Vollmer，2015）。而且，接种具有群体淬灭活性的珊瑚共生菌（Aiptasia pallida）能抑制黏质沙雷菌（Serratia marcescens）对珊瑚虫的降解能力（Alagely et al.，2011）。同样，一些研究黑带病的作者指出，由相关蓝细菌产生的林甘酸是一种强大的抗群体感应的化合物，能够选择性地抑制某些细菌物种的细胞间通信（Meyer et al.，2016）。总而言之，针对这两种不同珊瑚疾病的研究结果，强调了群体感应在健康或患病珊瑚中调控相关微生物群落的重要性（Hmelo et al.，2011；Teplitski et al.，2016）。

4.5 海绵相关群落

海绵（海绵动物门）微生物组是复杂且多样的。研究发现，一些海绵很少被细菌定殖，而其他一些则没有（Hentschel et al.，2003）。然而，定殖在海绵组织中的细菌可占海绵生物量的 35%（Vacelet and Donadey，1977）。同样，在这种类型的微生态位中，细菌可以频繁地达到足够的丰度，从而建立基于群体感应的通信，并产生 AIs 的多样化海绵共生体（Mohamed et al.，2008；Taylor et al.，2004）。此外，一些报告表明，海绵细菌共生体之间的这种通信是频繁的，在已研究的澳大利亚海绵中 77% 能激活基于 AHL 的生物传感器（Taylor et al.，2004），其中 46% 的海绵物种是在地中海和红海收集的（Britstein et al.，2017）。涉及这些微生物群落的 AHL 化学多样性已被阐明，发现其多样性包括短链和长链酰基侧链，碳链长度为 6~18 个碳（Saurav et al.，2017）。例如，在凯尔特海居蟹皮海绵中检测到 C6-HSL、C7-HSL 和 3-oxo-C12-HSL（Gardères et al.，2012），并且越来越多的来自海绵的 AHL 正在被描述（Bose et al.，2017；Britstein et al.，2016）。从草莓花瓶海绵中分离的鲁杰氏菌属 KLH11 菌株已被开发为模型，以更好地理解群体感应的细胞效应，并描述群体感应表达与细菌海绵共生体性状之间的关系。这些研究表明，KLH11 菌株模型中存在两对群体感应 luxI/R 基因和一个孤立的 luxI 基因，这些基因负调控生物膜形成和正调控基于鞭毛的运动（Zan et al.，2011，2013，2015）。这种调控可能会限制细菌在海绵内的聚集，并促进其在环境中的扩散和释放。海绵与其共生体之间的群体感应依赖关系似乎非常复杂。例如，细菌产生的 3-oxo-C12-HSL 改变了居蟹皮海绵基因的表达，并抑制了其先天免疫系统（Gardères et al.，2014）。此外，一些海绵化合物能干扰群体感应信号（Costantino et al.，2017）。

5 在海洋环境中由群体感应调控的原核生物功能多样性

在海洋环境中，原核生物的大量功能受群体感应调控，这些功能包括生物发光的产生、生物膜的形成和抑制以及色素合成等。图 3-7 总结这种功能的多样性，并提出了一种综合观点。

5.1 发光和色素的产生

关于群体感应的开创性研究详细描述了历史模型费氏弧菌中生物发光调控的机制。这些细节和参考文献在本章的第一部分提供给读者，并在此再次强调，生物发光是受群体感应调控的海洋细菌的一个重要特征。

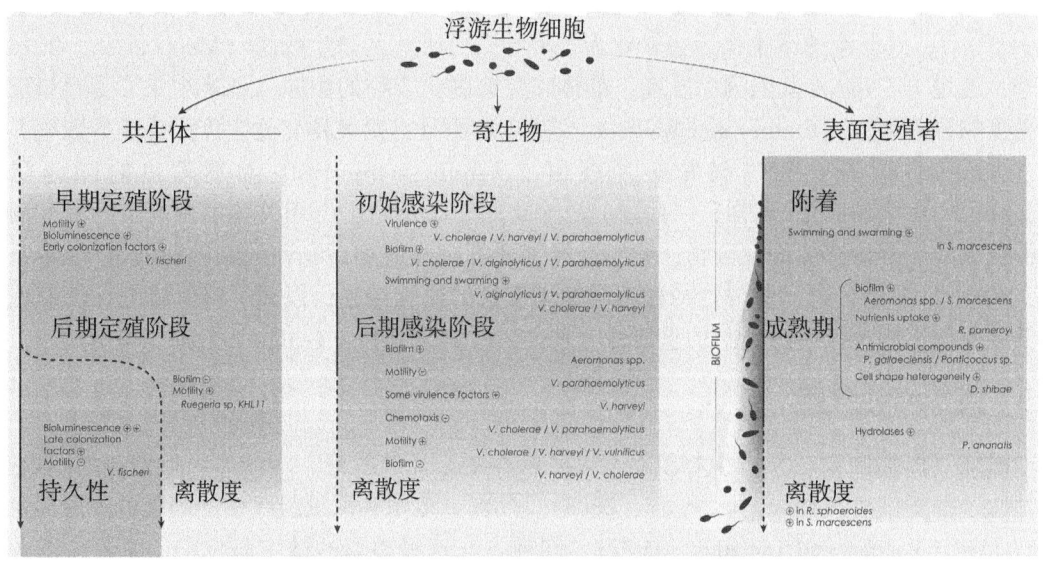

图3-7　合成和整理海洋环境和生物膜中群体感应调控功能多样性的合成
注：$V.$ = 弧菌；$S.$ = 沙雷菌；$R.\ pomeroyi$ = 玫瑰杆菌；$R.\ sphaeroides$ = 球形红杆菌。

色素的产生，尤其是紫色素，被认为是通过群体感应调控的（McClean et al.，1997）。在海洋环境中，依赖于群体感应的紫色素在海洋菌株假交替单胞菌（$Pseudoalteromonas\ ulvae$）TC14中的产生得到了证实（Ayé et al.，2015）。有趣的是，在浮游条件下，该菌株似乎不产生AHLs；然而，通过添加C6-HSL、C12-HSL、3-oxo-C8-HSL和3-oxo-C12-HSL可上调紫色素的产生，而添加3-oxo-C6-HSL可下调紫色素的产生。在固体培养条件下，发现3-oxo-C8-HSL能够上调紫色素的释放。

5.2　运动性

定殖的初始阶段，群体感应经常会在较低或平均细胞密度下激活运动性。例如，研究发现，费氏弧菌在夏威夷短尾乌贼上的定殖过程中，其运动性得到了上调（Lupp and Ruby，2005；Lupp et al.，2003）。AphA群体感应主调节因子参与了低细胞密度下的运动行为，如溶藻弧菌（$V.\ alginolyticus$）（Gu et al.，2016）和副溶血弧菌（$V.\ parahaemolyticus$）（Gu et al.，2016）。在哈维弧菌中也观察到了类似的结果（van Kessel et al.，2013）。然而，在霍乱弧菌中，运动性的上调是间接受群体感应控制的，并依赖于细胞内c-di-GMP池（Rutherford et al.，2011）。在高细胞密度下，主调节因子LuxR/HapR也能控制某些弧菌物种的运动性，如海洋病原体创伤弧菌（Lee et al.，2007）。相反，OpaR负调控副溶血弧菌和费氏弧菌的运动性（Lupp and Ruby，2005）。有趣的是，LuxR在高细胞密度下会激活霍乱弧菌的运动性，可能是为了促进细胞脱落和生物膜扩散（Reidl and Klose，2002）。

5.3　生物膜形成、成熟和散布的开始

医学领域的先驱研究揭示了群体感应表达与生物膜形成之间存在正相关（Davies et al.，1998）。然而，生物膜形成与群体感应化合物的释放之间的关系更为复杂。例如，在红杆菌科中，蓝湖硅酸杆菌（$S.\ lacuscaerulensis$）和波美罗氏硅酸杆菌都具有群体感应基因，但它

们在表面定殖的能力却存在差异（Slightom and Buchan，2009）。对于这种复杂性的一个可能解释是，生物膜的形成、成熟和扩散依赖于多种调控和信号通路下的多种步骤，这些步骤经常相互关联，并在细胞水平上以不同方式发挥作用。对水生病原体霍乱弧菌和哈维弧菌进行深入遗传分析，揭示了这种复杂性。在这些细胞中，群体感应、趋化性和双组分信号通路的转导途径彼此相互连接，并与细胞内的c-di-GMP池以及小RNA介导的信号通路相互作用。在这些菌株中，试验证明了该信号通路网络与生物膜之间的相关性（Hunter and Keener，2014；Srivastava et al.，2011；Srivastava and Waters，2012；Svenningsen et al.，2009；Tu and Bassler，2007）。

另一层解释是，生物膜形成、表面附着、生物膜成熟和扩散需要根据细菌生活策略进行不同类型的调控。研究发现，群体感应参与了黏质沙雷菌表面附着（Labbate et al.，2007），这是一种机会性病原体，有时可在海洋环境中发现（Alagely et al.，2011）。在某些弧菌物种中，低细胞密度下，AphA主调控因子也控制着生物膜的形成。这已在霍乱弧菌（Yang et al.，2010）、副溶血弧菌（Wang et al.，2013）和溶藻弧菌（Gu et al.，2016）中观察到。

在高细胞密度下，霍乱弧菌中群体感应LuxR型蛋白抑制生物膜的产生（Waters et al.，2008）。然而，在副溶血弧菌、创伤弧菌（Lee et al.，2007；Yildiz and Visick，2009）和鳗弧菌（Croxatto et al.，2002）中观察到相反的模式。许多细菌也通过群体感应正向调节生物膜的产生和成熟。例如，在鱼类病原体气单胞菌属中（Lynch et al.，2002），在成熟的生物膜中，群体感应还激活了其他功能，如抗菌化合物的产生或维持细胞形态异质性（见本章后文）。

在许多海洋细菌中，生物膜的扩散也受到群体感应的控制。在海绵共生体鲁杰氏菌属KLH11中（Zan et al.，2012），细胞运动性被群体感应激活，从而促进其传播。与此同时，群体感应在高细胞密度时抑制生物膜的产生，这可能也有利于鲁杰氏菌属KLH11细胞从其宿主中扩散（Zan et al.，2012）。这可能解释了为什么在病原体霍乱弧菌中，生物膜的产生在高细胞密度下受到负调控（Zhu and Mekalanos，2003）。水解酶的产生可能有利于这种动态。已有研究证明，群体感应正调控一些酶的表达，这与菠萝泛菌的情况类似（Jatt et al.，2015）。

综合来看，这些数据强调了群体感应在生物膜发展的每个阶段的重要作用，不论细菌生长在何种类型的海洋表面上：群体感应在海洋生物膜的形成、成熟和扩散过程中发挥着关键作用。此外，群体感应在不同菌株中以正向或负向方式调控类似的生物膜依赖表型，反映了细菌细胞多样的生活策略和适应能力。

5.4 毒力

毒力因子是细菌细胞释放的胞外产物，与致病性密切相关。毒力因子的产生对细胞来说代表重要的代谢成本，因此，是受到严格调控的（Yang and Defoirdt，2015）。毒力和群体感应化合物的表达之间的关系十分复杂：研究表明，无论是在低细胞密度还是高细胞密度下，群体感应可正向或负向调控毒力因子。这种复杂的调控机制被解释为细菌在不同的感染阶段需要以不同方式调节因子的产生（Natrah et al.，2011b）。例如，在哈维弧菌中，低细胞密度的主调控因子AphA和高细胞密度的主调控因子LuxR型共同控制着三型分泌系统1（T3SS1）的产生。这种调节机制使T3SS1在低细胞密度状态和高细胞密度状态之间有一个良好的表达峰值控制（van Kessel et al.，2013）。

在高细胞密度下，据报道哈维弧菌（Mok et al.，2003；Natrah et al.，2011b）、溶藻弧菌（Rui et al.，2008）和创伤弧菌（Shao and Hor，2001）中的酪蛋白酶、明胶酶和其他类型的蛋白酶呈现正调控。同样，金属蛋白酶（Mok et al.，2003）和细胞外毒素（Manefield et al.，2000）的产生也受到群体感应的正调控，鞭毛运动则被认为是一个重要的毒力因子（Yang and Defoirdt，2015）。相反，在哈维弧菌中，群体感应已被证明能下调几丁质酶 A（Defoirdt et al.，2010）和铁载体（Lilley and Bassler，2000）的产生。脂肪酶和溶血素作为毒力因子，其产生似乎与群体感应的表达无关（Natrah et al.，2011b）。

5.5　定殖因子的分泌

在海洋细菌中，群体感应调控在夏威夷短尾乌贼中寄生费氏弧菌的许多定殖因子。通过群体感应突变体的产生，结合对突变体和野生型培养物收集的微阵列数据的检测，提供了有趣的数据。因此，研究表明，*ainS* 基因表达有利于乌贼定殖的开始，并且在其他方面，正调节细胞运动性和胞外多糖的产生（Lupp and Ruby，2005）。在定殖的这个阶段，*lux* 基因没有完全表达。相反，*lux* 基因的表达参与后期定殖因子的产生，并在乌贼光器官的隐窝细胞中持续存在。总的来说，这些数据表明，在适度的细胞密度下，*ainS* 依赖的群体感应系统运行，而在非常高的细胞密度下，*lux* 依赖的系统活性达到峰值，如在光器官隐窝中观察到的细胞密度。

5.6　水平基因转移

在由混合的弧菌物种组成的生物膜中培养的霍乱弧菌中，已经证实存在群体感应依赖的水平基因转移。这些过程受到基因 *comEA* 的控制，其转录在 AI-1 和 AI-2 存在的情况下被诱导。由于试验设置是基于混合生物膜，因此，很可能霍乱弧菌中的这种水平转移是通过其他弧菌物种释放 AI 发生的（Antonova and Hammer，2011）。在霍乱弧菌中也有研究表明，IV 型分泌系统，允许细胞裂解和胞外 DNA 获取，受到群体感应的正调控（Papenfort and Bassler，2016）。

群体感应在基因转移中的重要性已在模式细菌荚膜红细菌（*R. capsulatus*）中得到了证明，其近亲在海洋环境中也被检测到。有趣的是，当 AIs 合成酶基因 *gtaI* 被 Sp^r 基因组插入元件静默时，研究人员观察到基因转移因子的产生减少，而添加 C16-HSL 后则恢复正常水平（Schaefer et al.，2002）。总的来说，所有这些数据表明，群体感应在控制基因转移中具有重要性，可能引起主要生态和进化影响。这些问题仍需在未来的研究中深入探讨。

5.7　营养获取

群体感应在细菌细胞中的作用最早的假设之一是，合作可能有利于营养物质的获取。在这个意义上，不同细菌释放的胞外水解酶是公共物品，有利于整个群落的生存。虽然这一假设并不特别适应海洋环境，在铜绿假单胞菌培养中，显示了蛋白酶分泌对种群水平具有积极影响（Darch et al.，2012）。

在海洋环境中，这个假设主要是通过研究细菌在微藻或海洋雪的藻际定殖得到验证。2012 年发表的一项研究表明，束毛藻的附生物依赖群体感应能产生碱性磷酸酶，从而上调其对磷酸盐的获取。这个过程似乎涉及 AHLs 的作用，而 AI-2 的产生导致磷酸盐的摄取减少（Van Mooy et al.，2012）。类似地，在二甲基硫代丙酸盐（DMSP）作为能源的情况下，玫瑰杆菌在培养时会过度产生3-oxo-C14-HSL。有趣的是，这个观察结果与细胞代谢组的剧烈变化相一致，表明当鲁杰氏菌属与藻类DMSP作为硫源培养时，会转变为一种合作的生活方式（Johnson et al., 2016）。

另一系列的研究揭示了腐烂藻类产生化学物质与群体感应表达之间存在有趣的联系。实际上，研究表明，p-香豆酸是腐烂浮游植物细胞释放的藻类木质素降解产物，同时也是沼泽红假单胞菌以及海洋波美罗氏硅酸杆菌DSS-3菌株群体感应中p-香豆酸-HSL前体（Schaefer et al., 2008）。在这个意义上，半化学物质的产生可能与浮游植物化合物释放有关，传达一些有关藻际内营养物质和外源性补充底物可用性的一些信息（Buchan et al., 2014; Schaefer et al., 2008）。有趣的是，藻类共生体什叶派二甲杆菌通过群体感应控制鞭毛生物合成（Patzelt et al., 2013），这可能使微藻能够趋化，从而有利于营养物质获取。

海洋细菌对营养物质获取的协同作用在全球范围内可能产生更广泛的影响，尽管这个假设仍未被充分探索。然而，Hmelo等（2011）和Jatt等（2015）的研究支持类似的结论。Hmelo等（2011）在温哥华岛附近收集并培养了海洋沉降颗粒样本，发现在添加AHLs后，水解酶活性增加。同样，Jatt等（2015）观察到最初从海洋雪中分离的菠萝泛菌培养物中添加C10-AHL时，碱性磷酸酶活性增强。总的来说，这些数据表明，群体感应可能调节海洋雪矿化动力学，然而，为了更好地描述和概括海水群体感应表达的生物地球化学意义，还需要进行更多的观察。

5.8 抗菌化合物的产生

群体感应参与调节许多化合物，这些化合物影响细胞密度和群体动态，如抗菌分子（Wood and Pierson, 1996）。在海洋环境中，观察到许多菌株会产生杀藻化合物。其中，TDA是一种抗菌化合物，能诱导自身合成，因此，也起到了AI的作用（Berger et al., 2011; Bruhn et al., 2005; Geng et al., 2008; Porsby et al., 2008）。有趣的是，TDA的产生也依赖于不同玫瑰杆菌属物种中的AHLs（Berger et al., 2011; Rao et al., 2007; Thole et al., 2012）。TDA的产生调控 *Phaeobacter gallaeciensis* BS107与微藻海洋球石藻之间有趣的群体特征。该细菌为藻类提供生长诱导剂，如在开花条件下的生长素，抗生素产生包括针对藻类病原体的TDA（Geng et al., 2008; Thiel et al., 2010）。在这种共生关系中，*P. gallaeciensis* 接收微藻产生的DMSP作为硫源（González et al., 1999; Newton et al., 2010）。

已从海洋细菌中鉴定出不同类型的杀藻剂，其可能受到群体感应化合物的调控（Nakashima et al., 2006; Paul and Pohnert, 2011; Skerratt et al., 2002）。例如，研究表明，藻酸杆菌（*Kordia algicida*）通过释放蛋白酶而显示出杀藻活性，其表达受群体感应调控（Paul and Pohnert, 2011）。此外，Skerratt等（2002）报道AI-2在对链状裸甲藻的杀藻活性起潜在作用。最近的研究表明，从微藻东海原甲藻（*Prorocentrum donghaiense*）藻际中分离的庞蒂球菌PD-2菌株，通过群体感应调控其杀藻活性。该菌株使用两个群体感应网络（zlaI/R和zlbI/R）产生3-oxo-C8-HSL和3-oxo-C10-HSL。使用β-环糊精抑制群体感应可将杀藻活性降低50%以上（Chi et al., 2017）。

5.9 细胞群体内表型异质性的诱导和维持

有趣的是，一些报告表明群体感应可能参与种群异质性的维持，这被假设认为在波动环境条件下的一种生存策略。在面对不断变化的条件时，这种表型异质性可能通过提高一个亚种群的生存率，从而增强整个种群的适应性。最初，这些观察是在哈维弧菌［现已重新分类为坎氏弧菌（*Vibrio campbellii*）］上进行的。这项研究发现，在一个生物发光的细胞群体中，只有69%的细胞能有效发光，且发光强度在细胞之间也存在差异。群体感应信号通路中的关键蛋白LuxO（详见本章前面的内容）似乎也在这些机制中发挥作用，因为*luxO*突变体的培养由发光细胞组成（Anetzberger et al., 2009）。

Patzelt et al.（2013）收集群体感应和表型异质性联系的有趣数据，并对海洋什叶派二硝酸杆菌进行遗传和转录组研究。这些试验揭示了*luxI*突变株中AHL的缺乏改变了344个基因的转录，这些基因参与了许多不同生理活动的调控，包括细胞分裂、鞭毛生物合成和sigma因子合成。有趣的是，这个*luxI*突变株培养物表现出二硝酸杆菌（*Dinoroseobacter*）细胞的单卵表型。相比之下，野生型表型以及添加饱和浓度的C18-AHL的突变株培养物中，观察到更多卵形、杆状和非常细长的细胞形态。在海洋细菌群体中，这种形态异质性的存在可能是一种生态优势。在浮游植物开花期间，捕食行为非常强烈，并已证明与细胞运动依赖性相关。因此，一个细菌种群可通过允许种群的一部分随机改变其表型来提高适应性，从而在环境波动期间确保至少有一部分群体得以生存（Acar et al., 2008）。例如，在季节性海洋浮游植物大量繁殖期间，这种现象尤为明显。这样的观察结果并不仅限于海洋环境，在微生物学的各个领域仍是一个研究主题（Grote et al., 2015）。

6 在海洋环境中寻找和应用群体感应抑制剂：水产养殖保护与绿色防污技术创新

6.1 一个模型故事：红色海洋藻类 *Delisea pulchra* 中发现防污化合物

在海水中首次发现抑制类群体感应的化合物。这一开创性研究常被作为一种鉴定潜在经济应用的抗群体感应化合物的模型过程。这个研究故事始于博物学家对 *D. pulchra* 的观察，其是一种小型的红色底栖大型藻类，叶子只有几厘米。首先，这些叶片有一个特点，就是很少被不同类型的生物所定殖。与大多数其他类型的海洋植物和藻类不同，叶片表面的细菌丰度显著低于其他海藻表面的细菌丰度（Maximilien et al., 1998）。然而，值得注意的是，在澳大利亚东海岸，这些藻类在夏季温度升高时会发生季节性漂白（Campbell et al., 2011）。有趣的是，*D. pulchra* 能产生卤代呋喃酮，其浓度在漂白事件期间会降低（Campbell et al., 2011）。

有趣的是，卤代呋喃酮与AHLs具有结构相似性，推测它们可能干扰群体感应信号通路（Givskov et al., 1996）。这些卤代呋喃酮能改变许多不同细菌的特性，如在液化沙雷菌（*Serratia liquefaciens*）（Givskov et al., 1996）和奇异变形杆菌（*Proteus mirabilis*）（Gram et al., 1996）中密集，多种细菌的胞外酶产生（Kjelleberg et al., 1997），以及生物发光和色素产生（Givskov et al., 1996；Kjelleberg et al., 1997）。对呋喃酮作用模式的分子和遗传机制进行了研究，证明这些化合物与LuxR受体结合并增加其降解周转率（Manefield et al., 2000, 2002）。

随后，对*Deliseas*漂白事件、温度升高和细菌群体感应之间的关系进行了深入研究。与浸泡在无菌海水中的藻类相比，在含有天然浮游细菌的水中接种 *D. pulchra* 孢子，增加了其对漂白的敏感性，这表明细菌在漂白事件中起到了一定的作用。此外，卤代呋喃酮的缺乏会增加 *D. pulcha* 对漂白的敏感性（Campbell et al., 2011），揭示了群体淬灭在限制这种疾病影响方面的重要性。其中，一种致病细菌为 *Nautella* sp. 已经被分离和鉴定出来，其能侵入藻类组织（Case et al., 2011）。其毒性机制仍有待阐明（Harder et al., 2014）；这些细菌似乎至少引起观察到的部分夏季漂白，且卤代呋喃酮的产生使得 *D. pulchra* 能控制病原体。

随后，研究了天然或合成的卤代呋喃酮类化合物对常见病原菌的影响，并取得了一定的成功（图3-8）。研究表明，这些化合物抑制铜绿假单胞菌生物膜中的群体感应过程（Hentzer et al., 2002），以及植物病原菌胡萝卜软腐欧文菌中碳青霉烯类抗生素的合成和外切酶毒力因子的产生（Manefield et al., 2001）。群体感应抑制化合物通常被怀疑是为了避免在目标细菌中出现耐药性。然而，这一结论需要结合最近的数据和观察来重新审视。研究发现，一些铜绿假单胞菌菌株对实验室大量使用的一些合成卤代呋喃酮类药物具有耐药性（García-Contreras et al., 2013）。耐药机制似乎与菌株增加呋喃酮类外排到细胞外的能力有关（Maeda et al., 2012）。

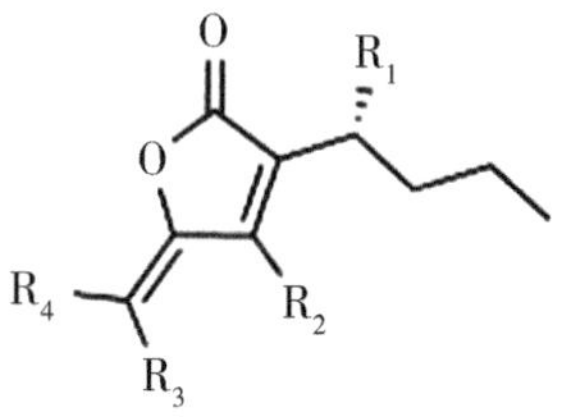

	R_1	R_2	R_3	R_4
呋喃酮1	H	Br	Br	Br
呋喃酮2	H	Br	H	Br
呋喃酮3	OAc	Br	H	Br
呋喃酮4	OH	Br	H	Br
呋喃酮5	OAc	Br	H	I
呋喃酮6	H	H	Br	Br

图3-8　其中部分卤代呋喃酮类化合物由 *Delisea pulchra* 生产

注：转引自 Manefield et al.（1999），化合物（6）是合成的。

6.2　从海洋生物中分离的群体感应淬灭化合物的多样性

从多种海洋生物（图3-9）中分离出了大量的群体淬灭化合物。这些生物包括细菌、蓝藻、真菌、海绵、海藻、刺胞动物和苔藓动物。最近发表的一篇综述详细介绍了目前从海水中分离得到的群体感应抑制化合物的目录，存在70种具有这种活性的海洋来源分子（Saurav et al., 2017）。

这篇综述揭示了海洋生物和具有这种生物活性的化合物的广谱性。在细菌中，苯乙酰胺类、环二肽类、酪醇类、酪醇乙酸酯类等化合物被发现具有群体淬灭活性（Abed et al., 2013; Martínez-Matamoros et al., 2016; Teasdale et al., 2011）。在海洋真菌中，发现了抗群体感应的化合物，如 aculenes C、D、E, penicitor, aspergillumarins A、B, meleagrin 和曲酸（Dobretsov et al., 2011; Kong et al., 2017; Li et al., 2003）。在蓝藻中，鞘丝藻酰胺C、8-表-鞘丝藻酰胺C、巨大鞘丝藻内酯和林比酸表现出抗群体感应的特性，以及肿瘤酸; honaucins A、B、C, pitinoic acid 和 microcolins A、B（Choi et al., 2012; Clark et al., 2008; Dobretsov et al., 2011; Kwan et al., 2010, 2011; Montaser et al., 2013）。从海绵中分离的甘露聚糖（Montaser et al., 2013），从刺胞动物中分离的类膜类物质（Tello et al., 2012），以及从苔藓动物类中分离的溴化生物碱（Peters et al., 2003）。

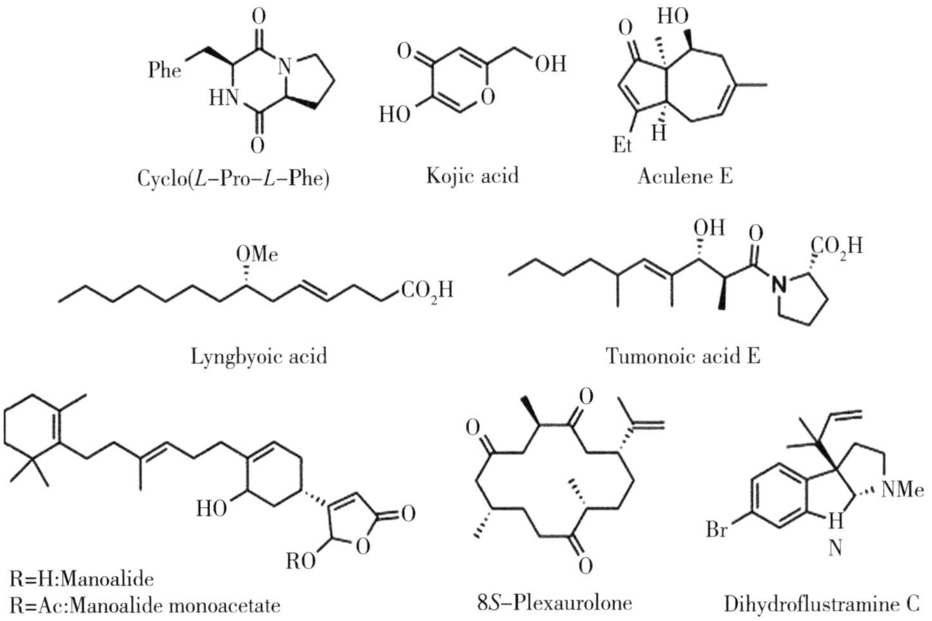

图 3-9 群体感应抑制化合物的多样性

6.3 海洋群体淬灭酶的多样性

大量的细菌能够产生群体淬灭酶，利用这些酶对抗依赖群体感应的细菌特性是具有很大潜力的。这些酶在细胞外起作用，降解 AIs，并可作为催化剂使用（Bzdrenga et al.，2017）。它们能够破坏细菌细胞附近环境中的群体感应信号，因此，不需要穿透细菌细胞，从而促进其行动方式和效果。尽管各种类型的酶都具有这种活性（Romero et al.，2015），但目前已描述两种主要类型的群体淬灭酶：内酯酶（水解 AHLs 的内酯环）和酰化酶（裂解 AHLs 中内酯环与酰基侧链之间的酰胺键）（Grandclement et al.，2016）。许多研究尝试使用群体淬灭酶来破坏病原体的毒力或生物膜形成。具体而言，群体淬灭酶已在植物病原体、膜生物反应器和医疗设备上的生物膜形成中进行了测试（Grandclement et al.，2016）。

许多海洋细菌是寻找新的群体淬灭酶的重要资源。Romero 等（2011）发表的一份报告筛选了 166 株海洋菌株，并鉴定出 24 株能显著降解 AHLs（Romero et al.，2011）。最近的研究表明，极地寡营养细菌（*Alteromonas stellipolaris*）是从 450 株菌株来自双壳类孵化场中分离筛选出的一种具有非常高效的内酯酶活性的细菌（Torres et al.，2016）。类似地，酶促群体淬灭活性也在珊瑚（Golberg et al.，2013；Tait et al.，2010）和海绵（Saurav et al.，2016a）中被检测到。在同样的视角下，宏基因组研究揭示了海洋微生物群落中丰富多样的群体淬灭基因（Muras et al.，2018a；Romero et al.，2012）。

毫不意外，一些研究小组尝试表达、表征和测试来自海洋环境的群体淬灭酶的潜力。例如，MomL 是一种从黄杆菌科（Flavobacteria）鼠尾菌属（*Muricauda olearia*）Th120 中分离出的一种内酯酶，其来自比目鱼黏液的菌株。有趣的是，该酶能够显著降低铜绿假单胞菌的毒力和生物膜形成（Tang et al.，2015）。类似地，从印度白虾的解剖肠道中分离出一株地衣芽孢杆菌（*Bacillus licheniformis*）菌株。这种酶对酸性条件具有很高的耐受性（可能是肠道中经历的自然条件的结果），其可降解多种 AHLs，并且似乎能够降低弧菌的生物膜（Vinoj et al.，2014）。同样，从海洋细菌黏着杆菌属（*Tenacibaculum* sp.）20J 中分离出一种广谱耐热的内酯酶（Mayer et al.，2015），其对变形链球菌（Muras et al.，2018b）和鱼病原菌迟缓爱德华氏菌（*Edwardsiella tarda*）具有抗生物膜活性（Romero et al.，2014）。

6.4 在海洋水下表面生物污垢的应用

生物污垢一词指的是生物在生物和非生物物质上的定殖。海洋生物在水下表面的定殖会引起诸多问题，例如生物腐蚀、船体摩擦力增加、水下仪器和设备，如螺旋桨或科学仪器的阻塞，以及海洋养殖中的问题（Dobretsov et al.，2011）。这种定殖始于细菌生物膜（微生物污垢物）的形成，其作为大型生物（藻类、多细胞生物：宏生物污损物）的定殖基质。因此，对抗生物膜的形成是许多行业当前面临的主要挑战，例如，水下电力生产、船舶建造和海军运输。大多数使用的防污化合物是有毒杀菌剂，它们释放到海水中，会对自然环境产生深远影响。例如，它们会沿着食物链积累，导致海洋哺乳动物体内积累高浓度的有毒物质。还可以直接影响过滤性生物，如牡蛎（Thomas and Brooks，2010）。克服这些困难的一种可能策略是在不影响细胞活力的情况下，靶向生物膜形成的关键机制，以选择毒性较低的化合物。在这个意义上，群体淬灭化合物的应用似乎符合这些标准，似乎是一个很有前景的抗生物污染的策略。因此，近年来，群体感应抑制化合物已引起越来越多的关注，尤其是考虑到之前描述的大型藻类模型 *D. pulchra* 上收集的有趣结果（Rasmussen et al.，2000）。

尽管已从各种各样的生物体中广泛分离出天然的群体感应抑制化合物，但它们对海洋生物膜的影响仍未得到充分的探索。其中一个原因是这些化合物在低浓度下产生，这限制了它们在环境中的应用。然而，现在可使用合成群体感应抑制化合物。在 2011 年一项专注于天然化合物的开创性研究中，Dobretsov 等（2011）从海洋生物和陆地植物中筛选分离出 78 种天然产物。其中，24% 能抑制经典报告菌株紫色色素杆菌 CV017 的群体感应而不引起毒性。研究表明，氢甲酸、去甲氧基胶囊、小结肠素 A 和 B 以及焦酸（一种氧吡咯酮）参与了这些过程，而焦酸能抑制玻璃片上生物膜的形成。此后，一些研究使用类似的方法，评估了群体感应抑制化合物对海洋生物膜的影响。例如，一些研究评估了三种群体感应抑制剂 [3,4-二溴-2(5)H-呋喃酮、4-硝基吡啶-*N*-氧化物和吲哚] 对海洋生物膜两种硅藻细柱藻属（*Cylindrotheca* sp.）和新月菱形藻（*Nitzschia closterium*）生长的潜在防污作用。该研究揭示了这些化合物对生物膜形成具有显著影响，报告了聚合物物质产量显著降低。4-硝基吡啶-*N*-氧化物似乎是最有效的化合物。此外，最近的研究表明，基于 AI-2 的群体感应也被用来限制生物膜的形成。据报道，青霉酸能够抑制太平洋盐单胞菌（*Halomonas pacifica*）中 AI-2 的产生以及生物膜的形成，而不影响细菌生长，且浓度高达 10 μmol/L（Liaqat et al.，2014）。

6.5 在水产养殖中的应用

水产养殖是另一个基于群体淬灭策略的行业，其具有潜在的重要应用（Grandclement et al.，2016）。密集的养殖条件以及这个经济部门的增长意味着鱼类疾病越来越普遍。预防措施的疗效很有限，而疫苗对一些水生病原体有效，但也会产生许多副作用。抗生素的使用限制了鱼类培养的商业化发展，因为过度使用抗生素导致许多病原体对这些化合物产生抗药性。因此，如果水产养殖经济要继续持续增长，就需要更新方法来对抗鱼类病原体。再次强调，群体感应是许多鱼类病原体毒力激活的关键机制；因此，群体淬灭策略似乎是控制鱼类疾病的一个非常有前景的研究方向（Defoirdt et al.，2011）。

在水产养殖中最早的策略之一是利用从（微）藻类中提取的化合物保护鱼类、贝类和甲壳类动物。研究表明，天然和合成的呋喃酮能保护虾类旧金山卤虫（*Artemia franciscana*）免受哈维弧菌、坎氏弧菌和副溶血弧菌的感染（Defoirdt et al.，2006；Givskov et al.，1996）。这些化合物也被报道可以保护虹鳟免受弧菌病的感染（Rasch et al.，2004）。然而，这些化合物对一些鱼类也有毒性，如虹鳟。使用像溴代硫酮这样毒性较低的合成化合物似乎很有前景（Defoirdt et al.，2012；Rasch et al.，2004）。同样，研究了多种微藻产生群体淬灭化合物的潜力，而嗜糖小球藻（*Chlorella saccharophila*）已被发现在水产养殖中具有重要的应用潜力（Natrah et al.，2011a）。

另一种策略是使用群体淬灭酶或提供能传递这种酶的益生菌。这些想法的应用已经取得了一些成功。例如，喂食群体淬灭 AiiA 重组蛋白的鲤鱼对水性气单胞菌（*Aeromonas hydrophila*）更具抵抗力（Cao et al.，2012）。此外，南美白对虾肠道微生物群落具有 AI-1 降解活性，在哈维弧菌存在下促进轮虫的生长。同样，来自欧洲鲈和尖吻鲈肠道的细菌群落被评估为保护大头虾幼虫（Nhan et al.，2010），而降解 AHL 的群落则提高了初食大比目鱼幼虫的存活率（Tinh et al.，2008）。这种选择群体淬灭产生菌株也为珊瑚养殖中的生物防治提供了有趣的初步数据。极地寡营养细菌菌株显著降低了地中海弧菌（*Vibrio mediterranei*）对珊瑚的致病性（Torres et al.，2016）。一些研究也尝试将能产生群体淬灭酶的芽孢杆菌菌株作为益生菌提供给鱼类。例如，从鲫鱼肠道分离出芽孢杆菌菌株和释放群体淬灭酶能够保护斑马鱼免受感染（Chu et al.，2014）。这一领域的研究已经促使一些公司开发了商业产品，例如，Biomin（奥地利因策尔斯多夫-盖策尔斯多夫），该公司销售益生菌"Aquastar"，该产品是一种用食物包裹的芽孢杆菌菌株，能产生群体淬灭酶，用于虾类养殖（Grandclement et al.，2016）。

7 总结

自 20 世纪 70 年代，在海洋环境中发现群体感应以来，相关研究主要集中于医学或农学领域感兴趣的菌株中。除弧菌菌株和一些重要的病原体外，群体感应在海洋菌株中的作用和重要性仍缺乏研究。因此，研究群体感应基因的多样性、群体感应化合物多样性特征以及在环境科学领域感兴趣菌株的生物学作用，仍有大量工作需要进行。此外，群体感应在生物地球化学循环中的影响尚未得到充分探索。未来研究的另一个重要领域是阐明群体感应在共生功能体微生物平衡中的作用。随着越来越多的研究揭示了基于群体感应化合物的排放、感知和抑制机制，细菌-真核生物之间的相互关系的重要性，尤其在海洋生态模型中。

对群体感应的了解带来重要的生物技术应用，尤其是在海洋环境中，水产养殖和防污工业均能从这些技术中受益。虽已发表了许多有趣的研究结果，但在实际条件下进行更大规模的试验或应用的数据仍很少。对于这样的结果，有多种可能的解释：（i）工业实验室内的研究正在进行，但尚未发表；（ii）在更大规模的试验中获得了负面结果或遇到了困难；（iii）这些工作的成果受到工业秘密的保护，无法向科学界公开。此外，目前在工业应用中只有少数群体淬灭剂进行了测试，并且可用的天然和合成化合物目录仍需扩展。无论是哪种情况，都需要更多的信息来更好地评估群体淬灭策略是否可以大规模地应用以解决工业问题。

术语表

克莱森缩合 克莱森缩合导致由两个酯（或一个酯和一个碳基化合物）形成碳-碳键的过程。该反应发生在强碱存在的情况下，导致 β-酮酯或 β-二酮生成。

浮游植物大量繁殖 浮游植物水华是指微藻在淡水或海水中的快速繁殖的现象。

小调控 RNA（小 RNA，sRNAs） 影响基因表达的 50~500 个核苷酸的 RNA。

缩写词

AHL	酰基高丝氨酸内酯
AI	自诱导物
CAI	霍乱病自诱导物
DPD	4,5-二羟基-2,3-戊二酮
GC-MS	气相色谱-质谱法
LC-MS	液相色谱-质谱法
NMR	核磁共振
RT-qPCR	定量逆转录 PCR
TDA	对二硫代酸
TLC	薄层色谱
UHPLC	超高效液相色谱法

致谢

作者非常感谢 Carole Petetin 和 Didier Stien 在数字绘制方面的帮助，也非常感谢 Sheree Yau 在英语语法和拼写方面的建议。

参考文献

ABED R M, DOBRETSOV S, AL-FORI M, et al., 2013. Quorum-sensing inhibitory compounds from extremophilic microorganisms isolated from a hypersaline cyanobacterial mat. J. Ind. Microbiol. Biotechnol., 40 (7): 759-772.

ACAR M, METTETAL J T, VAN OUDENAARDEN A, 2008. Stochastic switching as a survival strategy in fluctuating environments. Nat. Genet., 40 (4): 471-475.

ALAGELY A, KREDIET C J, RITCHIE K B, et al., 2011. Signaling-mediated cross-talk modulates swarming and biofilm formation in a coral pathogen *Serratia marcescens*. ISME J., 5 (10): 1609-1620.

ANETZBERGER C, PIRCH T, JUNG K, 2009. Heterogeneity in quorum sensing-regulated bioluminescence of *Vibrion harveyi*. Mol. Microbiol., 73 (2): 267-277.

ANTONOVA E S, HAMMER B K, 2011. Quorum-sensing autoinducer molecules produced by members of a multispecies biofilm promote horizontal gene transfer to *Vibrio cholerae*. FEMS Microbiol. Lett., 322 (1):

68-76.

AYÉ A M, BONNIN-JUSSERAND M, BRIAN-JAISSON F, et al., 2015. Modulation of violacein production and phenotypes associated with biofilm by exogenous quorum sensing N-acyl homoserine lactones in the marine bacterium *Pseudoalteromonas ulvae* TC14. Microbiology, 161 (10): 2039-2051.

BACHOFEN R, SCHENK A, 1998. Quorum sensing autoinducers: do they play a role in natural microbial habitats? Microbiol. Res., 153 (1): 61-63.

BERGER M, NEUMANN A, SCHULZ S, et al., 2011. Tropodithietic acid production in *Phaeobacter gallaeciensis* is regulated by N-acyl homoserine lactone-mediated quorum sensing. J. Bacteriol., 193 (23): 6576-6585.

BISWA P, DOBLE M, 2013. Production of acylated homoserine lactone by gram-positive bacteria isolated from marine water. FEMS Microbiol. Lett., 343 (1): 34-41.

BLANCHET E, PRADO S, STIEN D, et al., 2017. Quorum sensing and quorum quenching in the mediterranean seagrass *Posidonia oceanica* microbiota. Front. Mar. Sci., 4: 218.

BODOR A, ELXNAT B, THIEL V, et al., 2008. Potential for *luxS* related signalling in marine bacteria and production of autoinducer-2 in the genus *Shewanella*. BMC Microbiol., 8 (1): 13.

BOSE U, ORTORI C A, SARMAD S, et al., 2017. Production of N-acyl homoserine lactones by the sponge-associated marine actinobacteria *Salinispora arenicola* and *Salinispora pacifica*. FEMS Microbiol. Lett., 364 (2): fnx002.

BRITSTEIN M, DEVESCOVI G, HANDLEY K M, et al., 2016. A new N-acyl homoserine lactone synthase in an uncultured symbiont of the red sea sponge *Theonella swinhoei*. Appl. Environ. Microbiol., 82 (4): 1274-1285.

BRITSTEIN M, SAURAV K, TETA R, et al., 2017. Identification and chemical characterization of N-acyl-homoserine lactone quorum sensing signals across sponge species and time. FEMS Microbiol. Ecol., 94 (2): fix182.

BRUHN J B, NIELSEN K F, HJELM M, et al., 2005. Ecology, inhibitory activity, and morphogenesis of a marine antagonistic bacterium belonging to the *Roseobacter clade*. Appl. Environ. Microbiol., 71 (11): 7263-7270.

BUCHAN A, LECLEIR G R, GULVIK C A, et al., 2014. Master recyclers: features and functions of bacteria associated with phytoplankton blooms. Nat. Rev. Microbiol., 12 (10): 686-698.

BZDRENGA J, DAUDE D, REMY B, et al., 2017. Biotechnological applications of quorum quenching enzymes. Chem. Biol. Interact., 267: 104-115.

CAMPBELL A H, HARDER T, NIELSEN S, et al., 2011. Climate change and disease: bleaching of a chemically defended seaweed. Glob. Chang. Biol., 17 (9): 2958-2970.

CAO Y, HE S, ZHOU Z, et al, 2012. Orally administered thermostable N-acyl homoserine lactonase from *Bacillus* sp. strain AI96 attenuates *Aeromonas hydrophila* infection in zebrafish. Appl. Environ. Microbiol., 78 (6): 1899-1908.

CASE R J, LONGFORD S R, CAMPBELL A H, et al., 2011. Temperature induced bacterial virulence and bleaching disease in a chemically defended marine macroalga. Environ. Microbiol., 13 (2): 529-537.

CERTNER R H, VOLLMER S V, 2015. Evidence for autoinduction and quorum sensing in white band disease-causing microbes on Acropora cervicornis. Sci. Rep., 5: 11134.

CHARLTON T S, DE NYS R, NETTING A, et al., 2000. A novel and sensitive method for the quantification of N-3-oxoacyl homoserine lactones using gas chromatography-mass spectrometry: application to a model bacterial biofilm. Environ. Microbiol., 2 (5): 530-541.

CHEN X, SCHAUDER S, POTIER N, et al., 2002. Structural identification of a bacterial quorum-sensing signal containing boron. Nature, 415 (6871): 545-549.

CHI W, ZHENG L, HE C, et al., 2017. Quorum sensing of microalgae associated marine *Ponticoccus* sp. PD-2 andits algicidal function regulation. AMB Express, 7 (1): 59.

CHOI H, MASCUCH S J, VILLA, F A, et al., 2012. Honaucins A–C, potent inhibitors of inflammation and bacterial quorum sensing: synthetic derivatives and structure–activity relationships. Chem. Biol., 19 (5): 589-598.

CHU W, ZHOU S, ZHU W, et al., 2014. Quorum quenching bacteria *Bacillus* sp. QSI-1 protect zebrafish (*Danio rerio*) from *Aeromonas hydrophila* infection. Sci. Rep., 4: 5446.

CLARK B R, ENGENE N, TEASDALE M E, et al., 2008. Natural products chemistry and taxonomy of the marine cyanobacterium *Blennothrix cantharidosmum*. J. Nat. Prod., 71 (9): 1530-1537.

COSTANTINO V, DELLA SALA G, SAURAV K, et al., 2017. Plakofuranolactone as a quorum quenching agent from the Indonesian sponge *Plakortis* cf. *lita*. Mar. Drugs, 15 (3): 59.

CROXATTO A, CHALKER V J, LAURITZ J, et al., 2002. VanT, a homologue of *Vibrio harveyi* LuxR, regulates serine, metalloprotease, pigment, and biofilm production in *Vibrio anguillarum*. J. Bacteriol., 184 (6): 1617-1629.

CUDE W N, BUCHAN A, 2013. Acyl–homoserine lactone–based quorum sensing in the *Roseobacter clade*: complex cell-to-cell communication controls multiple physiologies. Front. Microbiol., 4: 336.

DARCH S E, WEST S A, WINZER K, et al., 2012. Density-dependent fitness benefits in quorum-sensing bacterial populations. Proc. Natl. Acad. Sci. U. S. A., 109 (21): 8259-8263.

DAVIES D G, PARSEK M R, PEARSON J P, et al., 1998. The involvement of cell-to-cell signals in the development of a bacterial biofilm. Science, 280 (5361): 295-298.

DECHO A W, VISSCHER P T, FERRY J, et al., 2009. Autoinducers extracted from microbial mats reveal a surprising diversity of *N*-acylhomoserine lactones (AHLs) and abundance changes that may relate to diel pH. Environ. Microbiol., 11 (2): 409-420.

DECHO A W, FREY R L, FERRY J L, 2011. Chemical challenges to bacterial AHL signaling in the environment. Chem. Rev., 111 (1): 86-99.

DEFOIRDT T, CRAB R, WOOD T K, et al., 2006. Quorum sensing–disrupting brominated furanones protect the gnotobiotic brine shrimp Artemia franciscana from pathogenic *Vibrio harveyi*, *Vibrio campbellii*, and *Vibrio parahaemolyticus* isolates. Appl. Environ. Microbiol., 72 (9): 6419-6423.

DEFOIRDT T, DARSHANEE RUWANDEEPIKA H A, KARUNASAGAr I, et al., 2010. Quorum sensing negatively regulates chitinase in *Vibrio harveyi*. Environ. Microbiol. Rep., 2 (1): 44-49.

DEFOIRDT T, SORGELOOS P, BOSSIER P, 2011. Alternatives to antibiotics for the control of bacterial disease in aquaculture. Curr. Opin. Microbiol., 14 (3): 251-258.

DEFOIRDT T, BENNECHE T, BRACKMAN G, et al., 2012. A quorum sensingdisrupting brominated thiophenone with a promising therapeutic potential to treat luminescent vibriosis. PLoS One, 7 (7): e41788.

DOBERVA M, SANCHEZ-FERANDIN S, TOULZA E, et al., 2015. Diversity of quorum sensing autoinducer synthases in the Global Ocean sampling metagenomic database. Aquat. Microb. Ecol., 74 (2): 107-119.

DOBERVA M, STIEN D, SORRES J, et al., 2017. Large diversity and original structures of acyl-homoserine lactones in strain MOLA 401, a marine Rhodobacteraceae bacterium. Front. Microbiol., 8: 1152.

DOBRETSOV S, TEPLITSKI M, BAYER M, et al., 2011. Inhibition of marine biofouling by bacterial quorum sensing inhibitors. Biofouling, 27 (8): 893-905.

EBERHARD A, BURLINGAME A L, EBERHARD C, et al., 1981. Structural identification of autoinducer of photobacterium fischeri luciferase. Biochemistry, 20 (9): 2444-2449.

ENGEBRECHT J, NEALSON K, SILVERMAN M, 1983. Bacterial bioluminescence: isolation and genetic analysis of functions from *Vibrio fischeri*. Cell, 32 (3): 773-781.

FUQUA W C, WINANS S C, GREENBERG E P, 1994. Quorum sensing in bacteria: the LuxR-LuxI family of cell density-responsive transcriptional regulators. J. Bacteriol., 176 (2): 269-275.

GANTNER S, SCHMID M, DURR C, et al, 2006. In situ quantitation of the spatial scale of calling distances and population densityindependent *N*-acylhomoserine lactone-mediated communication by rhizobacteria colonized on plant roots. FEMS Microbiol. Ecol., 56 (2): 188-194.

GARCIA-ALJARO C, VARGAS-CESPEDES G J, BLANCH A R, 2012. Detection of acylated homoserine lactones produced by *Vibrio* spp. and related species isolated from water and aquatic organisms. J. Appl. Microbiol., 112 (2): 383-389.

GARCIA-CONTRERAS R, MARTÍNEZ-VÁZQUEZ M, VELÁZQUEZ GUADARRAMA N, et al, 2013. Resistance to the quorum-quenching compounds brominated furanone C-30 and 5-fluorouracil in *Pseudomonas aeruginosa* clinical isolates. Pathog. Dis., 68 (1): 8-11.

GARDÈRES J, TAUPIN L, SAÏDIN JB, et al, 2012. *N*-Acyl homoserine lactone production by bacteria within the sponge *Suberites domuncula* (Olivi, 1792) (Porifera, Demospongiae). Mar. Biol., 159 (8): 1685-1692.

GARDÈRES J, HENRY J, BERNAY B, et al., 2014. Cellular effects of bacterial *N*-3-oxo-dodecanoyl-*L*-homoserine lactone on the sponge *Suberites domuncula* (Olivi, 1792): insights into an intimate inter-kingdom dialogue. PLoS One, 9 (5): e97662.

GENG H, BELAS R, 2010. Molecular mechanisms underlying *Roseobacter*-phytoplankton symbioses. Curr. Opin. Biotechnol., 21 (3): 332-338.

GENG H, BRUHN J B, NIELSEN K F, et al, 2008. Genetic dissection of tropodithietic acid biosynthesis by marine *Roseobacters*. Appl. Environ. Microbiol., 74 (5): 1535-1545.

GILSON L, KUO A, DUNLAP P V, 1995. AinS and a new family of autoinducer synthesis proteins. J. Bacteriol., 177 (23): 6946-6951.

GIRARD L, BLANCHET E, INTERTAGLIA L, et al., 2017. Characterization of *N*-acyl homoserine lactones in *Vibrio tasmaniensis* LGP32 by a biosensor-based UHPLC HRMS/MS method. Sensors, 17 (4): 906.

GIVSKOV M, DE NYS R, MANEFIELD M, et al., 1996. Eukaryotic interference with homoserine lactone-mediated prokaryotic signalling. J. Bacteriol., 178 (22): 6618-6622.

GOLBERG K, ELTZOV E, SHNIT-ORLAND M, et al., 2011. Characterization of quorum sensing signals in coral-associated bacteria. Microb. Ecol., 61 (4): 783-792.

GOLBERG K, PAVLOV V, MARKS R S, et al., 2013. Coral-associated bacteria, quorum sensing disrupters, and the regulation of biofouling. Biofouling, 29 (6): 669-682.

GONZÁLEZ J M, KIENE R P, MORAN M A, 1999. Transformation of sulfur compounds by an abundant lineage of marine bacteria in the α-subclass of the class *Proteobacteria*. Appl. Environ. Microbiol., 65 (9): 3810-3819.

GRAF J, RUBY E G, 1998. Host-derived amino acids support the proliferation of symbiotic bacteria. Proc. Natl. Acad. Sci. U. S. A., 95 (4): 1818-1822.

GRAM L, DE NYS R, MAXIMILIEN R, et al., 1996. Inhibitory effects of secondary metabolites from the red alga *Delisea pulchra* on swarming motility of *Proteus mirabilis*. Appl. Environ. Microbiol., 62 (11): 4284-4287.

GRAM L, GROSSART H P, SCHLINGLOFF A, et al., 2002. Possible quorum sensing in marine snow bacteria: production of acylated homoserine lactones by *Roseobacter* strains isolated from marine snow. Appl. Environ. Microbiol., 68 (8): 4111-4116.

GRANDCLEMENT C, TANNIERES M, MORERA S, et al., 2016. Quorum quenching: role in nature and applied developments. FEMS Microbiol. Rev., 40 (1): 86-116.

GREENBERG E, HASTINGS J, ULITZUR S, 1979. Induction of luciferase synthesis in *Beneckea harveyi* by other marine bacteria. Arch. Microbiol., 120 (2): 87-91.

GROTE J, KRYSCIAK D, STREIT W R, 2015. Phenotypic heterogeneity, a phenomenon that may explain why quorum sensing does not always result in truly homogenous cell behavior. Appl. Environ. Microbiol., 81 (16): 5280-5289.

GU D, LIU H, YANG Z, et al., 2016. Chromatin immunoprecipitation sequencing technology reveals global regulatory roles of low-cell-density quorum-sensing regulator AphA in the pathogen *Vibrio alginolyticus*. J. Bacteriol., 198 (21): 2985-2999.

HADFIELD M G, 2011. Biofilms and marine invertebrate larvae: what bacteria produce that larvae use to choose settlement sites. Ann. Rev. Mar. Sci., 3: 453-470.

HADFIELD M G, PAUL V J, 2001. Natural chemical cues for settlement and metamorphosis of marine invertebrate larvae. In: McClintock, J. B., Baker, W. (Eds.), Marine Chemical Ecology. In: 2001, CRC Press, Boca Raton, FL, pp: 431-461.

HARDER T, RICE S A, DOBRETSOV S, et al., 2014. Bacterial communication systems. In: La Barre, S., Kornprobst, J. M. (Eds.), Oustanding Marine Molecules: Chemistry, Biology, Analysis. Wiley-VCH, Weinheim, pp: 173-187.

HENTSCHEL U, FIESELER L, WEHRL M, et al., 2003. Microbial diversity of marine sponges. In: Müller, W. E. G. (Ed.), Sponges (Porifera). Progress in Molecular and Subcellular Biology., vol, 37. Springer, Berlin, pp: 59-88.

HENTZER M, RIEDEL K, RASMUSSEN T B, et al., 2002. Inhibition of quorum sensing in *Pseudomonas aeruginosa* biofilm bacteria by a halogenated furanone compound. Microbiology, 148 (Pt 1), 87-102.

HIGGINS D A, POMIANEK M E, KRAML C M, et al., 2007. The major *Vibrio cholerae* autoinducer and its role in virulence factor production. Nature, 450 (7171): 883-886.

HMELO L, VAN MOOY B A S, 2009. Kinetic constraints on acylated homoserine lactone-based quorum sensing in marine environments. Aquat. Microb. Ecol. 54 (2), 127-133.

HMELO L R, MINCER T J, VAN MOOY B A, 2011. Possible influence of bacterial quorum sensing on the hydrolysis of sinking particulate organic carbon in marine environments. Environ. Microbiol., Rep. 3 (6): 682-688.

HUANG Y L, KI J S, CASE R J, et al., 2008. Diversity and acyl-homoserine lactone production among subtidal biofilm-forming bacteria. Aquat. Microb. Ecol., 52 (2): 185-193.

HUANG Y L, KI J S, LEE O O, et al., 2009. Evidence for the dynamics of acyl homoserine lactone and AHL-producing bacteria during subtidal biofilm formation. ISME J., 3 (3): 296-304.

HUNTER G A, KEENER J P, 2014. Mechanisms underlying the additive and redundant Qrr phenotypes in *Vibrio harveyi* and *Vibrio cholerae*. J. Theor. Biol., 340: 38-49.

JATT A N, TANG K, LIU J, et al., 2015. Quorum sensing in marine snow and its possible influence on production of extracellular hydrolytic enzymes in marine snow bacterium *Pantoea ananatis* B9. FEMS Microbiol. Ecol., 91 (2): 1-13.

JOHNSON W M, KIDO SOULE M C, KUJAWINSKI E B, 2016. Evidence for quorum sensing and differential metabolite production by a marine bacterium in response to DMSP. ISME J., 10 (9): 2304-2316.

JOINT I, TAIT K, CALLOW M E, et al., 2002. Cell-to-cell communication across the prokaryote-eukaryote boundary. Science, 298 (5596): 1207.

KAUFMANN G F, SARTORIO R, LEE S H, et al., 2005. Revisiting quorum sensing: discovery of additional chemical and biological functions for 3-oxo-*N*-acylhomoserine lactones. Proc. Natl. Acad. Sci. U. S. A., 102 (2): 309-314.

KELLY R C, BOLITHO M E, HIGGINS D A, et al., 2009. The *Vibrio cholerae* quorum-sensing autoinducer CAI-1: analysis of the biosynthetic enzyme CqsA. Nat. Chem. Biol., 5 (12): 891-895.

KIØRBOE T, 2001. Formation and fate of marine snow: small-scale processes with large-scale implications. Sci. Mar., 65 (S2): 57-71.

KJELLEBERG S, STEINBERG P, GIVSKOV M, et al., 1997. Do marine natural products interfere with prokaryotic AHL regulatory systems? Aquat. Microb. Ecol., 13 (1): 85-93.

KONG F D, ZHOU L M, MA Q Y, et al., 2017. Metabolites with Gram negative bacteria quorum sensing inhibitory activity from the marine animal endogenic fungus *Penicillium* sp. SCS-KFD08. Arch. Pharm. Res., 40 (1): 25-31.

KRICK A, KEHRAUS S, EBERL L, et al., 2007. A marine *Mesorhizobium* sp. produces structurally novel long-chain *N*-acyl-L-homoserine lactones. Appl. Environ. Microbiol., 73 (11): 3587-3594.

KWAN J C, TEPLITSKI M, GUNASEKERA S P, et al., 2010. Isolation and biological evaluation of 8-epi-malyngamide C from the Floridian marine cyanobacterium *Lyngbya majuscula*. J. Nat. Prod., 73 (3): 463-466.

KWAN J C, MEICKLE T, LADWA D, et al., 2011. Lyngbyoic acid, a "tagged" fatty acid from a marine cyanobacterium, disrupts quorum sensing in *Pseudomonas aeruginosa*. Mol. Biosyst., 7 (4): 1205-1216.

LABBATE M, ZHU H, THUNG L, et al., 2007. Quorum-sensing regulation of adhesion in *Serratia marcescens* MG1 is surface dependent. J. Bacteriol., 189 (7): 2702-2711.

LANDA M, BURNS A S, ROTH S J, et al., 2017. Bacterial transcriptome remodeling during sequential co-culture with a marine dinoflagellate and diatom. ISME J., 11 (12): 2677-2690.

LAUE B E, JIANG Y, CHHABRA S R, et al., 2000. The biocontrol strain *Pseudomonas fluorescens* F113 produces the *Rhizobium* small bacteriocin, *N*-(3-hydroxy-7-cis-tetradecenoyl) homoserine lactone, via HdtS, a putative novel *N*-acylhomoserine lactone synthase. Microbiology, 146 (Pt 10): 2469-2480.

LEE J H, RHEE J E, PARK U, et al., 2007. Identification and functional analysis of *Vibrio vulnificus* SmcR, a novel global regulator. J. Microbiol. Biotechnol., 17 (2): 325-334.

LENZ D H, MOK K C, LILLEY B N, et al., 2004. The small RNA chaperone Hfq and multiple small RNAs control quorum sensing in *Vibrio harveyi* and *Vibrio cholerae*. Cell, 118 (1): 69-82.

LI X, JEONG J H, LEE K T, et al., 2003. γ-Pyrone derivatives, kojic acid methyl ethers from a marine-derived fungus *Altenaria* sp. Arch. Pharm. Res., 26 (7): 532-534.

LIAQAT I, BACHMANN R T, EDYVEAN R G, 2014. Type 2 quorum sensing monitoring, inhibition and biofilm formation in marine microrganisms. Curr. Microbiol., 68 (3): 342-351.

LILLEY B N, BASSLER B L, 2000. Regulation of quorum sensing in *Vibrio harveyi* by LuxO and sigma-54. Mol. Microbiol., 36 (4): 940-954.

LUPP C, RUBY E G, 2005. *Vibrio fischeri* uses two quorum-sensing systems for the regulation of early and late colonization factors. J. Bacteriol., 187 (11): 3620-3629.

LUPP C, URBANOWSKI M, GREENBERG E P, et al., 2003. The *Vibrio fischeri* quorum-sensing systems ain and lux sequentially induce luminescence gene expression and are important for persistence in the squid host. Mol. Microbiol., 50 (1): 319-331.

LYNCH M J, SWIFT S, KIRKE D F, et al., 2002. The regulation of biofilm development by quorum sensing in *Aeromonas hydrophila*. Environ. Microbiol., 4 (1): 18-28.

MAEDA T, GARCIA-CONTRERAS R, PU M, et al., 2012. Quorum quenching quandary: resistance to antivirulence compounds. ISME J., 6 (3): 493-501.

MANEFIELD M, DE NYS R, NARESH K, et al., 1999. Evidence that halogenated furanones from Delisea pulchra inhibit acylated homoserine lactone (AHL) -mediated gene expression by displacing the AHL signal from its receptor protein. Microbiology, 145 (2): 283-291.

MANEFIELD M, HARRIS L, RICE S A, et al., 2000. Inhibition of luminescence and virulence in the black tiger prawn (*Penaeus monodon*) pathogen *Vibrio harveyi* by intercellular signal antagonists. Appl. Environ. Microbiol., 66 (5): 2079-2084.

MANEFIELD M, WELCH M, GIVSKOV M, et al., 2001. Halogenated furanones from the red alga, *Delisea pulchra*, inhibit carbapenem antibiotic synthesis and exoenzyme virulence factor production in the phytopathogen *Erwinia carotovora*. FEMS Microbiol. Lett., 205 (1): 131-138.

MANEFIELD M, RASMUSSEN T B, HENZTER M, et al., 2002. Halogenated furanones inhibit quorum sensing through accelerated LuxR turnover. Microbiology, 148 (4): 1119-1127.

MARTÍNEZ-MATAMOROS D, LAITON FONSECA M, DUQUE C, et al., 2016. Screening of marine bacterial strains as source of quorum sensing inhibitors (QSI): first chemical study of *Oceanobacillus profundus* (RKHC-62B). Vitae, 23 (1): 30-47.

MAXIMILIEN R, DE NYS R, HOLMSTROM C, et al., 1998. Chemical mediation of bacterial surface colonisation by secondary metabolites from the red alga *Delisea pulchra*. Aquat. Microb. Ecol., 15 (3): 233-246.

MAYER C, ROMERO M, MURAS A, et al., 2015. Aii20J, a wide-spectrum thermostable N-acylhomoserine lactonase from the marine bacterium *Tenacibaculum* sp 20J, can quench AHL-mediated acid resistance in *Escherichia coli*. Appl. Microbiol. Biotechnol., 99 (22): 9523-9539.

MCCLEAN K H, WINSON M K, FISH L, et al., 1997. Quorum sensing and *Chromobacterium violaceum*: exploitation of violacein production and inhibition for the detection of N-acylhomoserine lactones. Microbiology, 143 (12): 3703-3711. Pt 12.

MCLEAN R J, WHITELEY M, STICKLER D J, et al., 1997. Evidence of autoinducer activity in naturally occurring biofilms. FEMS Microbiol. Lett., 154 (2): 259-263.

MEYER J L, GUNASEKERA S P, SCOTT R M, et al., 2016. Microbiome shifts and the inhibition of quorum sensing by black band disease cyanobacteria. ISME J., 10 (5): 1204-1216.

MILLER S T, XAVIER K B, CAMPAGNA S R., et al., 2004. *Salmonella typhimurium* recognizes a chemically distinct form of the bacterial quorum-sensing signal AI-2. Mol. Cell, 15 (5): 677-687.

MILLER S D, HADDOCK S H, ELVIDGE C D, et al., 2005. Detection of a bioluminescent milky sea from space. Proc. Natl. Acad. Sci. U. S. A., 102 (40): 14181-14184.

MOHAMED N M, CICIRELLI E M, KAN J, et al., 2008. Diversity and quorum-sensing signal production of *Proteobacteria* associated with marine sponges. Environ. Microbiol., 10 (1): 75-86.

MOK K C, WINGREEN N S, BASSLER B L, 2003. *Vibrio harveyi* quorum sensing: a coincidence detector for two autoinducers controls gene expression. EMBO J., 22 (4): 870-881.

MONTASER R, PAUL V J, LUESCH H, 2013. Modular strategies for structure and function employed by marine cyanobacteria: characterization and synthesis of pitinoic acids. Org. Lett., 15 (16): 4050-4053.

MURAS A, LÓPEZ-PÉREZ M, MAYER C, et al., 2018a. High prevalence of quorumsensing and quorum-quenching activity among cultivable bacteria and metagenomic sequences in the Mediterranean Sea. Genes, 9 (2): 100.

MURAS A, MAYER C, ROMERO M, et al., 2018b. Inhibition of *Steptococcus mutans* biofilm formation by extracts of *Tenacibaculum* sp. 20J, a bacterium with wide-spectrum quorum quenching activity. J. Oral Microbiol., 10 (1): 1429788.

NAKASHIMA T, MIYAZAKI Y, MATSUYAMA Y, et al., 2006. Producing mechanism of an algicidal compound against red tide phytoplankton in a marine bacterium gamma-proteobacterium. Appl. Microbiol. Biotechnol., 73 (3): 684-960.

NATRAH F M, KENMEGNE M M, WIYOTO W, et al., 2011a. Effects of microalgae commonly used in aquaculture on acyl-homoserine lactone quorum sensing. Aquaculture, 317 (1-4): 53-57.

NATRAH F M, RUWANDEEPIKA H A, PAWAR S, et al., 2011b. Regulation of virulence factors by quorum sensing in *Vibrio harveyi*. Vet. Microbiol., 154 (1-2): 124-129.

NEALSON K H, PLATT T, HASTINGS J W, 1970. Cellular control of the nsynthesis and activity of the bacterial luminescent system. J. Bacteriol., 104 (1): 313-322.

NEUMANN A, PATZELT D, WAGNER-DÖBLER I, et al., 2013. Identification of new *N*-acylhomoserine lactone signalling compounds of *Dinoroseobacter shibae* DFL-12T by overexpression of *luxI* genes. ChemBioChem, 14 (17): 2355-2361.

NEWTON R J, GRIFFIN L E, BOWLES K M, et al., 2010. Genome characteristics of a generalist marine bacterial lineage. ISME J., 4 (6): 784-798.

NG W L, PEREZ L J, WEI Y, et al., 2011. Signal production and detection specificity in *Vibrio* CqsA/CqsS quorum-sensing systems. Mol. Microbiol., 79 (6): 1407-1417.

NHAN D T, CAM D T, WILLE M, et al., 2010. Quorum quenching bacteria protect *Macrobrachium rosenbergii* larvae from *Vibrio harveyi* infection. J. Appl. Microbiol., 109 (3): 1007-1016.

PAGGI R A, MARTONE C B, FUQUA C, et al., 2003. Detection of quorum sensing signals in the haloalkaliphilic archaeon *Natronococcus occultus*. FEMS Microbiol. Lett., 221 (1): 49-52.

PAPENFORT K, BASSLER B L, 2016. Quorum sensing signal-response systems in Gram-negative bacteria. Nat. Rev. Microbiol., 14 (9): 576-588.

PAPENFORT K, SILPE J E, SCHRAMMA K R, et al., Seyedsayamdost, M. R., Bassler, B. L., 2017. A *Vibrio cholerae* autoinducer-receptor pair that controls biofilm formation. Nat. Chem. Biol., 13 (5): 551-557.

PATZELT D, WANG H, BUCHHOLZ I, et al., 2013. You are what you talk: quorum sensing induces individual morphologies and cell division modes in *Dinoroseobacter shibae*. ISME J., 7 (12): 2274-2286.

PAUL C, POHNERT G, 2011. Interactions of the algicidal bacterium *Kordia algicida* with diatoms: regulated protease excretion for specific algal lysis. PLoS One, 6 (6): e21032.

PEREZ-RODRIGUEZ I, BOLOGNINI M, RICCI J, et al., 2015. From deep-sea volcanoes to human pathogens: a conserved quorum-sensing signal in *Epsilonproteobacteria*. ISME J., 9 (5): 1222-1234.

PETERS L, KONIG G M, WRIGHT A D, et al., 2003. Secondary metabolites of *Flustra foliacea* and their influence on bacteria. Appl. Environ. Microbiol., 69 (6): 3469-3475.

PORSBY C H, NIELSEN K F, GRAM L, 2008. *Phaeobacter* and *Ruegeria* species of the *Roseobacter* clade colonize separate niches in a Danish Turbot (*Scophthalmus maximus*)-rearing farm and antagonize *Vibrio* anguillarum under different growth conditions. Appl. Environ. Microbiol., 74 (23): 7356-7364.

PUROHIT A A, JOHANSEN J A, HANSEN H, et al., 2013. Presence of acyl-homoserine lactones in 57 members of the *Vibrionaceae* family. J. Appl. Microbiol., 115 (3): 835-847.

RANSOME E, MUNN C B, HALLIDAY N, et al., 2014. Diverse profiles of *N*-acyl-homoserine lactone molecules found in cnidarians. FEMS Microbiol. Ecol., 87 (2): 315-329.

RAO D, WEBB J S, HOLMSTROM C, et al., 2007. Low densities of epiphytic bacteria from the marine alga *Ulva australis* inhibit settlement of fouling organisms. Appl. Environ. Microbiol., 73 (24): 7844-7852.

RASCH M, BUCH C, AUSTIN B, et al., 2004. An inhibitor of bacterial quorum sensing reduces mortalities caused by *Vibriosis* in rainbow trout (*Oncorhynchus mykiss*, Walbaum). Syst. Appl. Microbiol., 27

(3): 350-359.

RASMUSSEN T B, MANEFIELD M, ANDERSEN J B, et al., 2000. How Delisea pulchra furanones affect quorum sensing and swarming motility in *Serratia liquefaciens* MG1. Microbiology, 146 (12): 3237-3244.

RASMUSSEN B B, NIELSEN K F, MACHADO H, et al., 2014. Global and phylogenetic distribution of quorum sensing signals, acyl homoserine lactones, in the family of *Vibrionaceae*. Mar. Drugs 12 (11): 5527-5546.

REIDL J, KLOSE K E, 2002. *Vibrio* cholerae and cholera: out of the water and into the host. FEMS Microbiol. Rev., 26 (2): 125-139.

REZZONICO F, DUFFY B, 2008. Lack of genomic evidence of AI-2 receptors suggests a non-quorum sensing role for *luxS* in most bacteria. BMC Microbiol., 8 (1): 154.

RIVAS M, SEEGER M, JEDLICKI E, et al., 2007. Second acyl homoserine lactone production system in the extreme acidophile *Acidithiobacillus ferrooxidans*. Appl. Environ. Microbiol., 73 (10): 3225-3231.

ROLLAND J L, STIEN D, SANCHEZ-FERANDIN S, et al., 2016. Quorum sensing and quorum quenching in the phycosphere of phytoplankton: a case of chemical interactions in ecology. J. Chem. Ecol., 42 (12): 1-11.

ROMERO M, MARTIN-CUADRADO A B, ROCA-RIVADA A, et al., 2011. Quorum quenching in cultivable bacteria from dense marine coastal microbial communities. FEMS Microbiol. Ecol., 75 (2): 205-217.

ROMERO M, MARTIN-CUADRADO A-B, OTERO A, 2012. Determination of whether quorum quenching is a common activity in marine bacteria by analysis of cultivable bacteria and metagenomic sequences. Appl. Environ. Microbiol., 78 (17): 6345-6348.

ROMERO M, MURAS A, MAYER C, et al., 2014. In vitro quenching of fish pathogen *Edwardsiella tarda* AHL production using marine bacterium *Tenacibaculum* sp. strain 20J cell extracts. Dis. Aquat. Organ., 108 (3): 217-225.

ROMERO M, MAYER C, MURAS A, et al., 2015. Silencing bacterial communication through enzymatic quorumsensing inhibition. In: Kalia, V. C.(Ed.), Quorum Sensing vs. Quorum Quenching: A Battle With No End in Sight. Springer, Delhi, pp: 219-236.

ROSENBERG E, KOREN O, RESHEF L, et al., 2007. The role of microorganisms in coral health, disease and evolution. Nat. Rev. Microbiol., 5 (5): 355.

RUI H, LIU Q, MA Y, et al., 2008. Roles of LuxRin regulating extracellular alkaline serine protease a, extracellular polysaccharide and mobility of *Vibrio alginolyticus*. FEMS Microbiol. Lett., 285 (2): 155-162.

RUTHERFORD S T, VAN KESSEL J C, SHAO Y, et al., 2011. AphA and LuxR/HapR reciprocally control quorum sensing in *Vibrios*. Genes Dev., 25 (4): 397-408.

SAURAV K, BAR-SHALOM R, HABER M, et al., 2016a. In search of alternative antibiotic drugs: quorum-quenching activity in sponges and their bacterial isolates. Front. Microbiol., 7: 416.

SAURAV K, BURGSDORF I, TETA R, et al., 2016b. Isolation of marine *Paracoccus* sp. Ss63 from the sponge *Sarcotragus* sp. and characterization of its quorum-sensing chemical-signaling molecules by LC-MS/MS analysis. Isr. J. Chem., 56 (5): 330-340.

SAURAV K, COSTANTINO V, VENTURI V, et al., 2017. Quorum sensing inhibitors from the sea discovered using bacterial N-acyl-homoserine lactone-based biosensors. Mar. Drugs, 15 (3): 53.

SCHAEFER A L, HANZELKA B L, PARSEK M R, et al., 2000. Detection, purification, and structural elucidation of the acylhomoserine lactone inducer of *Vibrio fischeri* luminescence and other related molecules. Methods Enzymol., 305: 288-301.

SCHAEFER A L, TAYLOR T A, BEATTY J T, et al., 2002. Long-chain acyl-homoserine lactone quorum-sensing regulation of *Rhodobacter capsulatus* gene transfer agent production. J. Bacteriol., 184（23）：6515-6521.

SCHAEFER A L, GREENBERG E P, OLIVER C M, et al., 2008. A new class of homoserine lactone quorum-sensing signals. Nature, 454（7204）：595-599.

SHAO C P, HOR L I, 2001. Regulation of metalloprotease gene expression in *Vibrio vulnificus* by a *Vibrio harveyi* LuxR homologue. J. Bacteriol., 183（4）：1369-1375.

SHARIF D I, GALLON J, SMITH C J, et al., 2008. Quorum sensing in *Cyanobacteria*: N-octanoyl-homoserine lactone release and response, by the epilithic colonial cyanobacterium Gloeothece PCC6909. ISME J., 2（12）：1171-1182.

SIMON M, GROSSART H P, SCHWEITZER B, et al., 2002. Microbial ecology of organic aggregates in aquatic ecosystems. Aquat. Microb. Ecol., 28（2）：175-211.

SKERRATT J, BOWMAN J, HALLEGRAEFF G, et al., 2002. Algicidal bacteria associated with blooms of a toxic dinoflagellate in a temperate Australian estuary. Mar. Ecol. Prog. Ser., 244：1-15.

SLIGHTOM R N, BUCHAN A, 2009. Surface colonization by marine roseobacters: integrating genotype and phenotype. Appl. Environ. Microbiol., 75（19）：6027-6037.

SRIVASTAVA D, WATERS C M, 2012. A tangled web: regulatory connections between quorum sensing and cyclic di-GMP. J. Bacteriol., 194（17）：4485-4493.

SRIVASTAVA D, HARRIS R C, WATERS C M, 2011. Integration of cyclic di-GMP and quorum sensing in the control of vpsT and aphA in *Vibrio cholerae*. J. Bacteriol., 193（22）：6331-6341.

STEINDLER L, VENTURI V, 2006. Detection of quorum-sensing N-acyl homoserine lactone signal molecules by bacterial biosensors. FEMS Microbiol. Lett., 266（1）：1-9.

SURETTE M G, MILLER M B, BASSLER B L, 1999. Quorum sensing in *Escherichia coli*, *Salmonella typhimurium*, and *Vibrio harveyi*: a new family of genes responsible for autoinducer production. Proc. Natl. Acad. Sci. U. S. A., 96（4）：1639-1644.

SVENNINGSEN S L, TU K C, BASSLER B L, 2009. Gene dosage compensation calibrates four regulatory RNAs to control *Vibrio cholerae* quorum sensing. EMBO J., 28（4）：429-439.

TAIT K, HAVENHAND J, 2013. Investigating a possible role for the bacterial signal molecules N-acylhomoserine lactones in *Balanus improvisus* cyprid settlement. Mol. Ecol., 22（9）：2588-2602.

TAIT K, JOINT I, DAYKIN M, et al., 2005. Disruption of quorum sensing in seawater abolishes attraction of zoospores of the green alga Ulva to bacterial biofilms. Environ. Microbiol., 7（2）：229-240.

TAIT K, HUTCHISON Z, THOMPSON F L, et al., 2010. Quorum sensing signal production and inhibition by coral-associated vibrios. Environ. Microbiol. Rep., 2（1）：145-150.

TANG K, SU Y, BRACKMAN G, et al., 2015. MomL, a novel marine-derived N-acyl homoserine lactonase from *Muricauda olearia*. Appl. Environ. Microbiol., 81（2）：774-782.

TAYLOR M W, SCHUPP P J, BAILLIE H J, et al., 2004. Evidence for acyl homoserine lactone signal production in bacteria associated with marine sponges. Appl. Environ. Microbiol., 70（7）：4387-4389.

TEASDALE M E, DONOVAN K A, FORSCHNER-DANCAUSE S R, et al., 2011. Gram-positive marine bacteria as a potential resource for the discovery of quorum sensing inhibitors. Marine Biotechnol., 13（4）：722-732.

TELLO E, CASTELLANOS L, AREVALO-FERRO C, et al., 2012. Disruption in quorum-sensing systems and bacterial biofilm inhibition by cembranoid diterpenes isolated from the octocoral *Eunicea knighti*. J. Nat. Prod., 75（9）：1637-1642.

TEPLITSKI M, KREDIET C J, MEYER J L, et al., 2016. Microbial interactions on coral surfaces and within the coral holobiont. In: Goffredo, S., Dubinsky, Z. (Eds.), The Cnidaria, Past, Present and Future. 2016. Springer International Publishing, pp: 331-346.

THIEL V, BRINKHOFF T, DICKSCHAT J S, et al., 2010. Identification and biosynthesis of tropone derivatives and sulfur volatiles produced by bacteria of the marine *Roseobacter* clade. Org. Biomol. Chem., 8 (1): 234-246.

THOLE S, KALHOEFER D, VOGET S, et al., 2012. *Phaeobacter gallaeciensis* genomes from globally opposite locations reveal high similarity of adaptation to surface life. ISME J., 6 (12): 2229-2244.

THOMAS K V, BROOKS S, 2010. The environmental fate and effects of antifouling paint biocides. Biofouling, 26 (1): 73-88.

TINH N T N, YEN V H N, DIERCKENS K, et al., 2008. An acyl homoserine lactone-degrading microbial community improves the survival of first-feeding turbot larvae (*Scophthalmus maximus* L.). Aquaculture, 285 (1-4): 56-62.

TORRES M, RUBIO-PORTILLO E, ANTON J, et al., 2016. Selection of the N-acylhomoserine lactone-degrading bacterium *Alteromonas stellipolaris* PQQ-42 and of its potential for biocontrol in aquaculture. Front. Microbiol., 7: 646.

TU K C, BASSLER B L, 2007. Multiple small RNAs act additively to integrate sensory information and control quorum sensing in *Vibrio harveyi*. Genes Dev., 21 (2): 221-233.

VACELET J, DONADEY C, 1977. Electron microscope study of the association between some sponges and bacteria. J. Exp. Mar. Biol. Ecol., 30 (3): 301-314.

VAN KESSEL J C, RUTHERFORD S T, SHAO Y, et al., 2013. Individual and combined roles of the master regulators AphA and LuxR in control of the *Vibrio harveyi* quorum-sensing regulon. J. Bacteriol., 195 (3): 436-443.

VAN MOOY B A, HMELO L R, SOFEN L E, et al., 2012. Quorum sensing control of phosphorus acquisition in *Trichodesmium consortia*. ISME J., 6 (2): 422-429.

VINOJ G, VASEEHARAN B, THOMAS S, et al., 2014. Quorum-quenching activity of the AHL lactonase from *Bacillus licheniformis* DAHB1 inhibits *Vibrio* biofilm formation in vitro and reduces shrimp intestinal colonisation and mortality. Marine Biotechnol., 16 (6): 707-715.

WAGNER-DÖBLER I, THIEL V, EBERL L, et al., 2005. Discovery of complex mixtures of novel long-chain quorum sensing signals in free-living and host-associated marine *Alphaproteobacteria*. ChemBioChem, 6 (12): 2195-2206.

WANG L, LING Y, JIANG H W, et al., 2013. AphA is required for biofilm formation, motility, and virulence in pandemic *Vibrio parahaemolyticus*. Int. J. Food Microbiol., 160 (3): 245-251.

WATERS C M, LU W, RABINOWITZ J D, et al., 2008. Quorum sensing controls biofilm formation in *Vibrio cholerae* through modulation of cyclic di-GMP levels and repression of vpsT. J. Bacteriol., 190 (7): 2527-2536.

WEI Y, PEREZ L J, NG W L, et al., 2011. Mechanism of *Vibrio cholerae* autoinducer-1 biosynthesis. ACS Chem. Biol., 6 (4): 356-365.

WHITELEY M, DIGGLE S P, GREENBERG E P, 2017. Progress in and promise of bacterial quorum sensing research. Nature, 551 (7680): 313.

WIECZOREK S K, TODD C D, 1998. Inhibition and facilitation of settlement of epifaunal marine invertebrate larvae by microbial biofilm cues. Biofouling, 12 (1-3): 1-118.

WOOD D W, PIERSON L S, 1996. The phzI gene of *Pseudomonas aureofaciens* 30-84 is responsible for the production of a diffusible signal required for phenazine antibiotic production. Gene, 168 (1): 49-53.

YANG Q, DEFOIRDT T, 2015. Quorum sensing positively regulates flagellar motility in pathogenic *Vibrio harveyi*. Environ. Microbiol., 17 (4): 960-968.

YANG M, FREY E M, LIU Z, et al., 2010. The virulence transcriptional activator AphA enhances biofilm formation by *Vibrio cholerae* by activating expression of the biofilm regulator VpsT. Infect. Immun., 78 (2): 697-703.

YANG Q, HAN Y, ZHANG X H, 2011. Detection of quorum sensing signal molecules in the family *Vibrionaceae*. J. Appl. Microbiol., 110 (6): 1438-1448.

YATES E A, PHILIPP B, BUCKLEY C, et al., 2002. *N*-Acylhomoserine lactones undergo lactonolysis in a pH-, temperature-, and acyl chain length-dependent manner during growth of *Yersinia pseudotuberculosis* and *Pseudomonas aeruginosa*. Infect. Immun., 70 (10): 5635-5646.

YILDIZ F H, VISICK K L, 2009. *Vibrio* biofilms: so much the same yet so different. Trends Microbiol., 17 (3): 109-118.

ZAN J, FRICKE W F, FUQUA C, et al., 2011. Genome sequence of *Ruegeria* sp. strain KLH11, an *N*-acylhomoserine lactone-producing bacterium isolated from the marine sponge Mycale laxissima. J. Bacteriol., 193 (18): 5011-5012.

ZAN J, CICIRELLI E M, MOHAMED N M, et al., 2012. A complex LuxR-LuxI type quorum sensing network in a roseobacterial marine sponge symbiont activates flagellar motility and inhibits biofilm formation. Mol. Microbiol., 85 (5): 916-933.

ZAN J D, HEINDL J E, LIU Y, et al., 2013. The CckA-ChpT-CtrA phosphorelay system is regulated by quorum sensing and controls flagellar motility in the marine sponge symbiont *Ruegeria* sp. KLH11. PLoS One, 8 (6): 1932-6203.

ZAN J, CHOI O, MEHARENA H, et al., 2015. A solo *luxI*-type gene directs acylhomoserine lactone synthesis and contributes to motility control in the marine sponge symbiont *Ruegeria* sp. KLH11. Microbiology, 161 (Pt 1): 50-56.

ZHAI C, ZHANG P, SHEN F, et al., 2012. Does *Microcystis aeruginosa* have quorum sensing? FEMS Microbiol. Lett., 336 (1): 38-44.

ZHU J, MEKALANOS J J, 2003. Quorum sensing-dependent biofilms enhance colonization in *Vibrio cholerae*. Dev. Cell, 5 (4): 647-656.

ZIMMER B L, MAY A L, BHEDI C D, et al., 2014. Quorum sensing signal production and microbial interactions in a polymicrobial disease of corals and the coral surface mucopolysaccharide layer. PLoS One, 9 (9): e108541.

第4章　极端微生物中的群体感应

Gennaro Roberto Abbamondi[*], Margarita Kambourova[†], Annarita Poli[*], Ilaria Finore[*], Barbara Nicolaus[*]

[*] National Research Council of Italy—Institute of Biomolecular Chemistry, Pozzuoli, Italy

[†] Institute of Microbiology—Bulgarian Academy of Sciences, Sofia, Bulgaria

1 极端微生物和极端生态系统简介

从物理和化学的角度来看，极端微生物是指能在极其恶劣的条件下生存的生物，地球上的大多数生物都会在这种条件下艰难地生存或灭亡。这些微生物能在一个或多个化学-物理胁迫的环境中执行任何重要的功能，这对于科学界来说是特别有趣，因为它们在不同的工业领域和生物修复过程中具有潜在的应用价值（Antranikian et al., 2005; Canganella and Wiegel, 2014; Poli et al., 2017）。而且，它们对于理解进化非常有用，因为它们可能是我们星球上生命起源时的第一批"居民"。极端微生物隶属于古菌、细菌和真核三个领域，可在热液喷口和泉水、咸水湖、石盐晶体、极地冰和湖泊、火山区、沙漠或厌氧条件下的环境生态位中被发现。地球以外生命形式的存在需要对经典生命极限的延伸：嗜极生物在温度、盐、pH值、压力、干旱和干燥等恶劣条件下的抵抗能力，使这些生命体成为评估其他星球可居住性的良好候选者（Di Donato et al., 2018; Mastascusa et al., 2014）。

随着先进基因技术的不断发展，鉴定新的代谢物和酶可应用于医疗、食品和工业领域，以至于现在将微生物认为是细胞工厂（Dalmaso et al., 2015; van den Burg, 2003）。在极端条件下（pH值、温度、压力、盐和营养物）生存和繁殖的能力导致产生各种生物技术上有用的分子，例如脂质、酶、多糖和可溶性物质，其在多个工业过程中被使用（Poli et al., 2017）。许多极端嗜热酶和内源性化合物在食品工业、洗涤剂制备、药物应用和遗传学研究中都取得了成功的应用。尤其是来源于嗜热微生物的酶，被称为嗜热酶，是工业上感兴趣的新型催化剂的良好来源（Finore et al., 2016; Poli et al., 2010）。极端微生物包括但并不局限于嗜热菌（thermophiles）（高温）、嗜冷菌（psychrophiles）（低温）、嗜酸菌（acidophiles）（低pH值）、嗜碱菌（alkaliphiles）（高pH值）、嗜压菌（piezophiles）（高压，以前称为嗜压微生物）、嗜盐菌（halophiles）（高盐浓度）、嗜高渗菌（osmophiles）（高浓度有机溶质）、寡营养菌（oligotrophs）（低浓度溶质和/或营养物质）和嗜干菌（xerophiles）（非常干燥的环境）。极端微生物也包括能在高金属浓度或高剂量辐射条件下生长的微生物。

研究最多的极端环境是那些受到高温影响的环境，其中的微生物被命名为嗜热菌（最适温度>45℃）或超嗜热菌（>70℃和高达113℃）。温度极高的热场所是温泉、深海喷口和深层地下区域。温泉分布在陆地地壳上，温热的地热水在这里涌出。而且，地下水的输送引起水的矿物质含量因温度较高而富集，使泉水具有治疗性。深海喷口是深海海底热液的排放口。新出现的水温度在 60~400℃；地下水所输送的矿物质与寒冷的海水接触形成沉淀物，这些沉淀物负责构建烟囱结构，其可以是白色或黑色的排烟口，根据矿物的性质：分别是钡、钙、硅或硫化物。这些生态位的特征是冷热水接近、酸性水（嗜酸菌）、无太阳能、高压（嗜压微生物）、无氧（厌氧菌）（Reith，2011）。

与炎热地区相对应的是寒冷和冰冷的地方，这些地方遍布全球，如高山和北极的土壤、高纬度和深海水域、极地冰、冰川和雪原。这里的微生物被称为嗜冷微生物。普遍认为，从深海喷口分离出的细菌代表了一种宝贵资源，可用于生物技术的特殊分子，同时对产生这些特殊分子的作用机制也提供了重要见解。大多数深海环境受到高压、低温和低营养浓度的影响。然而，许多深海耐冷细菌生活在深海生态群落中，并得到了广泛研究，因为其揭示细菌在不断变化的深海生态系统中的形态、生理和代谢适应的重要生态特征。特别是，许多嗜冷菌能产生胞外多糖（EPSs）从而保护细胞免受冻结；事实上，高浓度的 EPSs 已在南极和北极海洋细菌中被发现（Poli et al.，2010）。有趣的是，颗粒聚集物在世界海洋中是无处不在且含量丰富。海洋细菌从聚集体中受益，因为它们接近其他细胞和表面并提供相互作用和营养吸收的机会。由细菌分泌的 EPSs 提供了一个网络来将这些结构固定在一起，为微生物的生存提供了一个合适的微生境。

另一类极端生境是以高盐浓度为特征的微生物，生活在这里的微生物被称为嗜盐菌。盐环境被定义为盐浓度比海水观察到的浓度高五倍的环境；这些环境可以是自然形成的，也可以是人工创造的。例如，盐湖、内陆海、海水蒸发池、腌制盐水、盐渍食品和盐碱土壤。所有这些生态系统都含有大量的盐，且趋于饱和；通常 pH 值介于中性和碱性之间（嗜盐碱菌）。高盐环境存在于各种水生和陆生生态系统中，在这种环境下中栖息着耐盐微生物和嗜盐微生物（最适盐浓度在 0.5~2.5 mol/L NaCl）以及极端嗜盐微生物（超过 2.5 mol/L NaCl）。中度和极端嗜盐菌不仅从高盐生态系统（盐湖、海盐场和盐碱土壤）中分离出来，而且还从碱性生态系统（碱性湖泊）中分离出来。为了生存，嗜盐微生物在高盐条件下形成了各种生化策略，包括合成相溶性物质，以维持细胞结构和功能。它们的产物如赤藓糖、细菌视黄醇蛋白、胞外多糖、水解酶和生物表面活性剂，具有明显的工业价值（Oren，2010；Poli et al.，2017）。

此外，有一种多极端生物（即至少需要三种不同的压力参数才能生长的微生物），例如高盐浓度、高温和碱性 pH 值（Karan et al.，2013）。事实上，许多自然环境都具有两个或更多的极端参数。举几个例子，许多温泉同时具有酸性或碱性，且通常富含金属成分；一些高盐度湖泊表现出非常强的碱性；深海环境通常很寒冷、寡营养（养分含量非常低）且承受高压。有趣的是，从这些环境中分离出越来越多的物种和菌株，被发现能耐受多种极端条件（Seckbach and Rampelotto，2015）。

最初对极端生境中的极端微生物的研究主要依赖于经典培养法进行菌株分离。通过这种方法，才能在实验室中分离出代谢和生理需要的微生物。为了克服这一限制，最近研发了宏基因组方法来探索和获取未培养的微生物群落（Finore et al.，2014；Hedlund et al.，2014；Leon et al.，2014）。

生态系统的生物多样性成为深入研究的对象，从而获得关于世界各地微生物群落分布的丰富信息。从这些天然来源中分离新化合物仍是主要目标，并结合对化合物与生物系统之间关系的研究。尤其是来自极端海洋微生物的酶广泛应用于化学工业中，它们作为生物催化剂在其他行业中也是必不可少的。例如，这些酶被应用于啤酒和生物燃料的生产，以及生物洗涤剂和纸张工业。海洋生物如细菌、真菌、海绵和藻类被确定为一种未被探索出来的酶的来源，但它们仍然未被完全研究。事实上，只有极少部分的海洋酶达到了商业化阶段。此外，寻找高产胞外多糖（EPS）的菌株是一个持续进行的过程，改善发酵条件及其后续步骤对EPS的回收仍在进行中。通过遗传和代谢工程生产具有明确性的聚合物，以及探索低成本底物的生产，对于广泛使用微生物来源的生物聚合物是至关重要的（Finore et al., 2014）。

2 极端微生物的群体感应

群体感应是指几种生物通过群体依赖的信号来调节基因表达的分子通信机制。这种集体行为的证据已在生命的三个领域中被描述：细菌、古细菌和真核生物中。以细菌和古细菌为重点，QS 系统调控着细菌和古细菌的多种生理活动，包括运动性、生物被膜发育、生长抑制、毒力表达、质粒接合等（Williams, 2007）。在 QS 基础上，微生物合成小分子可扩散化合物，称为自诱导物（AIs），其可在胞外环境中释放。研究最多的 QS 机制是基于 N-酰基高丝氨酸内酯（AHLs）（AI-1 QS）的合成，但在不同微生物中也发现了其他不同信号分子的 QS 系统。在 AI-2 体系中，微生物的集体行为受 4,5-二羟基-2,3-戊二酮（DPD）、（$2R$,$4S$）-2-甲基-2,3,3,4-四羟基四氢呋喃（R-THMF）和（$2S$,$4S$）-2-甲基-2,3,3,4-四羟基四氢呋喃硼酸盐（S-THMF）控制；CAI-1 系统是基于霍乱弧菌自诱导物-1（CAI-1）和（Z）-3-氨基十一烯-2-酮（Ea-C8-CAI-1）合成；而且，多肽作为二酮哌嗪（DKPs）参与基因表达的调控（Tommonaro et al., 2015；Abbamondi et al., 2014）。

这些分子在周围环境中积累，当它们的浓度达到临界阈值时，菌株开始协调基因表达并调节特定的表型结果（Fuqua et al., 1994）。

在极端环境中的分子通信系统尚不清楚，但 QS 可能在嗜极微生物的生存策略中起着关键作用。而且，对这些恶劣生境中 QS 的深入理解，有助于开发合成"极端生物分子"的创新策略及其在生物技术过程中的应用。本章介绍了对嗜极微生物 QS 研究的最新进展。

2.1 嗜盐菌

嗜盐菌是一群能在高盐条件下繁衍的生物。在这种极端环境中，盐浓度超过海水，氯化钠甚至可达 20%~30%。实际上，嗜盐菌需要在这些异常的高盐浓度栖息地中生存。液态水是每个活细胞的新陈代谢反应必不可少的溶剂，其缺乏会导致这群生物进化出不同的策略来适应这种极端条件。细菌主要通过产生高浓度的相溶性物质（小有机分子）来平衡这些恶劣环境的渗透压，其并不参与细胞代谢（氨基酸、糖类、多醇等）。细胞内高浓度盐（氯化钾）的积累是另一种策略，主要采用古菌和一些嗜盐厌氧细菌（Saum and Müller, 2008）。

高盐环境特点通常具有碱性（高 pH 值）和/或高温；因此，嗜盐菌株也可能是嗜热菌和/或嗜碱菌（Montgomery et al., 2013；Pikuta et al., 2007）。

据文献报道，QS 可能在复杂环境生态位上如高盐场所参与调节细菌适应和生长的过程（Tommonaro et al., 2015）。在嗜盐菌中，QS 的研究主要来自对盐单胞菌属（*Halomonas*）。Llamas 等（2005）分析了盐单胞菌属不同中度嗜盐细菌产生参与 QS 机制的信号分子。分析了 11 株胞外多糖的菌株：广盐盐单胞菌（*H. eurihalina*）（3 株）、摩尔盐草胞菌（*H. maura*）（4 株）、樊氏盐单胞菌（*H. ventosae*）（1 株）和安替卡尔盐单胞菌（*H. anticariensis*）（3 株）。通过指示菌株根瘤农杆菌 NTL4（pZLR4）和紫色色素杆菌 CV026

检测到了 N-酰基-高丝氨酸内酯（AHLs）的存在（Luo et al., 2001；McClean et al., 1997）。所有盐单胞菌菌株中都检测到了 AHL 信号分子；这一发现使作者推测 QS 可能在恶劣的生境中发挥关键作用，特别是在生物膜形成和 EPS 产生方面。EPSs 作为防止干燥的保护层具有保护功能，此外，EPSs 促进细胞间的化学通信（Decho, 2000）。通过质谱（GC-MS 和 ESI 串联质谱）也鉴定了来自安替卡尔盐单胞菌 FP35（T）中的一些 AHLs：C4-HSL、C6-HSL、C8-HSL 和 C12-HSL（Llamas et al., 2005）。

同一课题组对安替卡尔盐单胞菌 FP35(T)的 QS 进行了深入的研究。发现其 QS 系统由 *luxR/luxI* 同源物：*hanR*（转录调控基因）和 *hanI*（AHL 合成酶基因）组成（Tahrioui et al., 2011）。该伽马变形菌的基因组序列草图随后在《基因组公告》上报道（Tahrioui et al., 2013a）。Tahrioui 等（2013b）对盐单胞菌科（Halomonadaceae）家族的 43 株菌进行 AHLs 的筛选。在这种情况下，所有被检测的细菌都激活了指示菌株（用于中长酰基链）。PCR 和 DNA 测序分析表明，在大部分物种中都存在 *luxI* 同源物，并检测到 C6-HSL 是最主要的 AHL 分子（Tahrioui et al., 2013b）。PCR 和 DNA 测序分析表明，在大多数研究的物种中存在 *luxI* 同源物，并且检测到 C6-HSL 作为最主要的 AHL 分子。嗜盐单胞菌 AAD6 是从土耳其爱琴海地区的盐田地区分离出来的一种中度嗜盐性、产胞外多糖的细菌（Poli et al., 2013）。通过 LC-MS 分析，在固定相培养物中提取二氯甲烷萃取物鉴定出一种未取代的酰基侧链长度为 C16 的 AHL。推测 EPS 的生长阶段依赖性产生可能受到 QS 的调控（Abbamondi et al., 2016）。

嗜盐芽孢杆菌属（*Halobacillus halophilus*）是从德国海岸一片盐沼中分离出来的一株中度嗜盐菌。其以生长阶段依赖的方式产生自诱导物-2（AI-2），高度依赖 Cl 离子的存在。更具体地说，lux S 同源蛋白受阴离子浓度上调。QS 可能在盐沉淀（氯离子调节）中发挥重要作用，但在细胞运动中也发挥重要作用（Averhoff and Müller, 2010; Sewald et al., 2007）。表 4-1 报道了嗜盐菌中 QS 证据的实例。

表 4-1 嗜盐菌群体感应（QS）

菌株	QS 信号分子/已确定的基因	生物报告菌株	调节活动	参考文献
摩尔盐单胞菌（4 株）、樊氏盐单胞菌（1 株）和安替卡尔盐单胞菌（3 株）	N-酰基高丝氨酸内酯（AHLs）	根瘤农杆菌 NTL4（pZLR4）和紫色色素杆菌 CV026	生物膜形成和胞外多糖（EPS）产生的调节	Lamas et al., 2005
安替卡尔盐单胞菌 FP35（T）	C4-HSL、C6-HSL、C8-HSL 和 C12-HSL *hanR/hanI*（*luxR/luxI* 同系物）	根瘤农杆菌 NTL4（pZLR4）和紫色色素杆菌 CV026		Lamas et al., 2005; Tahrioui et al., 2011
盐单胞菌科家族（43 株）	C6-HSL（最主要的 AHL）LuxI 同源物	根瘤农杆菌 NTL4（pZLR4）		Tahrioui et al., 2013b
嗜盐单胞菌 AAD6	C16-HSL	根瘤农杆菌 NTL4（pCF218；pCF372）	推测调控 EPS 的产生	Abbamondi et al., 2016

续表

菌株	QS 信号分子/已确定的基因	生物报告菌株	调节活动	参考文献
喜盐芽孢杆菌	自诱导物-2（AI-2）		盐沉淀的调节（氯离子调节和细胞运动）	Averhoff and Müller, 2010; Sewald et al., 2007

2.2 嗜热菌

嗜热微生物已成功地在极端炎热的陆地和海洋地区定殖，这些地区通常对地球上的大多数生物都是不利的。在生命的三个领域（古菌、细菌和真核）中可找到嗜热菌，但主要是细菌和古菌生活在如此恶劣的环境中。嗜热微生物实际上能在至少 60℃ 的异常高温下生长和增殖。根据最佳生长温度（OGT），可以将其分为两类：一类嗜热微生物在 60~80℃ 范围内生长，而另一类嗜热微生物能在极热温度（80~110℃）下生长。

在过去的几年中，越来越多的研究人员出于不同的原因对嗜热菌进行了研究。首先，为了更好地理解生命如何在极端温度下增殖，深入研究这些生物的生存机制是非常重要的。此外，这些知识可以为我们提供关于地球上已知最早的生命形式的信息。此外，嗜热微生物的独特代谢物使其非常适用于生物技术应用。具体来说，嗜热菌的胞外多糖和胞外酶即使在极高的温度下也能有效地发挥作用，这为它们在生物技术过程中的应用提供了显著优势（Raddadi et al., 2015）。

嗜热菌通过 QS 机制进行交流的能力是最近被发现。事实上，天然 AHLs 的一个限制是它们的相对热不稳定性；因此，在极高温度下预计不会发现活性信号分子（Schopf et al., 2008）。

Johnson 等（2005）在湿热环境中发现 QS 机制的第一个证据。作者观察到海栖热袍菌（*Thermotoga maritima*）与甲烷球菌（*Methanococcus jannaschii*）共培养时，细胞密度和胞外多糖产量显著增加。在共培养中，还检测到一个编码多肽的基因（TM0504）上调，该基因包含一个参与 QS 机制的多肽共同基序。当在培养基中添加合成的外源肽（基于 TM0504）时，胞外多糖的产生在较低的细胞密度下被诱导。这一发现使作者推测，基于肽的 QS 系统可能与湿热环境中微生物生态策略有关，尤其是在种群密度依赖的胞外多糖形成中。

为了研究基于 QS 的自诱导物-2（AI-2）在热液环境中的作用，海栖热袍菌和激烈火球菌（*Pyrococcus furiosus*）被选为高温嗜热模型，其被认为是嗜中温细菌中一种普遍存在的种间化学通信系统。尽管在这两株菌株中均检测到 AI-2 的产生，但 AI-2（和其他呋喃酮）在热液生境中的重要性需要进一步研究。事实上，在模型菌株中并未观察到 QS 表型的明显变化（Nichols et al., 2009）。

在伊朗西北省的温泉中分离出栖热菌属（*Thermus* sp.）GH5 菌株并检测到了 AHL 信号传导系统的存在（Yousefi-Nejad et al., 2011）。作者研究了这种嗜热细菌温度从 75℃（最适生长温度）下降到 45℃（持续 2/5 h）后的蛋白质组。在早期冷激条件下，乙酰转移酶和 S-腺苷甲硫氨酸合成酶的过表达，证明 AHL 信号传导在嗜热微生物对冷激反应生存机制中起到了作用。这些蛋白质实际上参与了 AHL 合成循环。尤其，将这些 AHL 前体的合成与生物膜的积累关联起来。

嗜热脂肪地芽孢杆菌（*Geobacillus stearothermophilus*）T-6 是一种产生胞外木聚糖酶（Xyn10A）的嗜热土壤细菌。据观察，该酶在早期指数生长阶段的活性比晚期高出 50 倍。此外，向低细胞密度培养物中添加煮沸处理的培养基也增加了 Xyn10A 的表达。这一发现表明，细胞密度调控（QS）机制可能控制嗜热脂肪地芽孢杆菌 T-6 木聚糖酶的产生（Shulami et al.，2014）。

来自深海热液喷口的两株 ε-变形菌门细菌：嗜无机物硫卵菌和嗜热硫氧化菌也检测到了 QS 活性。更确切地说，在生长过程中检测到了 *luxS* 基因的表达和 QS 信号的产生。而且，*luxS* 的转录在活跃的深海热液喷口中以 ε-变形菌为主的生物膜中被鉴定出来（Pérez-Rodríguez et al.，2015）。

在嗜热厌氧的热纤梭菌（*Clostridium thermocellum*）中检测到一个小型新基因 *Cthe_3383*。该基因高度表达，并编码一个可能涉及 QS 机制的多肽（Wilson et al.，2013）。

破坏 QS 信号的过程，如 QS 抑制剂（QSIs）或 QQ 酶，已被描述为群体淬灭（QQ）（Grandclément et al.，2016）。尤其是 QQ 酶，作为一种抑制不良细菌性状的创新策略，在医学、农业、水产养殖和生物污损的生物技术应用中是非常有前途的（Bzdrenga et al.，2017）。

研究发现，一株属于地芽孢杆菌属的菌株热解木糖地芽孢杆菌（*Geobacillus caldoxylosilyticus*）能产生一种参与 QQ 机制的金属 β-内酰胺酶；事实上，该酶能水解不同的内酯，包括参与 QS 系统的 AHLs（Bergonzi et al.，2016）。

在极端嗜热菌（*Thermaerobacter marianensis*）中也检测 QQ。该菌株是从马里亚纳海沟中分离得到的，其最适生长温度为 75℃。极端嗜热菌的基因组中存在一个 AHL 降解基因同源物（*aiiT*）。AiiT 表现出高耐热的 AHL 内酯酶活性，在 60~80℃ 时活性最高（Morohoshi et al.，2015）。表 4-2 总结了嗜热菌中 QS/QQ 更有代表性的发现。

表 4-2　嗜热菌群体感应（QS）和群体淬灭（QQ）

菌株	QS-QQ 分子/已确定的基因	调节活性/活性	参考文献
海栖热袍菌（与甲烷球菌共培养）	多肽 TM0504 的基因编码（基于肽的 QS 系统）	推测在胞外多糖（EPS）形成中的作用	Johnson et al.，2005
海栖热袍菌，激烈火球菌	AI-2		Nichols et al.，2009
栖热菌属 GH5	AHL 信号系统	推测生物膜的积累和产生在应对冷激的生存机制中发挥重要作用	Yousefi-Nejad et al.，2011
嗜热脂肪地芽孢杆菌 T-6		推测调控胞外木聚糖酶（Xyn10A）的产生	Shulami et al.，2014
热解木糖地芽孢杆菌	金属 β-内酰胺酶（QQ）		Bergonzi et al.，2016
极端嗜热菌	*aiiT*（AHL-降解基因同源物），其编码高耐热性 AHL-内酯酶（QQ）的 *luxS* 基因		Morohoshi et al.，2015

菌株	QS-QQ 分子/已确定的基因	调节活性/活性	参考文献
嗜无机物硫卵菌，嗜热硫氧化菌，热纤梭菌	Cthe-3383（编码可能参与 QS 机制的肽段的新基因）		Perez-Rodriguez et al., 2015；Wilson et al., 2013

2.3 嗜酸菌

嗜酸菌的术语是用来描述能够在强酸性环境中栖息的生物，它们可以适应很低的 pH 值。酸性环境在天然（如酸性硫磺泉）和人造（如来自采矿金属和煤炭活动）环境中都可以找到（Oren，2010）。酸性环境往往重金属元素含量高，而嗜酸微生物往往对这些潜在毒性元素具有抗性。此外，也有一些嗜酸菌能抵抗高温。这些微生物能在如此恶劣的环境中生存，因为它们将细胞内 pH 值维持在接近中性的策略，实际上它们的细胞质膜可能支持极低的 pH 值梯度。与嗜中性和嗜碱性微生物相反，它们通常保持反向（外负）的膜电位。

关于这些嗜酸微生物的 QS 机制报道较少。极端嗜酸菌嗜酸氧化亚铁硫杆菌（*Acidithiobacillus ferrooxidans*）报道了一种 Lux-like QS 系统（Rivas et al.，2007）。嗜酸氧化亚铁硫杆菌是一株化能无机自养型菌株，能在极低 pH 值（pH 值 1~2）和高浓度金属条件下增殖。基因组分析鉴定到 2 个基因（*afeI* 和 *afeR*），其编码的蛋白与 LuxI 和 LuxR 家族蛋白相似度极高。而且，利用色谱/质谱分析检测到一个长度为 C14 且未被取代的酰基侧链的 AHL。2007 年，同一课题组对嗜酸氧化亚铁硫杆菌的基因组进行了更深入的研究（Rivas et al.，2007）。根据 GC-MS 结果，鉴定了一个编码酰基转移酶（*act*）的基因，该基因可产生 AHLs 主要以 C14 酰基链为主。结果表明，两种 QS 系统对不同的外界刺激均有响应。具体而言，Act-based QS 系统在含铁培养基中高表达，而在 Lux-like QS 系统中，编码 LuxI 和 LuxR-like 蛋白的发散定向基因在含硫培养基中表达更多。在嗜酸氧化亚铁硫杆菌 ATCC 23270（T）中研究了 QS 与生物膜形成之间的关系。研究发现，QS 网络至少代表了其基因组的 141 个基因（4.5%）。有趣的是，这些基因中有 60 个（42.5%）与生物被膜的形成有关（Mamani et al.，2016）。对 5 株嗜酸氧化亚铁硫杆菌、3 株嗜酸氧化硫酸硫杆菌（*A. thiooxidans*）和 2 株氧化亚铁钩端螺旋菌（*Leptospirillum ferrooxidans*）进行了筛选，以评估它们产生 AHL 的能力。除 2 株氧化亚铁钩端螺旋菌外，所有受试菌株都能激活生物载体根瘤农杆菌 NTL4（pZLR4）（Ruiz et al.，2008）。据报道，嗜酸氧化亚铁硫杆菌能产生 9 种不同类型的 AHLs，其酰基链在 8~16 个碳之间且 C-3 位上有氧代和羟基取代（Farah et al.，2005）。Ruiz 等（2008）从 3 株嗜酸氧化硫酸硫杆菌中鉴定出两种 AHLs。具体而言，所有嗜酸氧化硫酸硫杆菌菌株均具有合成 3-oxo-C8-AHL 的能力，而只有一株菌株（DSMZ 11478）能合成 3-oxo-C6-HSL。对另一株钩端螺菌属的嗜铁钩端螺旋菌^{T}DSM 14647 进行多组学分析。检测到参与 QS 过程的代谢系统的存在（Christel et al.，2018）。最后，通过对嗜酸铁氧化菌属（*Ferrovum sp.*）JA12 基因组序列进行筛选，发现存在一种嗜热类烯还原酶（FOYE-1）。推测 FOYE-1 可能参与 QS 过程（Scholtissek et al.，2017）。

2.4 古菌

古菌是一组独立的原核生物，主要生活在恶劣的环境中。用来区分这些微生物和细菌的主要特征之一是细胞壁的组成不同。事实上，肽聚糖不是古菌细胞壁的组成成分。这些微生物具有各种类型的细胞壁，但共同的细胞壁结构是由蛋白质表面（S）层形成的。在相对有限的情况下，也有可能发现类似于假肽聚糖的多聚糖等。这些细胞结构可以由额外的S层支撑，但也可能存在完全缺乏细胞壁的古菌物种（Albers et al., 2017; Meyer and Albers, 2001）。

尽管人们对微生物通信系统的科学兴趣日益增加，但对古菌中的QS机制仍然缺乏了解。Paggi等（2003）首次描述了古菌中存在基于AHL的QS系统。研究报道了在嗜盐碱古菌隐藏嗜盐碱球菌（*Natronococcus occultus*）中检测到QS分子，可能属于AHL家族。更详细地，无细胞培养基的乙酸乙酯提取物能激活根瘤农杆菌NTL4生物传感器中的QS，该传感器对广泛的AHLs都有响应。此外，研究还描述了向低密度培养物中添加隐藏嗜盐球菌晚期指数条件培养基，可诱导蛋白酶的早期产生，这可能为控制该酶活性的细胞密度调节机制的存在提供了证据。

在古菌中，第一个具有化学特征的信号分子属于AHL家族。具体而言，研究表明，产甲烷古菌（*Methanosaeta harundinacea*）6Ac通过羧化AHL的QS系统，调节细胞组装和碳代谢通量。研究人员纯化了3个分子，分别为N-羧基-癸酰基-高丝氨酸内酯、N-羧基-十二酰基-高丝氨酸内酯和N-羧基-丁酰基-高丝氨酸内酯（Zhang et al., 2016），在火山盐古菌（*Haloferax volcanii*）DS2的培养提取物中也检测到C4和C6酰基高丝氨酸内酯样信号分子的存在；该提取物实际上可诱导其与QS生物报告物根瘤ATCC BAA-2240的反应（Megaw and Gilmore, 2017）。

从西班牙丰德彼德拉盐湖中分离出的盐陆生菌属（*Haloterrigena*）伊斯巴尼亚盐陆生菌（*Haloterrigena hispanica*）菌株中也进行了QS调查。实际上，从嗜盐古菌中第一次也是唯一一个发现二酮哌嗪（DKPs）的产生。Tommonaro等（2012）纯化了5个具有化学特征的DKPs：cyclo-(D-Pro-L-Tyr), cyclo-(L-Pro-L-Tyr), cyclo-(L-Pro-L-Val), cyclo-(L-Pro-L-Phe), cyclo-(L-Pro-L-isoLeu)。在所有分离的DKPs中，只有cyclo-(L-Pro-L-Val)刺激了两种生物报告菌株根癌杆菌根瘤农杆菌NTL4（pCF218；pCF372）和鳗弧菌DM27。

在另一株盐陆生菌属中检测到QQ活性。从阿曼沙漠高盐蓝藻垫中分离到一株顾菌，其与嗜盐陆生菌（*Haloterrigena saccharevitans*）的系统发育相似性为99.6%。非极性（二氯甲烷）和极性（水）组分均抑制紫色色素杆菌CV017生物报告物中紫罗兰酸的产生（Abed et al., 2013）。生物膜的形成受QS的控制；因此，在海洋环境中具有QQ能力的嗜盐菌株可能是一种潜在的防污剂。

最后，在两个属于硫化叶菌（*Sulfolobus*）的嗜热古菌中检测到QS，分别为硫黄矿硫化叶菌（*Sulfolobus solfataricus*）和冰岛硫化叶菌（*Sulfolobus islandicus*）作为内酯酶活性（AHL水解）的抑制作用（Hiblot et al., 2012; Ng et al., 2011）。表4-3总结了有关古菌中QS/QQ发现的主要研究报告。

表4-3 古菌的群体感应（QS）和群体淬灭（QQ）

菌株	QS-QQ分子/已确定的基因	生物报告菌株	调节活动	参考文献
隐藏嗜盐球菌	假定的AHL信号分子	根瘤农杆菌NTL4（pCF218）（pCF372）	对蛋白酶生产的假定调节	Paggi et al., 2003

续表

菌株	QS-QQ 分子/已确定的基因	生物报告菌株	调节活动	参考文献
产甲烷古菌 6Ac	N-羧基-C10-HSL, N-羧基-C12-HSL, N-羧基-C14-HSL	根瘤农杆菌 NTL4 (pZLR4)	细胞组装和碳代谢通量的调控	Zhang et al., 2012
火山盐古菌 DS2	C4-和 C6-HSL 类信号分子	根瘤农杆菌 ATCC BAA-2240		Megaw and Gilmore, 2017
嗜盐陆生菌（系统发育相似性为 99.6%）	QQ 活性	紫色色素杆菌 CV017		Abed et al., 2013
伊斯巴尼亚盐陆生菌	Cyclo-(L-Pro-L-Val)	根瘤农杆菌 NTL4 (pCF218; pCF372) 和鳗弧菌 DM27		Tommonaro et al., 2012
硫黄矿硫化叶菌和冰岛硫化叶菌	QQ 活性（内酯酶-AHL-水解）			Hiblot et al., 2012; Ng et al., 2011

2.5 其他极端细菌

目前，在其他类群的极端微生物中 QS 机制的存在仍缺乏了解。嗜冷菌可在极低温环境下成功定殖。实际上，寒冷地区占据了地球生物圈的大部分（Margesin and Miteva, 2011）。只有少数几篇文章对嗜冷菌的 QS 进行了研究。对海冰细菌隐单胞菌（*Psychromonas ingrahamii*) 37 的基因组序列进行筛选，发现 HapR（LuxR）同源物的存在，可能参与生物膜形成（Riley et al., 2008）。此外，嗜冷菌洛氏另类弧菌（*Aliivibrio logei*）生物发光受 QS 控制。更详细地，对 16 株洛氏另类弧菌的 *lux* 操纵子结构的分析表明，存在具有两个拷贝的 *LuxR* 基因（*luxR*1 和 *luxR*2）的 LuxI/LuxR QS 系统。实验观察表明，在洛氏另类弧菌中 *lux* 操纵子的转录受到 LuxR1 和 LuxR2 蛋白的协同调节（Khrulnova et al., 2016; Konopleva et al., 2016）。

嗜压微生物（或适压生物）是能够在相对高压（最高可达 110 MPa）下生活的生物体，通常栖息于深海沉积物、高静岩压力的地下岩石或热液喷口等极端环境中。这些环境也与异常高的（高达 375℃）或低（1~2℃）温度，以及缺乏光照和养分有效性相关（Kato, 1999）。一大类群的嗜压菌属于希瓦氏菌属。Bodor 等（2008）首次描述了希瓦氏菌属中 AI-2 的产生。在所研究的 10 种希瓦氏菌属中，均检测了 *luxS* 基因的存在和 AI-2 的产生。Tait 等（2009）在希瓦氏菌属晚期指数期提取物中鉴定出 3 种主要的 AHLs（OC4-HSL、OC10-HSL 和 OC12-HSL）。他们还发现 AHL 的产生与大型海藻浒苔 *Ulva* 的游动孢子定殖之间存在有趣的联系，证明 QS 的跨界生态效应。中度嗜压菌对明亮发光杆菌（*Photobacterium phosphoreum*）ANT-2200 在 30℃ 和 10 MPa 条件下表现出最佳的生长条件。据报道，与 0.1 MPa 相比，明亮发光杆菌在 22 MPa 下能产生更高的生物发光强度。该菌通过形成细胞聚集模拟更高的细胞密度以响应高压；因此，研究者推测存在一个细胞密度依赖的信号系统可调节生物发光的产生（Martini et al., 2013）。

3 极端微生物生物分子合成：QS 调控

极端微生物可产生不寻常生物分子的能力，作为适应原始环境条件的一种适应机制之一，其中，热稳定酶和胞外多糖是它们在极端生态位中发挥功能所必需的。QS 通过有机小分子或肽类调控它们的产生，被定义为"自诱导物"。QS 机制为微生物菌株适应极端环境提供了一种策略，并涉及参与极端菌群种内和种间的相互作用。

3.1 极端酶

生物修复是一种环境友好的方法，其原理是利用微生物对越来越多工业过程中释放的石油、重金属和多环芳烃等有害物质进行破坏或转化为无害的产物。细菌烃降解物自诱导物类似物高丝氨酸内酯（AHL）多基因的存在以及重度污染后烃降解菌群结构的不同，表明 QS 调控环境解毒过程（Huang et al., 2013）。来自氨基水解酶家族的酶能作为杀虫剂、除草剂和化学药剂对有机磷类物质进行解毒。从超嗜热古菌硫黄矿硫化叶菌（Ng et al., 2011）和岩化火山鬃菌（*Vulcanisaeta moutnovskia*）（Kallnik et al., 2014）中分离得到水解有机磷的极端耐热磷酸三酯酶类内酯酶。在铜绿假单胞菌中表达的硫黄矿硫化叶菌酶可以减弱毒力因子产生的 QS 信号（Ng et al., 2011）。由于甲烷生成途径中初始酶的上调，羧化的 AHLs 改变了碳代谢，使其更有利于乙酸转化为甲烷（Zhang et al., 2012）。许多细菌群落的细胞密度往往受到 QS 的调控。内酯酶、氧化还原酶、酰基转移酶和对氧磷酶等 QQ 酶干扰了微生物群落中的 QS 系统，并通过抑制 QS 酶的表达来限制细胞生长（Chen et al., 2013）。当培养物达到较高的细胞密度时，自诱导分子的积累刺激胞外水解酶（包括蛋白酶）的产生（Nakayama et al., 2001）。在古菌隐藏嗜盐球菌中观察到通过 QS 在细菌中进行细胞间通信，且存在类似的密度依赖性诱导蛋白酶基因表达（Paggi et al., 2003）。当细胞密度达到较高时，在后期指数增长中观察到高丝氨酸内酯分子的积累。

3.2 胞外多糖

胞外多糖（EPS）创造了一种环境，使细胞在生物膜中黏附，其稳定性由 EPS 决定。聚合物将生物膜中的细菌维持在一起，保护它们在极端环境条件下免受恶劣环境的影响。EPS 基质通过与可生物降解化合物的结合和积累，来增加物质的吸收速率，这在营养缺乏环境（如许多极端环境）中是一个特别重要的机制，同时确保酶与其底物之间的密切接触。QS 通过酰基高丝氨酸内酯调控生物膜的形成，允许从快速细胞生产到大量 EPS 产生的快速开关，以抵御环境威胁（Frederick et al., 2011）。

由两个 *N*-酰基高丝氨酸内酯自诱导因子 C4-HSL 和 OC12-HSL 介导的信号系统，控制着铜绿假单胞菌在细菌生物膜形成中分化为多细胞微菌落所必需的基因（Whiteley et al., 1999）。当菌株在两种多环芳烃上生长，在细菌生物膜中形成分化的多细胞微菌落，在生物膜形成的铜绿假单胞菌 N6P6 中建立了编码 C4-HSL 和 OC12-HSL 的两个 QS 基因 *lasI* 和 *rhlI* 的差异表达（Mangwani et al., 2015）。

4 极端生物分子：潜在的生物技术应用

对嗜极端微生物的持续研究兴趣在于它们能够产生生物分子，这些生物分子在高温或低温、pH值、压力等不同且同时存在的环境压力下，表现出高度稳定性。正是由于这些生物分子及其适应的细胞机制，极端微生物在特殊生态位中以任何生命形式繁殖都是可以想象到的（Poli et al., 2017）。事实上，从药理学到生态学，从分子生物学到医学，极端分子的性质在许多领域对可能的生物技术应用具有吸引力。因此，来自嗜极微生物的分子，如酶、脂质和胞外多糖，已在许多领域得到应用，显著提高相应工业生产过程的效率和/或可持续性（Finore et al., 2015）。

为了改善基于生物催化的工业流程，新稳定的酶是必需的（Woodley, 2013）。极端酶是由微生物在极端温度（$-12 \sim 122℃$）、pH值（$0 \sim 12$）、压力（高达100 MPa）和盐度（高达饱和浓度）的极端条件下合成的，它们在许多制造过程的极端条件下发挥重要作用（Coker, 2016）。到目前为止，大规模的生物技术应用中，只有少量的极端酶被发现使用。其中，引用DNA聚合酶是必不可少的，其基本活性依赖于脱氧核糖核苷酸链的合成。在复制过程中它们参与了生理基因组修复和细胞机制的维持。从嗜热微生物水生栖热菌（*Thermus aquaticus*）、极端嗜热菌激烈火球菌和嗜热高温球菌海岸热球菌（*Thermococcus litoralis*）中纯化的耐热DNA聚合酶，通常被认为是Taq22、Pfu23和Vent24（Ishino and Ishino, 2014）。这些聚合酶在分子生物学中的聚合酶链式反应（PCR）技术中具有重要作用。这种实践的发展和效率不仅限于研究领域，还为整个人类社会带来了许多好处；在任何情况下，基于PCR技术的应用在生物样本数量有限的情况下都是至关重要的，例如在法医科学中（Gunn et al., 2014）。

糖基水解酶负责复合糖中糖苷键的水解。它们在自然界中非常普遍，被划分为100多个家族。随着时间的推移，发现这些酶在生物技术上的应用越来越广泛。

无法消化乳糖的现象在全球人口中普遍存在，并会产生令人尴尬的影响。这是由于乳糖不耐受和缺乏β-半乳糖苷酶所致（Coker, 2016）；因此，为解决这一不便，许多情况下，食用不含乳糖的膳食产品是首选，通过使用乳酸克鲁维酵母（*Kluyveromyces lactis*）β-半乳糖苷酶对产品进行预处理可以有效降低乳糖含量（Messia et al., 2007）。然而，酶催化水解的反应温度高于食品维持的温度（$5 \sim 25℃$），意味着存在较大的污染和变质风险。一种解决方案可能是利用来自嗜冷微生物的β-半乳糖苷酶，这些酶在较低的温度下具有活性。目前，许多嗜冷酶被研究，其产生的效果与目前使用的中温酶相似，其最适工作温度范围为$15 \sim 37℃$（Coker and Brenchley, 2006）。

类似地，纤维素、半纤维素和淀粉水解过程涉及使用酶池，这些酶均属于糖基水解酶。先前提到的聚合物在高温下的酶促降解更为有利，因为随着寡糖和单糖浓度逐渐富集，污染风险降低。因此，使用嗜热酶成为首选。针对耐热纤维素酶、α-葡萄糖苷酶、木聚糖酶、β-木糖苷酶、泡菜糖酶、葡萄糖淀粉酶和β-淀粉酶的研究也在不断进行中。在许多工业领域中，水解产品得到广泛应用，例如，洗衣液和瓷器清洁剂、烘焙中的防腐剂，以及从可再生农业废弃物中生产生物乙醇和有附加值的有机化合物（Dornez et al., 2011; Finore et al., 2011; Finore et al., 2015; Finore et al., 2016; Mohammad et al., 2017; van der Maarel et al., 2002）。

此外，令人惊讶的是，极端微生物也是氢源，作为厌氧发酵的代谢产物。气火菌（*Aeropyrum*）、解糖热纤维素菌（*Caldicellulosiruptor*）、嗜热厌氧细杆菌（*Thermoanaerobacterium*）（Ren et al.，2008）和火球菌（*Pyrococcus*）（Baker et al.，2009）均显示出巨大的潜力。

另一个应用领域生物采矿是利用嗜极端微生物酶的催化特性，这是一种利用微生物进行金属回收的生物技术和环保过程。事实上，与传统的堆浸法不同，该法需要许多化学物质来结合和分离金属，利用嗜酸和嗜热微生物能够氧化铁（Ⅱ）或硫化物（Coker，2016）。嗜酸硫杆菌、钩端螺旋菌属、铁质浆菌（*Ferroplasma*）、硫化叶菌属、金属球菌（*Metallosphaera*）、酸性菌（*Acidianus*）和硫黄球形菌（*Sulfurisphaera*）在铜、金、钴、镍、锌和铀的生物采矿中得到了应用（Podar and Reysenbach，2006；Vera et al.，2013）。显然，这种微生物应用可以扩展到处理残渣或矿山废料堆（矿尾矿）（Chen et al.，2016），以及从工业废料残留物中提取金属（Mishra and Rhee，2014）。此外，生物冶金微生物群落的 DNA 测序技术进步是可取的，因为，此技术提供了前所未有的机会来获取有关基因组的信息，从而构建代谢潜力和生态系统水平相互作用的预测模型（Cárdenas et al.，2016）。

除这些具有商业价值的酶外，还有用于肽键水解的蛋白酶。其中，碱性蛋白酶在洗涤剂中表现出较高的稳定性；广泛应用于洗衣粉、皮革加工、食品工业和制药等多个生物工业领域（Mohammad et al.，2017；Sarmiento et al.，2015；Sellami-Kamoun et al.，2008）。通常，商业上使用的蛋白酶主要来自嗜中温微生物，尤其是自芽孢杆菌属，这些产品由诺维信和杰能科等公司销售。然而，针对嗜冷蛋白酶改善冷水洗涤的研究刚刚起步。大多数被测试的嗜冷酶在室温下的稳定性较低，显示出不适合的特性。尽管如此，设计一种工程化的嗜冷/嗜中温蛋白酶，以提高其在冷水洗涤过程中性能，仍然是非常有前景的（Tindbaek et al.，2004）。

脂肪酶是催化脂肪水解的酶。由于其广泛的底物范围、高特异性和稳定性，脂肪酶在工业应用中非常有吸引力（Coker，2016；Hasan et al.，2006；Sarmiento et al.，2015）。许多嗜温脂肪酶通常来自类似于芽孢杆菌和曲霉（*Aspergillus*）的生物，它们在高温下表现出活跃的催化能力。因此，嗜极端脂肪酶往往被忽视。然而，来自嗜热芽孢杆菌的脂肪酶被证明比目前使用的酶更高效（Imamura and Kitaura，2000）。最近，研究显示，嗜冷微生物的脂肪酶具有催化效率和低热稳定性，这与其嗜中温和嗜热的对应物存在显著区别（Maiangwa et al.，2015）。

EPS 是一种由糖残基组成的高分子量聚合物，由微生物分泌到周围环境中。根据其储存位置和/或功能，EPS 可分为细胞内多糖、结构性多糖和细胞外多糖。它们通常由单糖和一些取代基（如醋酸盐、丙酮酸盐、琥珀酸盐和磷酸盐）组成。由于 EPS 组成的广泛多样性，意味着其化学结构和物理性质也各不相同，因此，在不同领域具有不同的潜在应用（Nicolaus et al.，2010）。

许多研究表明，通过调整生长条件、温度、压力、盐浓度和营养成分，可以诱导和提高 EPS 的产生。此外，基因组学和统计学研究可选择合适的底物，以支持聚合物的产生（Finore et al.，2014）。研究证明，废弃材料可作为微生物的碳源，用于 EPS 的产生（Kucukasik et al.，2011）。

据报道，嗜极端微生物通常能产生多种分子以应对环境压力（Poli et al.，2017）。在工业上应用 EPS 的一个例子是在烘焙业中使用右旋糖酐（Kothari et al.，2014）。一般来说，工业用途的 EPSs 是由嗜温微生物产生的。然而，利用嗜极端原核生物产生的多糖是有限的，但它们的非典型特性对开发新生物技术应用具有吸引力（Nicolaus et al., 2010）。事实上，有许多案例研究旨在证明来自极端微生物的 EPSs 有巨大潜力。例如，嗜盐单胞菌菌株 AAD6^T 产生高水平的果聚糖（Poli et al., 2009; Poli et al., 2013）。对这种聚合物的深入研究表明，在两种不同细胞系统（成骨细胞和小鼠巨噬细胞）中，它不会影响细胞存活和增殖，显示出较高的生物相容性。此外，其对阿瓦醇的毒性具有保护作用，阿瓦醇是从海洋海绵贪婪掘海绵（*Dysidea avara*）中分离的重排倍半萜对苯二酚，显示出不同的生物活性（例如抗病毒、抗微生物、抗氧化、细胞毒性、抗炎和抗银屑病）（Minale et al., 1974; Crispino et al., 1989）。从嗜盐单胞菌菌株 AAD6^T 中分离得到具有细胞保护作用的果聚糖，表明其可作为一种抗细胞毒性药物的额外用途。果聚糖作为血浆扩张剂、工业胶、稳定剂、增稠剂、乳化剂、甜味剂、配方助剂、表面处理剂、包埋剂以及风味和香料的载体已被众所周知（Beine et al., 2008; Shih et al., 2005）。然而，盐单胞菌属 AAD6 在确定的培养基上生长时，可作为果聚糖的另一来源，假设其作为一种致病微生物在工业应用中具有更广泛的潜力（Sam et al., 2011; Sezer et al., 2011）。

对极端古菌和细菌细胞膜的脂质组成进行了深入研究；这成为其最独特的分类因素之一，研究结果本质上受到微生物暴露于温度、压力和盐浓度的影响（Poli et al., 2017; Siliakus et al., 2017）。古菌脂类在生物、医学和生物技术应用展现出很大的潜力。为了产生超分子结构，如脂质膜或脂质体，四醚脂质能自组装，并用作药物和基因传递系统和疫苗的载体（Jacquemet et al., 2009）。

此外，Müller 等（2006）利用天然四醚衍生物的超薄薄膜覆盖纳米多孔氧化铝膜，以改变相应的过滤性能，应用于生物领域。研究结果表明，新膜的渗透性较低，且易于消毒。通过设计特殊的膜纳米系统，四醚脂质成为生物纳米技术和材料科学的合适系统（Chugunov et al., 2014）。然而，对这些脂类的理解仍需进一步研究，以及扩大其生物技术应用的范围（Jacquemet et al., 2009）。

来自极端微生物的另一个有趣的应用与海藻糖的生产有关，海藻糖可作为疫苗和抗体的稳定剂（Guo et al., 2000）。对极端嗜热古细菌硫黄矿硫化叶菌产生的海藻糖进行了研究，发现其能够毫无困难地替代目前使用的来自节杆菌属（*Arthrobacter* sp.）Q36 的中温酶（Schiraldi et al., 2002）。

另一个微生物产品是四氢嘧啶，它属于渗透调节类物质，能平衡细胞所受的渗透压。研究发现，四氢嘧啶可作为一种皮肤保护剂，防止 UVA 引起的损伤。天然细胞保护因子（RonaCare Ectoin）是由默克集团（德国达姆施塔特）生产，作为一种保湿剂可从嗜盐微生物中提取（DasSarma et al., 2009）。

目前，微生物在极端环境中的作用和机制仍然复杂且不清楚。因此，似乎所有这些生物分子的产生和 QS 之间可能存在关联并相互影响（Tommonaro et al., 2015）。实际上，通信机制可能会影响这些生物分子的产生和/或抑制。由于在极端环境中产生的生物分子可被视为对物理-化学环境胁迫因子的响应，因此，QS 涉及适应机制，允许这些特殊的生命形式在不适宜居住的生态位中增殖。

5 结论

在"极端"环境中发现能增殖的生物体，表明生命可以存在于非典型的化学和物理条件下。事实上，极端微生物能够在高温或低温，异常高或低的 pH 值，高盐度，压力或毒性，甚至高辐射水平下生长。出乎意料的是，这些极端环境参数并不是限制因素，而是它们的最适生长条件。对极端环境中生命的研究，特别关注极端微生物为对抗恶劣条件而形成的机制，这是研究人员特别感兴趣的领域。这些生物的研究并不仅局限于研究其生物过程的基本特性，而且还集中在"极端生物分子"在工业生物技术中的潜在应用。极端微生物的代谢产物，特别是胞外酶，实际上在极端环境的恶劣条件下是稳定和活跃的。与不稳定的嗜中温分子相比，极端生物分子独特的抗性特征，表明其在生物技术应用过程中具有明显优势。在各种微生物中 QS 是一种被广泛描述的细胞间通信策略，调控各种社会行为，如发光产生、生物膜形成、EPS 产生、运动性等。尽管 QS 系统在微生物中普遍存在，但在极端环境下细胞间通信机制仍有待进一步研究。新技术的出现，如生物信息学和基因组测序，对许多极端微生物中参与 QS 机制的基因进行了鉴定。利用质粒生物传感器，在恶劣栖息地中不同菌株的无细胞培养基中检测不同的信号分子，主要是 AHLs。然而，要全面了解对 QS 在极端微生物中的作用，仍需要进一步研究。对自诱导物产生的遗传基础，以及对这些信号分子调节嗜极端微生物不同表型机制的深入研究，将有助于理解这些生物如何在极端条件下生存。而且，正像在不同的微生物模型中所描述的那样，QS 可以在一些"极端分子"（特别是胞外多糖和胞外酶）的合成中发挥作用，这些分子在工业应用中具有潜在价值。因此，在恶劣环境中研究 QS 可能对生物技术过程改进具有重要贡献。

术语表

古菌 含有原核微生物的生命领域，在遗传上与细菌不同，通常生活在极端环境中。

自诱导物 参与群体感应（QS）的信号分子是在细胞群体达到阈值时合成的信号分子。

噬菌调理素 古菌使用的光合蛋白质。

生物催化 由一种酶或酶复合物组成或来源于生物体或细胞培养物（无细胞或全细胞形式），可催化生物体中的代谢反应和/或各种化学反应中的底物转化。

生物膜 微生物通过自我产生的基质主要由胞外多糖和蛋白质组成聚集在一起。

生物修复 利用生物体来处理污染物或修复受污染的土壤、水或空气。

生物表面活性剂 在活的表面上产生两亲性化合物。

催化剂 一种能够增加化学反应速率而不发生任何永久化学变化的物质。

接合 一个细菌（供体）将遗传物质，通常是一个质粒或转座子，转移到另一个细胞（受体）的机制。

胞外多糖 由糖残基组成的高分子量聚合物，由微生物分泌到周围的环境中。

嗜极微生物 在物理上或地球化学上的极端条件下生长的生物体。

矿盐 氯化钠（NaCl）的矿物（天然）形式。

嗜温微生物 任何生物，特别是某些细菌和真菌，在 25~40℃ 的中等温度下生长。

宏基因组学 通过 DNA 剪切和测序的方法对微生物进行基因组分析，并使用生物信息学工具进行数据处理

肽聚糖 一种由氨基酸和碳水化合物组成的生物聚合物，形成大多数细菌的细胞壁。

多羟基化合物 低分子量、水溶性聚合物和含有大量羟基的低聚物的通称。

多聚糖 由糖残基组成的高分子量聚合物。

群体淬灭 破坏群体感应（QS）信号传导的一组过程。

群体感应 通过几种生物种群依赖的信号通路调节基因表达的分子通信机制。调控基因：一组由同一调控分子调控的基因。

调节子 由同一调控分子调控的一组基因。

紫色杆菌素 由某些细菌产生的紫色色素。

游走孢子 一种由某些藻类和真菌组成的能够独立运动的微小无性生殖细胞。

缩写词

R-THMF	$(2R,4S)$-2-Methyl-2,3,3,4-tetrahy droxytetrahydrofuran
S-THMF	$(2S,4S)$-2-Methyl-2,3,3,4-tetrahydroxytetrahydrofuran-borate
DPD	(Z)-3-aminoundec-2-en-4-one(Ea-C8-CAI-1);4,5-Dihydroxy-2,3-pentanedione
AI	自诱导物
AI-2	自诱导物-2
CAI-1	霍乱弧菌自诱导物-1
DKP	二酮哌嗪
ESI tandem MS	电喷雾电离质谱法
EPS	胞外多糖
GC-MS	气相色谱-质谱法
LC-MS	液相色谱-质谱法
OC4-HSL	N-(3-氧-丁酰)-高丝氨酸内酯
OC10-HSL	N-(3-氧-癸酰)-高丝氨酸内酯
OC12-HSL	N-(3-氧-十二酰)-高丝氨酸内酯
AHL	N-酰基高丝氨酸内酯
C4-HSL	N-丁酰高丝氨酸内酯
C12-HSL	N-十二酰-高丝氨酸内酯
C6-HSL	N-己酰-高丝氨酸内酯
C8-HSL	N-辛酰-高丝氨酸内酯
OGT	最佳生长温度
PAHs	多环芳香烃
QQ	群体淬灭
QS	群体感应
FOYE-1	嗜热烯还原酶

致谢

作者感谢 BAS/CNR 联合项目 2016-18 对这项工作的支持。

参考文献

ABBAMONDI G R, DE ROSA S, IODICE C, et al., 2014. Cyclic dipeptides produced by marine spongeassociated bacteria as quorum sensing signals. Nat. Prod. Commun., 9 (2): 229-232.

ABBAMONDI G R, SUNER S, CUTIGNANO A, et al., 2016. Identification of N-Hexadecanoyl-L-homoserine lactone (C16-AHL) as signal molecule in halophilic bacterium *Halomonas smyrnensis* AAD6. Ann. Microbiol., 66 (3): 1329-1333.

ABED R M M, DOBRETSOV S, AL-FORI M, et al., 2013. Quorum-sensing inhibitory compounds from extremophilic microorganisms isolated from a hypersaline cyanobacterial mat. J. Ind. Microbiol., Biotechnol. 40 (7): 759-772.

ALBERS S, EICHLER J, AEBI M, 2017. Archaea. In: Varki A, Cummings R D, Esko J D, et al., (Eds), Essentials of Glycobiology, third ed. Cold Spring Harbor, NY.

ANTRANIKIAN G, VORGIAS C E, BERTOLDO C, 2005. Extreme environments as a resource for microorganisms and novel biocatalysts. In: Ulber R, Le Gal Y (Eds), Marine Biotechnology I. Springer Berlin Heidelberg, Berlin, Heidelberg, pp: 219-262.

AVERHOFF B, MÜLLER V, 2010. Exploring research frontiers in microbiology: recent advances in halophilic and thermophilic extremophiles. Res. Microbiol., 161 (6): 506-514.

BAKER S E, HOPKINS R C, BLANCHETTE C D, et al., 2009. Hydrogen production by a hyperthermophilic membrane-bound hydrogenase in water-soluble nanolipoprotein particles. J. Am. Chem. Soc., 131 (22): 7508-7509.

BEINE R, MORARU R, NIMTZ M, et al., 2008. Synthesis of novel fructooligosaccharides by substrate and enzyme engineering. J. Biotechnol., 138 (1): 33-41.

BERGONZI C, SCHWAB M, ELIAS M, 2016. The quorum-quenching lactonase from Geobacillus caldoxylosilyticus: Purification, characterization, crystallization and crystallographic analysis. Acta Crystallogr. Sect. F., 72 (9): 681-686.

BODOR A, ELXNAT B, THIEL V, et al., 2008. Potential for luxS related signalling in marine bacteria and production of autoinducer-2 in the genus Shewanella. BMC Microbiol., 8: 13.

BZDRENGA J, DAUDE D, REMY B, et al., 2017. Biotechnological applications of quorum quenching enzymes. Chem. Biol. Interact., 267: 104-115.

Canganella F, Wiegel J, 2014. Anaerobic thermophiles. Life, 4 (1): 77-104.

CÁRDENAS J P, QUATRINI R, HOLMES D S, 2016. Genomic and metagenomic challenges and opportunities for bioleaching: A mini-review. Res. Microbiol., 167 (7): 529-538.

CHEN F, GAO Y, CHEN X, et al., 2013. Quorum quenching enzymes and their application in degrading signal molecules to block quorum sensing-dependent infection. Int. J. Mol. Sci., 14 (9): 17477-17500.

CHEN L, HUANG L, MÉNDEZ-GARCÍA C, et al., 2016. Microbial communities, processes and functions in acid mine drainage ecosystems. Curr. Opin. Biotechnol., 38: 150-158.

CHRISTEL S, HEROLD M, BELLENBERG S, et al., 2018. Multi-omics reveals the lifestyle of the acidophilic, mineral-oxidizing model species *Leptospirillum ferriphilum*^{T}. Appl. Environ. Microbiol, 84 (3): e02091-17.

CHUGUNOV A O, VOLYNSKY P E, KRYLOV N A, et al., 2014. Liquid but durable: Molecular dynamics simulations explain the unique properties of archaeal-like membranes. Sci. Rep., 4: 7462.

COKER J A, 2016. Extremophiles and biotechnology: current uses and prospects. F1000Research 5, F1000 Faculty Rev-1396.

COKER J A, BRENCHLEY J E, 2006. Protein engineering of a cold-active β-galactosidase from Arthrobacter sp. SB to increase lactose hydrolysis reveals new sites affecting low temperature activity. Extremophiles, 10 (6): 515-524.

CRISPINO A, DE GIULIO A, DE ROSA S, et al., 1989. A new bioactive derivative of avarol from the marine sponge Dysidea avara. J. Nat. Prod., 52 (3): 646-648.

DALMASO Z G, FERREIRA D, VERMELHO B A, 2015. Marine extremophiles: a source of hydrolases for biotechnological applications. Mar. Drugs, 13 (4).

DASSARMA P, COKER J A, HUSE V, et al., 2009. Halophiles, Industrial Applications, Encyclopedia of Industrial Biotechnology. John Wiley & Sons, Inc. Decho, A. W, 2000. Microbial biofilms in intertidal systems: An overview. Cont. Shelf Res., 20 (10): 1257-1273.

DI DONATO P, ROMANO I, MASTASCUSA V, et al., 2018. Survival and adaptation of the thermophilic species Geobacillus thermantarcticus in simulated spatial conditions. Origins Life Evol. Biospheres, 48 (1): 141-158.

DORNEZ E, CUYVERS S, HOLOPAINEN U, et al., 2011. Inactive fluorescently labeled xylanase as a novel probe for microscopic analysis of arabinoxylan containing cereal cell walls. J. Agric. Food Chem., 59 (12): 6369-6375.

FARAH C, VERA M, MORIN D, et al., 2005. Evidence for a functional quorum-sensing type ai-1 system in the extremophilic bacterium Acidithiobacillus ferrooxidans. Appl. Environ. Microbiol., 71 (11): 7033-7040.

FINORE I, DI DONATO P, MASTASCUSA V, et al., 2014. Fermentation technologies for the optimization of marine microbial exopolysaccharide production. Mar. Drugs, 12 (5): 3005-3024.

FINORE I, KASAVI C, POLI A, et al., 2011. Purification, biochemical characterization and gene sequencing of a thermostable raw starch digesting alphaamylase from Geobacillus thermoleovorans subsp stromboliensis subsp nov. World J. Microbiol. Biotechnol., 27 (10): 2425-2433.

FINORE I, LAMA L, POLI A, et al., 2015. Biotechnology implications of extremophiles as life pioneers and wellspring of valuable biomolecules. In: Kalia, V. C. (Ed.), Microbial Factories: Biodiversity, Biopolymers, Bioactive Molecules: Volume 2. Springer India, New Delhi, pp: 193-216.

FINORE I, POLI A, DI DONATO P, et al., 2016. The hemicellulose extract from Cynara cardunculus: a source of value-added biomolecules produced by xylanolytic thermozymes. Green Chem., 18 (8): 2460-2472.

FREDERICK M R, KUTTLER C, HENSE B A, et al., 2011. A mathematical model of quorum sensing regulated EPS production in biofilm communities. Theor. Biol. Med. Modell., 8: 8.

FUQUA W C, WINANS S C, GREENBERG E P, 1994. Quorum sensing in bacteria: the LuxR-LuxI family of cell density-responsive transcriptional regulators. J. Bacteriol., 176 (2): 269-275.

GRANDCLEMENT C, TANNIÈRES M, MORÉRA S, et al., 2016. Quorum quenching: role in nature and applied developments. FEMS Microbiol. Rev., 40 (1): 86-116.

GUNN P, WALSH S, ROUX C, 2014. The nucleic acid revolution continues—will forensic biology become forensic molecular biology? Front. Genet., 5: 44.

GUO N, PUHLEV I, BROWN D R, et al., 2000. Trehalose expression confers desiccation tolerance on human cells. Nat. Biotechnol., 18, 168.

HASAN F, SHAH A A, HAMEED A, 2006. Industrial applications of microbial lipases. Enzym. Microb. Technol., 39 (2): 235-251.

HEDLUND B P, DODSWORTH J A, MURUGAPIRAN S K, et al., 2014. Impact of single-cell genomics and metagenomics on the emerging view of extremophile "microbial dark matter". Extremophiles, 18 (5):

865-875.

HIBLOT J, GOTTHARD G, CHABRIERE E, et al., 2012. Structural and enzymatic characterization of the lactonase SisLac from *Sulfolobus islandicus*. PLoS One, 7 (10): e47028.

HUANG Y L, ZENG Y, YU Z, et al., 2013. In silico and experimental methods revealed highly diverse bacteria with quorum sensing and aromatics biodegradation systems - a potential broad application on bioremediation. Bioresour. Technol., 148: 311-316.

IMAMURA S, KITAURA S, 2000. Purification and characterization of a monoacylglycerol lipase from the moderately thermophilic *Bacillus* sp. H-257. J. Biochem., 127 (3): 419-425.

ISHINO S, ISHINO Y, 2014. DNA polymerases as useful reagents for biotechnology—the history of developmental research in the field. Front. Microbiol., 5: 465.

JACQUEMET A, BARBEAU J, LEMIÈGRE L, et al., 2009. Archaeal tetraether bipolar lipids: structures, functions and applications. Biochimie, 91 (6): 711-717.

JOHNSON M R, MONTERO C I, CONNERS S B, et al., 2005. Population densitydependent regulation of exopolysaccharide formation in the hyperthermophilic bacterium *Thermotoga maritima*. Mol. Microbiol., 55 (3): 664-674.

KALLNIK V, BUNESCU A, SAYER C, et al., 2014. Characterization of a phosphotriesterase-like lactonase from the hyperthermoacidophilic crenarchaeon *Vulcanisaeta moutnovskia*. J. Biotechnol., 190: 11-17.

KARAN R, CAPES M D, DASSARMA P, et al., 2013. Cloning, overexpression, purification, and characterization of a polyextremophilic β - galactosidase from the Antarctic haloarchaeon *Halorubrum lacusprofundi*. BMC Biotechnol., 13: 3.

KATO C, 1999. Barophiles (Piezophiles). In: Horikoshi K, Tsujii K (Eds), Extremophiles in Deep-Sea Environments. Springer Japan, Tokyo, pp: 91-111.

KHRULNOVA S A, BARANOVA A, BAZHENOV S V, et al., 2016. Lux-operon of the marine psychrophilic bacterium Aliivibrio logei: A comparative analysis of the LuxR1/LuxR2 regulatory activity in *Escherichia coli* cells. Microbiology, 162 (4): 717-724.

KONOPLEVA M N, KHRULNOVA S A, BARANOVA A, et al., 2016. A combination of luxR1 and luxR2 genes activates Pr-promoters of psychrophilic *Aliivibrio logei* luxoperon independently of chaperonin GroEL/ES and protease Lon at high concentrations of autoinducer. Biochem. Biophys. Res. Commun., 473 (4): 1158-1162.

KOTHARI D, DAS D, PATEL S, et al., 2014. Dextran and food application. In: Ramawat K G, Merillon J-M (Eds), Polysaccharides: Bioactivity and Biotechnology. Springer International Publishing, Cham, pp: 1-16.

KUCUKASIK F, KAZAK H, GUNEY D, et al., 2011. Molasses as fermentation substrate for Levan production by Halomonas sp. Appl. Microbiol. Biotechnol., 89 (6): 1729-1740.

LEÓN M J, FERNÁNDEZ A B, GHAI R, et al., 2014. From metagenomics to pure culture: Isolation and characterization of the moderately halophilic bacterium *Spiribacter salinus* gen. Nov, sp. nov. Appl. Environ. Microbiol., 80 (13): 3850-3857.

LLAMAS I, QUESADA E, MARTÍNEZ - CÁNOVAS M J, et al., 2005. Quorum sensing in halophilic bacteria: detection of N - acyl - homoserine lactones in the exopolysaccharide - producing species of *Halomonas*. Extremophiles, 9 (4): 333-341.

LUO Z Q, CLEMENTE T E, FARRAND S K, 2001. Construction of a derivative of agrobacterium tumefaciens C58 that does not mutate to tetracycline resistance. Molecular plant-microbe interactions. Journal, 14 (1): 98-103.

MAIANGWA J, ALI M S M, SALLEH A B, et al., 2015. Adaptational properties and applications of cold-active lipases from psychrophilic bacteria. Extremophiles, 19 (2): 235-247.

MAMANI S, MOINIER D, DENIS Y, et al., 2016. Insights into the quorum sensing regulon of the acidophilic *Acidithiobacillus ferrooxidans* revealed by transcriptomic in the presence of an acyl homoserine lactone superagonist analog. Front. Microbiol., 7 (1365).

MANGWANI N, KUMARI S, DAs S, 2015. Involvement of quorum sensing genes in biofilm development and degradation of polycyclic aromatic hydrocarbons by a marine bacterium *Pseudomonas aeruginosa* N6P6. Appl. Microbiol. Biotechnol., 99 (23): 10283-10297.

MARGESIN R, MITEVA V, 2011. Diversity and ecology of psychrophilic microorganisms. Res. Microbiol., 162 (3): 346-361.

MARTINI S, AL ALI B, GAREL M, et al., 2013. Effects of hydrostatic pressure on growth and luminescence of a moderately-piezophilic luminous bacteria *Photobacterium phosphoreum* ANT-2200. PLoS One, 8 (6): e66580.

MASTASCUSA V, ROMANO I, DI DONATO P, et al., 2014. Extremophiles survival to simulated space conditions: an astrobiology model study. Orig. Life Evol. Biosph., 44 (3): 231-237.

MCCLEAN K H, WINSON M K, FISH L, et al., 1997. Quorum sensing and *Chromobacterium violaceum*: Exploitation of violacein production and inhibition for the detection of *N*-acyl homoserine lactones. Microbiology, 143: 3703-3711.

MEGAW J, GILMORE B F, 2017. Archaeal persisters: Persister cell formation as a stress response in *Haloferax volcanii*. Front. Microbiol., 8: 1589.

MESSIA M C, CANDIGLIOTA T, MARCONI E, 2007. Assessment of quality and technological characterization of lactose-hydrolyzed milk. Food Chem., 104 (3): 910-917.

MEYER B H, ALBERS S V, 2001. Archaeal Cell Walls, eLS. John Wiley & Sons, Ltd.

Minale L, Riccio R, Sodano G, 1974. Avarol a novel sesquiterpenoid hydroquinone with a rearranged drimane skeleton from the sponge Dysidea avara. Tetrahedron Lett., 38 (15): 3401-3404.

MISHRA D, RHEE Y H, 2014. Microbial leaching of metals from solid industrial wastes. J. Microbiol., 52 (1): 1-7.

MOHAMMAD B T, AL DAGHISTANI H I, JAOUANI A, et al., 2017. Isolation and characterization of thermophilic bacteria from jordanian hot springs: *Bacillus licheniformis* and *Thermomonas hydrothermalis* isolates as potential producers of thermostable enzymes. Int. J. Microbiol., 2017: 12.

MONTGOMERY K, CHARLESWORTH J C, LEBARD R, et al., 2013. Quorum sensing in extreme environments. Life, 3 (1): 131-148.

MOROHOSHI T, TOMINAGA Y, SOMEYA N, et al., 2015. Characterization of a novel thermostable *N*-acylhomoserine lactonase from the thermophilic bacterium *Thermaerobacter marianensis*. J. Biosci. Bioeng., 120 (1): 1-5.

MÜLLER S, PFANNMÖLLER M, TEUSCHER N, et al., 2006. New method for surface modification of nanoporous aluminum oxide membranes using tetraether lipid. J. Biomed. Nanotechnol., 2 (1): 16-22.

NAKAYAMA J, CAO Y, HORII T, et al., 2001. Gelatinase biosynthesis-activating pheromone: A peptide lactone that mediates a quorum sensing in *Enterococcus faecalis*. Mol. Microbiol., 41 (1): 145-154.

NG F S W, WRIGHT D M, SEAH S Y K, 2011. Characterization of a phosphotriesterase-like lactonase from *Sulfolobus solfataricus* and its immobilization for disruption of quorum sensing. Appl. Environ. Microbiol., 77 (4): 1181-1186.

NICHOLS J D, JOHNSON M R, CHOU C J, et al., 2009. Temperature, not LuxS, mediates AI-2 formation in hydrothermal habitats. FEMS Microbiol. Ecol., 68 (2): 173-181.

NICOLAUS B, KAMBOUROVA M, ONER E T, 2010. Exopolysaccharides from extremophiles: From fundamentals to biotechnology. Environ. Technol., 31 (10): 1145-1158.

OREN A, 2010. Industrial and environmental applications of halophilic microorganisms. Environ. Technol., 31 (8-9): 825-834.

PAGGI R A, MARTONE C B, FUQUA C, et al., 2003. Detection of quorum sensing signals in the haloalkaliphilic archaeon *Natronococcus occultus*. FEMS Microbiol. Lett., 221 (1): 49-52.

PEREZ-RODRÍGUEZ I, BOLOGNINI M, RICCI J, et al., 2015. From deep-sea volcanoes to human pathogens: a conserved quorum-sensing signal in Epsilonproteobacteria. ISME J., 9 (5): 1222-1234.

PIKUTA E V, HOOVER R B, TANG J, 2007. Microbial extremophiles at the limits of life. Crit. Rev. Microbiol., 33 (3): 183-209.

PODAR M, REYSENBACH A L, 2006. New opportunities revealed by biotechnological explorations of extremophiles. Curr. Opin. Biotechnol., 17 (3): 250-255.

POLI A, ANZELMO G, NICOLAUS B, 2010. Bacterial exopolysaccharides from extreme marine habitats: Production, characterization and biological activities. Mar. Drugs, 8 (6): 1779.

POLI A, FINORE I, ROMANO I, et al., 2017. Microbial diversity in extreme marine habitats and their biomolecules. Microorganisms, 5 (2): 25.

POLI A, KAZAK H, GÜRLEYENDAĞ B, et al., 2009. High level synthesis of Levan by a novel *Halomonas* species growing on defined media. Carbohydr. Polym., 78 (4): 651-657.

POLI A, NICOLAUS B, DENIZCI A A, et al., 2013. *Halomonas smyrnensis* sp. nov, a moderately halophilic, exopolysaccharide-producing bacterium. Int. J. Syst. Evol. Microbiol., 63 (1): 10-18.

RADDADI N, CHERIF A, DAFFONCHIO D, et al., 2015. Biotechnological applications of extremophiles, extremozymes and extremolytes. Appl. Microbiol. Biotechnol., 99 (19): 7907-7913.

REITH F, 2011. Life in the deep subsurface. Geology, 39 (3): 287-288.

REN N, CAO G, WANG A, et al., 2008. Dark fermentation of xylose and glucose mix using isolated *Thermoanaerobacterium thermosaccharolyticum* W16. Int. J. Hydrog. Energy, 33 (21): 6124-6132.

RILEY M, STALEY J T, DANCHIN A, et al., 2008. Genomics of an extreme psychrophile, *Psychromonas ingrahamii*. BMC Genomics, 9: 210.

RIVAS M, SEEGER M, JEDLICKI E, et al., 2007. Second acyl homoserine lactone production system in the extreme acidophile *Acidithiobacillus ferrooxidans*. Appl. Environ. Microbiol., 73 (10): 3225-3231.

RUIZ L M, VALENZUELA S, CASTRO M, et al., 2008. AHL communication is a widespread phenomenon in biomining bacteria and seems to be involved in mineral-adhesion efficiency. Hydrometallurgy, 94 (1): 133-137.

SAM S, KUCUKASIK F, YENIGUN O, et al., 2011. Flocculating performances of exopolysaccharides produced by a halophilic bacterial strain cultivated on agro-industrial waste. Bioresour. Technol., 102 (2): 1788-1794.

SARMIENTO F, PERALTA R, BLAMEY J M, 2015. Cold and hot extremozymes: Industrial relevance and current trends. Front Bioeng Biotechnol, 3: 148.

SAUM S H, MÜLLER V, 2008. Regulation of osmoadaptation in the moderate halophile *Halobacillus halophilus*: chloride, glutamate and switching osmolyte strategies. Saline Systems, 4 (1): 4.

SCHIRALDI C, GIULIANO M, DE ROSA M, 2002. Perspectives on biotechnological applications of archaea. Archaea, 1 (2): 75-86.

SCHOLTISSEK A, ULLRICH S R, MÜHLING M, et al., 2017. A thermophilic-like enereductase originating from an acidophilic iron oxidizer. Appl. Microbiol. Biotechnol., 101 (2): 609-619.

SCHOPF S, WANNER G, RACHEL R, et al., 2008. An archaeal bi-species biofilm formed by *Pyrococcus furiosus* and *Methanopyrus kandleri*. Arch. Microbiol., 190 (3): 371-377.

SECKBACH J, RAMPELOTTO P H, 2015. Chapter 8: Polyextremophiles. In: Bakermans, C. (Ed.), Microbial Evolution under Extreme Conditions. De Gruyter.

SELLAMI-KAMOUN A, HADDAR A, ALI N E H, et al., 2008. Stability of thermostable alkaline protease from *Bacillus licheniformis* RP1 in commercial solid laundry detergent formulations. Microbiol. Res., 163 (3): 299-306.

SEWALD X, SAUM S H, PALM P, et al., 2007. Autoinducer-2-producing protein LuxS, a novel salt- and chloride-induced protein in the moderately halophilic bacterium *Halobacillus halophilus*. Appl. Environ. Microbiol., 73 (2): 371-379.

SEZER A D, KAZAK H, ÖNER E T, et al., 2011. Levan-based nanocarrier system for peptide and protein drug delivery: Optimization and influence of experimental parameters on the nanoparticle characteristics. Carbohydr. Polym., 84 (1): 358-363.

SHIH I L, YU Y T, SHIEH C J, et al., 2005. Selective production and characterization of Levan by *Bacillus subtilis* (natto) Takahashi. J. Agric. Food Chem., 53 (21): 8211-8215.

SHULAMI S, SHENKER O, LANGUT Y, et al., 2014. Multiple regulatory mechanisms control the expression of the *Geobacillus stearothermophilus* gene for extracellular xylanase. J. Biol. Chem., 289 (37): 25957-25975.

SILIAKUS M F, VAN DER OOST J, KENGEN S W M, 2017. Adaptations of archaeal and bacterial membranes to variations in temperature, pH and pressure. Extremophiles, 21 (4): 651-670.

TAHRIOUI A, QUESADA E, LLAMAS I, 2011. The hanR/hanI quorum-sensing system of *Halomonas anticariensis*, a moderately halophilic bacterium. Microbiology, 157 (12): 3378-3387.

TAHRIOUI A, QUESADA E, LLAMAS I, 2013a. Draft genome sequence of the moderately halophilic gammaproteobacterium *Halomonas anticariensis* FP35 (T). Genome Announc., 1 (4): e00497-00413.

TAHRIOUI A, SCHWAB M, QUESADA E, et al., 2013b. Quorum sensing in some representative species of *Halomonadaceae*. Life, 3 (1): 260-275.

TAIT K, WILLIAMSON H, ATKINSON S, et al., 2009. Turnover of quorum sensing signal molecules modulates cross-kingdom signalling. Environ. Microbiol., 11 (7): 1792-1802.

TINDBAEK N, SVENDSEN A, OESTERGAARD P R, et al., 2004. Engineering a substrate-specific cold-adapted subtilisin. Protein Eng. Des. Sel., 17 (2): 149-156.

TOMMONARO G, ABBAMONDI G R, IODICE C, et al., 2012. Diketopiperazines produced by the halophilic archaeon, *Haloterrigena hispanica*, activate AHL bioreporters. Microb. Ecol., 63 (3): 490-495.

TOMMONARO G, ABBAMONDI G R, TOKSOY ONER E, et al., 2015. Investigating the quorum sensing system in halophilic bacteria. In: Maheshwari D K, Saraf M (Eds), Halophiles: Biodiversity and Sustainable Exploitation. Springer International Publishing, Cham, pp. 189-207.

VAN DEN BURG B, 2003. Extremophiles as a source for novel enzymes. Curr. Opin. Microbiol., 6 (3): 213-218.

VAN DER MAAREL M J E C, VAN DER VEEN B, UITDEHAAG J C M, et al., 2002. Properties and applications of starch-converting enzymes of the α-amylase family. J. Biotechnol., 94 (2): 137-155.

VERA M, SCHIPPERS A, SAND W, 2013. Progress in bioleaching: fundamentals and mechanisms of bacterial metal sulfide oxidation-Part A. Appl. Microbiol. Biotechnol., 97 (17): 7529-7541.

WHITELEY M, LEE K M, GREENBERG E P, 1999. Identification of genes controlled by quorum sensing in *Pseudomonas aeruginosa*. Proc. Natl. Acad. Sci. U. S. A., 96 (24): 13904-13909.

WILLIAMS P, 2007. Quorum sensing, communication and cross-kingdom signalling in the bacterial world. Microbiology, 153 (12): 3923-3938.

WILSON C M, RODRIGUEZ M, JOHNSON C M, et al., 2013. Global transcriptome analysis of *Clostridi-

um thermocellum ATCC 27405 during growth on dilute acid pretreated Populus and switchgrass. Biotechnol. Biofuels, 6: 179.

WOODLEY J M, 2013. Protein engineering of enzymes for process applications. Curr. Opin. Chem. Biol., 17 (2): 310-316.

YOUSEFI-NEJAD M, MANESH H N, KHAJEH K, 2011. Proteomics of early and late cold shock stress on thermophilic bacterium, *Thermus* sp. GH5. J. Proteome, 74 (10): 2100-2111.

ZHANG G, ZHANG F, DING G, et al., 2012. Acyl homoserine lactone-based quorum sensing in a methanogenic archaeon. ISMEJ., 6 (7): 1336-1344.

ZHANG X, YU S, GONG Y, et al., 2016. Optimization design for turbodrill blades based on response surface method. Adv. Mech. Eng., 8 (2): 1-12.

第2部分
跨界通信

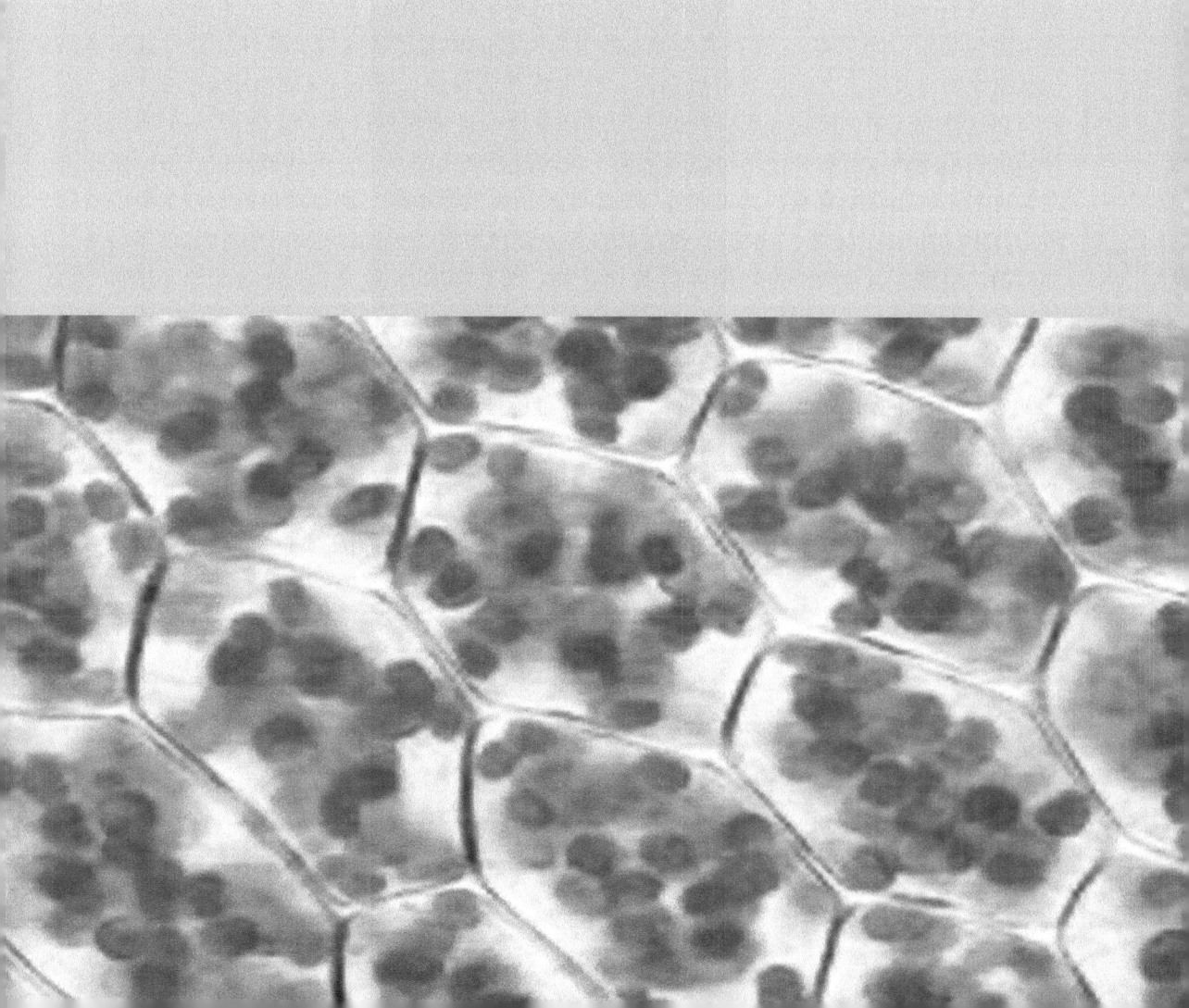

第5章 群体感应在植物致病性上的作用

Onur Kırtel[*,a], Maxime Versluys[†,a], Wim Van den Ende[†],
Ebru Toksoy Öner[*]

[*] Industrial Biotechnology and Systems Biology Research Group, Marmara University, Bioengineering Department, Istanbul, Turkey, [†] Laboratory of Molecular Plant Biology, KU Leuven, Leuven, Belgium

1 前言

微生物无处不在，虽然它们的体积很小，但它们与环境或宿主发生无数的相互作用，从而引起重大变化。例如，近年来人类微生物组一直是一个热门的研究领域，因为微生物组的特征和作用与人类健康之间存在着强烈的相互作用（Althani et al., 2016）。就像动物一样，植物也是多种多样的共生微生物群的宿主。无论微生物对植物是有益（互惠关系）还是有害（寄生关系；病原体），这些共生体不断参与复杂的种间和种内信号事件。共生这一术语包括微生物与其宿主之间的所有致病性、共生或互惠关系（Martin and Schwab, 2013）。因此，在这篇综述中，植物发病机制被视为这些共生相互作用的一部分。如图5-1所示，这种共生关系是通过植物和微生物之间交换大量的信号来巩固。在一系列事件中，根际微生物在含有根系的土壤中定殖，不仅通过与根系定殖引发信号串联，而且通过调节宿主免疫发挥核心作用（Mendes et al., 2013）。微生物在根系定殖被激活是通过根系分泌物包含多种分子如单糖、多糖、氨基酸、蛋白质及酚类化合物、维生素和激素（Lareen et al., 2016）等。趋化反应被激活是通过特异性未知的受体。此外，根表面离子和质子运输会产生电流，从而吸引游动的孢子。然而，趋化性和趋电性在塑造植物微生物组中的意义尚不清楚（Mendes et al., 2013）。除了根际土壤相关的微生物组外，叶际微生物也会被茎和叶的渗出物包括挥发性有机化合物或激素所吸引（Vorholt, 2012）。

尽管没有根际微生物那么丰富，但已知细菌也存在于植物器官中，如果实、种子、花和花蜜（Compant et al., 2011）。

根据生活方式，植物病原菌被分为三组。第一组，生物营养体，从活细胞中获取必要的代谢物，因此，它们不会迅速杀死宿主；第二组，坏死营养体是腐生的，它们能迅速杀死宿主，然后从死亡的宿主细胞中获取营养；第三组，半生物营养体，在感染初期表现出生物营养特征，然后在后期转向坏死性营养（Moore et al., 2011）。植物病原菌存在于不同的生命群体中如细菌、真菌、病毒和线虫（Chagas et al., 2018）。虽然一些植物致病性物种如丁香假单胞菌（*P. syringae*）（Baltrus et al., 2017），解淀粉欧文菌（*E. amylovora*）（McNally et al., 2015）或稻瘟病菌（*Magnaporthe oryzae*）（Martin-Urdiroz et al., 2016）都被认为是"模式生物"，多年来对它们进行了非常充分的研究。近年来，组学方法和对这些共生体及其宿主之间信号事件的鉴定表明，植物发病的复杂机制远比目前所知的要复杂得多。

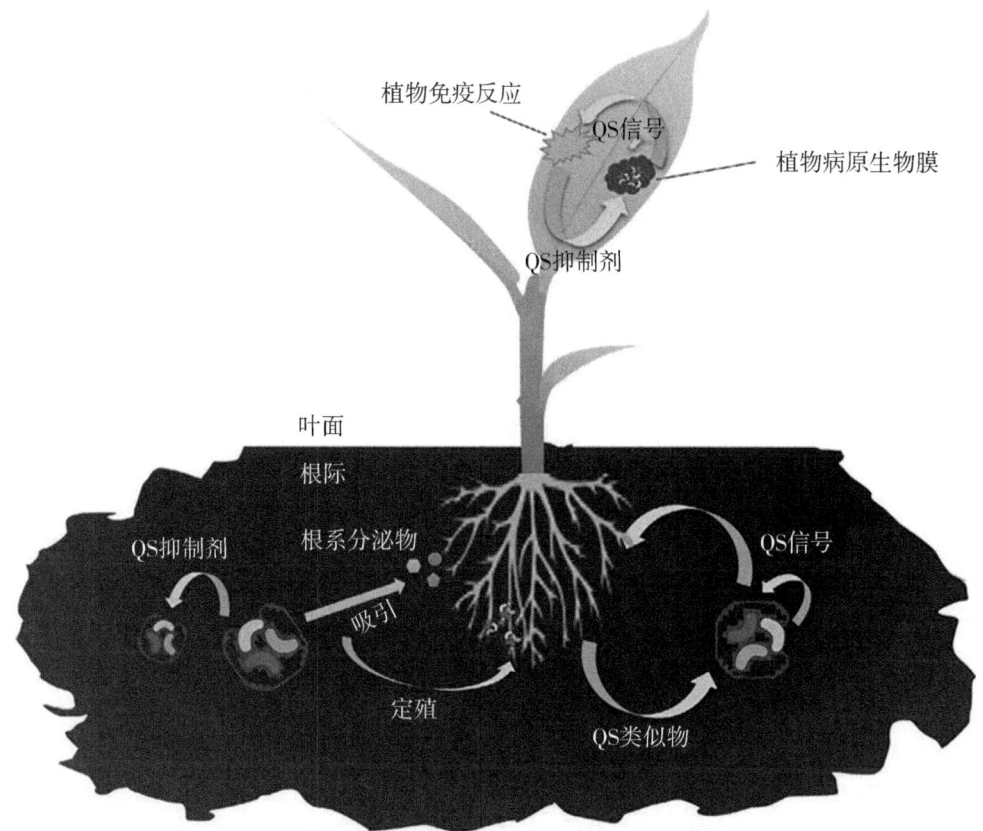

图 5-1 由群体感应（QS）系统介导的植物和微生物之间的互惠和寄生关系

注：植物根系分泌物以及模拟 QS 吸引互惠共生的微生物到根际定殖。互惠共生微生物和植物病原微生物都依赖 QS 系统进行各种活动，如胞外多糖（EPS）的合成、毒力因子的分泌、逃避植物的识别等。病原体的存在可能产生病原体相关分子模式（PAMPs）或损伤相关分子模式（DAMPs），从而触发植物的免疫反应途径，导致 QS 抑制剂的分泌。当与其他物种竞争资源时，细菌也会分泌各种具有 QS 抑制作用的酶。

植物病原菌摧毁了世界各地的农作物，给农业和食品工业造成了巨大的损失。仅在美国，每年因植物病原菌造成的作物损失价值约为 330 亿美元（Pimentel，2011）。水稻稻瘟病菌（*M. grisea*）造成全球水稻损失 10%～30%，据估计，这些损失足以养活约 6 000 万人（Talbot，2003）。很明显人类仍无法有效地与这些病原菌作斗争。使用化学杀菌剂可能在短期内有效，但不仅会对环境和人类健康产生有害影响，还会对病原菌造成巨大的选择压力，为耐药菌株的出现或被其他病原菌迅速取代铺平了道路。因此，了解植物发病的潜在作用模式对于克服这些损失至关重要。近年来，科学界对这些病原菌及其群落之间的信号有更深入的了解，如主要存在于革兰氏阴性细菌中的 AHLs；革兰氏阳性细菌中的几种寡肽；酵母中的酪醇和法尼醇；以及许多细菌中的种间信号分子自诱导物呋喃酮酰硼酸二酯（AI-2）（Chagas et al.，2018）。对这些信号分子的鉴定和表征，使得了解和形成抵抗植物病害的新策略成为可能。

2 群体感应现象

群体感应（QS）是植物与微生物间互作中的一种重要现象。通过与寄主植物、微生物间和微生物内的信号通信，QS 信号分子在微生物定殖和共生关联中发挥重要作用。微生物致病的主要驱动力依赖于种群密度的化学信号分子，其构成了整个群体感应系统。QS 系统最早是在海洋细菌费氏弧菌中发现的（Nealson and Hastings，1979），该细菌寄居在海洋真核生物如夏威夷短尾乌贼的光器官上。当细菌在光器官中达到一定的细胞密度后，费氏弧菌通过荧光素酶启动生物发光。这种互惠关系使费氏弧菌能够在营养贫乏的海水中定殖在其表面，同时在夜间费氏弧菌通过去除水柱上的阴影为短尾乌贼提供有效的伪装。基本上，荧光素酶的合成始于细胞外 AHL 分子的浓度（3-oxo-C6-HSL）达到一定阈值浓度（100~200 nmol/L）后，与 LuxR 蛋白结合。随后，AHL-LuxR 复合物结合到荧光素酶启动子区域，启动荧光素酶和 LuxI 蛋白的转录，形成一个前馈环，使 AHL 的合成呈指数增长。荧光素酶将一个还原的黄素单核苷酸和一个长链脂肪酸分别氧化为黄素单核苷酸和脂肪酸，在此过程中产生生物发光现象（Verma and Miyashiro，2013）。一旦在群落中 AHL（或任何其他自诱导物）的浓度与合成它的细胞的浓度成正比，那么，QS 系统可被视为一个细胞密度依赖的化学信号级联，导致差异基因表达，进而产生各种表型特征，如生物发光、生物膜形成或毒力因子分泌。

2.1 植物相关的微生物群体感应

植物与多种微生物之间存在着密切的关系。叶际是由植物地上部分的微生物群落组成，根际是位于植物根周围的狭窄区域，包含大量微生物如附生菌（附着在植物上）、内生菌（居住在植物体内）。微生物物种的组成不仅依赖于植物种类，而且也依赖于非生物和生物环境因子的影响。根际微生物群影响植物的生长发育，地下微生物物种丰富度甚至可以用来预测植物的生产力。除土壤病原菌外，对有益的根际微生物包括菌根真菌、植物生长促进的根际细菌及固氮菌，已进行了广泛而深入的研究，这些微生物显著提高植物的生长动态（Mendes et al.，2013）。叶际微生物群也参与许多植物过程，如免疫反应、营养获取和生长发育。植物激素的产生和分泌是植物有利于微生物生存的一种常见策略。胞外多糖（EPS）的产生是植物相关细菌耐受干旱等逆境的最重要的适应性之一（Chagas et al.，2018）。

与植物密切相关的高丰度微生物与 QS 机制密切相关。大约 20 年前，首次发现 AHL QS 分子，包括不同酰基链长度的 3-oxo-和 3-羟基 AHL，这些在植物相关细菌中广泛存在，包括农杆菌、欧文菌、泛菌属、假单胞菌和根瘤菌（Cha et al.，1998）。此外，黄单胞菌和伯克霍尔德菌主要产生一种不同类型的 QS 分子，即 DSFs（扩散信号因子）。这些 QS 分子在调控生物膜形成和物种内和物种间的通信中起着非常重要的作用。研究表明，AHL QS 在根际和叶际有益微生物和致病微生物中都很常见。在植物致病菌中，QS 也影响毒力因子的表达。例如，在胡萝卜软腐果胶杆菌胡萝卜亚属（*Pectobacterium carotovorum* subsp. *carotovorum*），以前被称为胡萝卜软腐欧文菌胡萝卜亚属（*E. carotovora* subsp. *carotovora*）中，QS 影响细胞外酶的表达和抗生素的生产（von Bodman et al.，2003）。有趣的是，根瘤菌物种产生的 AHL 结构多样性最大，这清楚地表明了 QS 在这些固氮细菌中的重要性及其与宿主植物建立良好的相互作用（Cha et al., 1998；Sanchez-Contreras et al., 2007）。

2.1.1 植物致病细菌的群体感应

在费氏弧菌中发现 AHL 依赖的信号系统后，在许多其他细菌中也发现了类似的系统，其中，很大一部分属于植物革兰氏阴性病原菌，如根瘤农杆菌（Lang and Faure, 2014；Subramoni et al, 2014）、胡萝卜软腐果胶杆菌（Pirhonen et al., 1993）、斯氏泛菌（*P. stewartii*）（von Bodman and Farrand, 1995）以及丁香假单胞菌（Quiñones et al., 2005）。尽管这些细菌的 QS 系统表现出经典种群密度依赖信号的模式，但所涉及的分子和基因表现出巨大的变异性，从而表现出截然不同的致病性策略。

农杆菌是一种革兰氏阴性菌，会引起许多植物冠瘿病。农杆菌致病性是基于从肿瘤诱导（Ti）质粒转移 DNA（T-DNA）并整合到植物细胞的细胞核中，诱导植物生长素和细胞分裂素的表达，从而产生冠瘿碱。在根瘤农杆菌 QS 系统中，*TraR* 基因是转录激活因子，位于 Ti 质粒上，与费氏弧菌中 *LuxR* 同源，因此，在基因表达水平上受冠瘿碱的控制。有冠瘿病的植物通过 T-DNA 插入基因和上调 *TraR* 的表达，从而产生冠瘿碱。尽管其他 3-oxo-AHL 在一定程度上会诱导 *TraR* 的表达，但 *TraR* 的共诱导物为 3-oxo-C8-HSL。AHL 是由 3-氧-辛酰基载体蛋白和 S-腺苷甲硫氨酸合成的，由 TraI3 产生。通过启动子区域的 tra 盒子调控 QS 依赖性基因，激活 TraR-3-oxo-C8-HSL 二聚体。这些 QS 调控的基因也位于 Ti 质粒上。*tra*、*rep* 和 *traM* 基因的上调可诱导 Ti 质粒的整合。TraM 通过直接相互作用抑制 TraR 活性，从而抑制 TraR 与 DNA 的结合。激活 TraR 也能上调 *traM*，从而调控 TraR 水平。此外，QS 信号还受到内酯酶（降解 AHL 的酶）的负调控，这些酶是由植物信号诱导的（在"植物 QS 抑制剂"部分讨论）（Lang and Faure, 2014；Subramoni et al., 2014）。

解淀粉欧文菌是导致苹果和梨等蔷薇类植物火疫病的病原菌，其通过一种类似菌毛的结构入侵宿主，这种结构属于Ⅲ型分泌系统的一部分（T3SS；Piqué et al., 2015）。解淀粉欧文菌还拥有 5~25 个基因组成Ⅵ型分泌物（T6SS）。尽管 T6SS 的确切功能尚不清楚，但一些研究表明，T6SS 缺失的突变体表现出运动性受损（Kamber et al., 2017）或 EPS 产生受损（Tian et al., 2017），二者都是细菌存活和致病的关键因素。解淀粉欧文菌可产生由葡萄糖、半乳糖和葡萄糖醛酸组成的杂多糖以及由果糖聚合而成的果聚糖，作为 EPSs。Koczan 等（2009）研究表明，苹果接种杂多糖缺失的菌株无致病性且快速死亡，而接种果聚糖缺失的菌株表现出运动性降低，且无法侵染木质部导管。果胶杆菌属含有 LuxI 同源物 ExpI，在一定的 AHL 浓度下，能抑制 T3SS 等多种毒力因子、运动性、黏附性以及植物细胞壁降解酶的转录（Moleleki et al., 2017）。因此，只有当群落达到一定的细胞密度后，才能保证病原菌对植物的成功入侵。斯氏泛菌是玉米细菌性枯萎病的致病因子，主要通过被称为 *stewartan* 的荚膜多糖（CPS）阻断木质部中水分的运输，*stewartan* 是由一种由葡萄糖、半乳糖和葡萄糖醛酸残基组成的支链酸性杂多糖（Nimtz et al., 1996）。当 AHL 浓度低于阈值时，斯氏泛菌中的 LuxR 同源物 EsaR 抑制 CPS 生物合成的转录因子 *rcsA*，激活控制细菌表面运动的 *lrhA*（Duong and Stevens, 2017）。丁香假单胞菌在引起大豆褐斑病之前大量聚集在叶片表面，也依赖于 AHL 介导的 QS 系统，调节氧化应激耐受性、EPS 产生和运动（Quiñones et al, 2005）。

尽管大多数革兰氏阴性菌依赖 AHL 介导的 QS 系统，但伯克霍尔德菌属和黄单胞菌分别使用 DSF，如顺式-2-十二烯酸和顺式-11-甲基-2-十二烯酸（Carlier et al., 2014）。在黄单胞菌属中，DSF 是由 RpfF 同源物合成的，该物质由一种来自烯酰水合酶亚家族的酶组成。

DSFs 在细胞外空间达到一定浓度后，膜上的传感激酶 RpfC 会感知到这一变化，进而通过激活细胞质蛋白 RpfG 降解 bis-(3′,5′)-环二聚鸟苷单磷酸（环二鸟苷酸）。环二鸟苷酸作为黄单胞菌和伯克霍尔德菌毒力因子的抑制子。与黄单胞菌不同的是，伯克霍尔德菌的 DSF 是通过胞质 RpfR 直接感知的，RpfR 也能降解环二鸟苷酸（Dow，2017）。值得注意的是，伯克霍尔德菌也有经典的 AHL 介导的 QS 系统，并且有证据表明这两个系统之间存在相互作用（Udine et al.，2013）。

与革兰氏阴性细菌相比，革兰氏阳性细菌中致病细菌较少。最典型的菌种是密执安棒状杆菌（*Clavibacter michiganensis*）和带化红球菌（*R. fascian*），它们分别引起细菌性溃疡病和叶瘿病。一些链霉菌属（*Streptomyces* spp.），如酸性链霉菌（*S. acidiscabies*）、疥疮链霉菌（*S. scabies*）和充血链霉菌（*S. turgidiscabies*）会导致马铃薯结痂病（Kers et al.，2005）。在革兰氏阳性 QS 系统中，截至目前，尚未发现 AHL 介导的信号通路。相反，革兰氏阳性细菌使用一些翻译后修饰的寡肽作为信号分子。分泌的寡肽要么通过膜表面的双组分系统检测，要么通过寡肽运输系统进入细胞质，然后与不同的调节因子相互作用，从而改变与毒力、能力或偶联相关基因的表达（Monnet and Gardan，2015）。

2.1.2 植物病原真菌中的群体感应

植物病原真菌和卵菌占全球每年由微生物引起的作物损失 70%~80%（Oerke，2006）。植物病原真菌/卵菌及其引起的疾病包括蜜环菌（*Armillaria* spp.）（蜂蜜真菌病）、小麦白粉菌（*Blumeria graminis*）（白粉病）、灰霉病菌（*Botrytis cinerea*）（灰霉病）、尾孢菌属（*Cercospora* spp.）（叶斑病）、尖孢镰孢菌（*Fusarium oxysporum*）（枯萎病）、稻瘟病菌（稻瘟病）、亚麻锈菌（*Melampsora lini*）（亚麻锈病）、新榆枯萎病菌（*Ophiostoma novo-ulmi*）（荷兰榆树病）、致病疫霉（*Phytophthora infestans*）（马铃薯晚疫病，这些病害导致了 1845—1946 年爱尔兰饥荒，导致爱尔兰人口中 1/8 的人口死亡）、小麦锈病菌（*Puccinia graminis tritici*）（茎锈病）和黑粉病菌属（*Ustilago* spp.）（黑粉真菌病；Moore et al.，2011）。真菌是极具侵略性的病原菌，利用不同的策略入侵植物宿主。蜜环菌属利用高度保护的地下多根茎结构，被称为根茎形态，在森林地面上运输水分和营养物质。一个生活在俄勒冈州森林里的口耳蜜环菌（*A. ostoyae*）无性系杀死了 30%的黄松，覆盖了 9.65 km² 的森林地面，被认为是目前现存的最大生物体（Ferguson et al.，2003）。在土耳其黑海地区的一片针叶林中也发现了一个面积为 450 m² 的类似森林（Lehtijärvi et al.，2017）。像亚麻锈菌和小麦白粉菌真菌产生被称为吸器的入侵结构，这些结构能穿透植物细胞获取营养并抑制宿主的防御系统。相似的，黑粉病菌属为了同样的目的会形成细胞内菌丝（Moore et al.，2011）。

正如 Oliveira-Garcia 和 Valent 所述（2015），寄生真菌已经发展出高度专门化的策略，以避免被植物免疫系统检测到。第一个策略是在吸器/胞内菌丝和植物细胞之间建立一个凝胶-基质界面。这种基质中含有大量来自真菌和植物的碳水化合物和蛋白质，以及植物所表达的糖和氨基酸转运体。这些转运体有效地将碳源转移到入侵的真菌中，从而最大限度地减少触发参与植物防御机制的外质体信号形成的风险。第二种策略是合成操纵宿主代谢的特殊效应因子。这些效应因子是小蛋白质，能够通过抑制宿主溶解酶活性或宿主免疫反应途径，防止几丁质低聚物从真菌细胞壁释放，或将已经释放的低聚物进行隔离。同时，它们还抑制植物的过氧化物酶，从而保护入侵菌丝免受活性氧（ROS）的伤害，或直接操纵宿主代谢以合成辅助致病性的毒力因子。最后，第三种策略包括对真菌细胞壁结构进行修饰，以躲避寄主的识别。为此，入侵真菌可以将细胞壁几丁质脱酰化为壳聚糖，产生 α-1,3-葡聚糖，以保护菌丝免受植物裂解酶的影响，或调节 β-1,3/1,6-葡聚糖的含量，以逃避病原体相关分子模式（PAMP）引发的免疫反应。

真菌和卵菌利用的 QS 信号分子与细菌完全不同，其中最著名的包括来自白假丝酵母菌（*Candida albicans*）的法尼醇、法尼酸和酪醇；来自单胞锈菌（*Uromyces phaseoli*）的二甲氧基肉桂酸；来自接合菌纲（Zygomycetes）的三孢酸；以及酿酒酵母菌（*Saccharomyces cerevisiae*）的一个因子（Hogan，2006）。与细菌相比，真菌 QS 系统是最近才被发现的，并影响多态真菌在不同生命阶段的形态，间接影响其致病力。与细菌类似，信号分子在细胞内产生并分泌到细胞外，当其浓度达到阈值浓度后，便与细胞膜上的信号受体蛋白结合，从而改变各种基因的表达。在玉米黑粉病菌（*U. maydis*）中，短法尼基化肽作为交配信息素，在相反交配类型的细胞上诱导从出芽到丝状双核体的形态发生（Spellig et al.，1994）。环磷酸单腺苷（cAMP）在这种真菌中也发挥类似的作用（Gold et al.，1994）。寄生疫霉（*P. parasitica*）通过细胞密度依赖的细胞外信号分子的作用形成生物膜（Galiana et al.，2008）。在小麦白粉菌中，检测到与革兰氏阳性细菌相似的短链肽信号分子（Rajput et al.，2015）。在自然界中，真菌总是与细菌密切接触，细菌的 QS 信号分子也会影响真菌，诱导其形态变化或触发参与 AHL 降解的酶（Hartmann and Schikora，2012）。大多数文献中涉及 QS 和真菌的研究仅限于通过细菌 QS 分子诱导植物对病原真菌的系统抗性，而植物病原真菌本身 QS 系统的特性研究在很大程度上是缺乏的。

3 细菌群体感应对植物免疫和生理的影响

3.1 植物先天免疫

植物与动物一样，在环境胁迫条件下，包括生物相互作用，具有一个先天免疫系统来维持体内平衡。植物通过识别微生物相关分子模式（MAMPs）来应对微生物的存在，并将其作为其免疫反应的一部分。最初，Jones 和 Dangl（2006）根据锯齿型模型中提出了植物免疫系统，并将其分为两个阶段：PTI（模式触发免疫）和 ETI（效应触发免疫）。近年来，为了更好地理解植物的免疫反应，相关理论已进行相应的调整和改进。ETI 依赖于病原菌 Avr 蛋白和宿主 R 蛋白之间特定的相互作用，导致程序性细胞死亡途径。另外，PTI 是一种基于宿主受体识别 MAMPs 的更古老的进化机制。对 MAMPs 的感知是由位于植物质膜上 PRRs（模式识别受体）进行的，受体激活后诱导下游信号级联反应，涉及 ROS 动态变化、MAPK 级联反应和激素信号。LRR（富亮氨酸重复序列）或 LysM（赖氨酸基序）是 PRRs 中最常见的结构域，后者参与了多糖分子的识别。研究表明，MAMPs 中研究最为深入的是 flg22，它是细菌鞭毛蛋白的保守表位，能被植物中 FLS2 受体识别。在 flg22 存在的情况下，异二聚体受体与 BAK1 激酶结合，从而诱导下游信号转导（Bigeard et al，2015）。在真菌 MAMPs 中，几丁质是最具有特征的，由基于 *N*-乙酰葡萄糖胺低聚物片段的植物 LysM 受体识别。其他的 MAMPs 包括 EF-Tu、肽聚糖和果聚糖等。在致病微生物存在的情况下，由于侵染可能导致细胞渗漏和/或细胞壁降解，因此，对 MAMPs 的识别可能伴随着 DAMP（损伤相关分子模式）的感知（Versluys et al.，2016）。

在 MAMP 受体的下游，一个信号级联在感知时被激活。最早的信号反应涉及 Ca^{2+} 和 ROS 信号。Ca^{2+} 依赖的信号涉及 Ca^{2+} 结合蛋白，如钙依赖蛋白激酶，这导致参与生物胁迫响应的转录因子被激活。在生物胁迫中，ROS 的主要来源是 NADPH 氧化酶 RBOHs，它可以长距离传导 ROS 波（Choi et al.，2017）。MAPKs（丝裂原活化蛋白激酶）在 MAMP 受体下游的磷酸化级联反应中起核心作用。在其他 MAPKs 的 Thr-X-Tyr 激活基序中磷酸化丝氨酸/苏氨酸残基。尽管植物中含有大量的 MAPKs，但只有一小部分参与植物免疫反应的激活，尤其是 MPK3 和 MPK6。研究表明，MPK3 在 PTI 信号转导中更为重要（Meng and Zhang，2013）。在微生物攻击的情况下，这些途径诱导下游适当的植物防御反应。

植物拥有几种形式的组成型防御，其中包括物理屏障，如细胞壁和角质层，以及抗菌化合物，如皂苷和硫代葡萄糖苷。后者是一种非活性的储存化合物，在细胞损伤过程中降解，释放有毒的异硫氰酸盐，这是十字花目（Brassicales）中常见的机制（Bigeard et al.，2015）。如果以上防御不足以阻止病原菌，那么 MAMP 感知可引起一系列的防御，包括木质素和胼胝质沉积、程序性细胞死亡，以及抗菌次生代谢产物和酶的产生（Taiz and Zeiger，2012）。苯丙氨酸解氨酶是木质素和黄酮类化合物合成的重要酶。此外，它还参与了水杨酸（SA）的合成，SA 是一种参与植物免疫应答的植物激素（Gao et al.，2015）。

除先天免疫外，植物还可诱导系统免疫系统。在病原菌攻击过程中，抗性机制在植物的远端部分被诱导，导致一种系统获得抗性（SAR），从而降低了进一步侵染的敏感性。对于植物有益细菌的相互作用，也会发生类似的机制，被称为诱导系统抗性（ISR；Gao et al.，2015）。这些系统途径需要可移动的信号分子，如壬二酸和丙戊酸，这些信号分子能在整个植物中运输（Shah and Zeier，2013）。参与 SAR 的植物激素是 SA，它在感染部位增加，并刺激远距离信号以产生系统抗性。ISR 的作用机制受到植物激素茉莉酸（JA）和乙烯的调控。在诱导不同防御反应过程中，通常 SA 和 JA 是拮抗的信号分子（Gimenez-Ibanez and Solano，2013）。

3.2 群体感应分子与植物免疫

QS 分子在细菌中广泛存在，因此，推测其可能被植物中的 PRRs 识别，从而作为 MAMPs。然而，迄今为止，只有 MAMP Ax21 肽被提出作为 QS 分子。这种硫酸肽与水稻的 XA21 PRR 结合，XA21 是一种受体激酶，当被激活时，将细胞内激酶结构域转移到细胞核进行下游免疫应答（Park and Ronald，2012）。最近的研究表明，这种 Ax21 肽在植物和动物的病原菌均有发现，并通过 I 型分泌系统分泌。尽管一些研究暗示了 Ax21 在 QS 中的作用，但截至目前，这一点仍没有得到充分的证实。一直存在的争论是，由于 QS 分子具有多样性，它们可能并没有被 MAMPs 恰当的识别。尽管如此，下面的部分将讨论 QS 分子对植物免疫影响的几项研究。

之前已有研究表明，通过动物系统检测 AHL，其对巨噬细胞、白细胞介素产生和炎症反应等具有免疫调节作用。在植物中，Mathesius 等（2003）研究表明，苜蓿根能够检测"nmol"浓度的细菌 AHL，并改变 150 多种蛋白质的积累，其中包括胁迫响应因子如 PR 蛋白、过氧化物酶和超氧化物歧化酶，以及参与类黄酮代谢的酶和植物激素信号的蛋白（Hartmann and Schikora，2012）。You 等（2006）研究表明，因为叶面应用 QS 分子的检测效率较低，因此大多数研究是在植物根应用 QS 分子。对拟南芥叶片喷施来自根瘤农杆菌的 3-oxo-C8-HSL（QS 信号分子）时，其基因表达变化是非常有限的。

在番茄中，Schuhegger 等（2006）报道了一种来自液化沙雷菌的 C6-HSL，能够诱导番茄对交替链格氏菌（*Alternaria alternata*）的抗病性。von Rad 等（2008）研究了 C6-HSL 对拟南芥中的作用，发现其仅引起防御相关基因在转录水平上的有限变化，表明植物物种之间的检测存在差异。在后续的研究中，Schikora等（2011）证明寄主植物的响应取决于酰基部分的长度和在AHLγ位置上的官能团。拟南芥用3-oxo-C14-HSL处理后，增强了拟南芥对生物营养真菌丁香花球菌（*Golovinomyces orontii*）和半生物营养细菌丁香假单胞菌番茄致病变种DC3000的抗性。有趣的是，通过对atmpk6突变体的研究发现，在AHL诱导的免疫中，AtMPK6的需求在植物免疫信号通路中扮演着核心角色。

此外，Schikora等（2011）研究3-oxo-C14-HSL对于大麦根对小麦白粉菌的抗性诱导，发现其与拟南芥有相似的作用。在后来的一项研究中，在拟南芥与病原细菌丁香假单胞菌番茄致病变种DC3000中的互作系统中，发现较长的AHL酰基链具有更高的免疫反应诱导能力，从而抵抗细菌的侵染（Schenk et al., 2012）。在后续的试验中，研究人员利用3-oxo-C14-HSL在拟南芥中进一步验证了AHL上调的特异性防御和信号通路。植物通过木质化、胼胝质沉积以及酚类化合物的产生增强细胞壁，从而提高对丁香假单胞菌的抵抗能力（Schenk and Schikora, 2015）。

AHL检测不仅能诱导局部防御，而且似乎还能在宿主植物中诱导系统抗性途径。在番茄根际与来自液化沙雷菌的C6-HSL相互作用后，其叶片SA水平升高。因此，来自该物种的QS分子可以诱导系统抗性反应，这一过程可通过SA-和乙烯-依赖基因的系统诱导得到证明（Schuhegger et al., 2006）。在同一品系中，Pang等（2009）发现大豆和番茄的AHL基于ISR反应对灰霉病菌侵染表现出抗性，而Schenk和Schikora（2015）发现，C14-HSL处理拟南芥根后，拟南芥表现出系统抗病性。随后观察到，拟南芥用3-oxo-C14-HSL处理后，氧脂素如JA及其相关代谢物和SA的变化。尽管SA和JA在系统抗性途径中表现出拮抗作用，但在AHL启动系统中，它们似乎具有协同效应，表明存在一种新的AHL特异性信号通路（Schenk and Schikora, 2015）。

3.3 群体感应分子影响植物生理

为了被植物系统地检测到，QS分子必须被宿主植物吸收。QS分子通过根部的吸收和运输取决于AHL的长度以及植物的种类。Götz等（2007）研究了大麦根系如何同时吸收C6-和C8-HSLs，而豆薯根只运输可检测到的C6-HSL水平。在拟南芥中，von Rad等（2008）研究表明，AHL的运输取决于AHL酰基链的长度。虽然C6-HSL很容易从根运输到地上部，但C10-HSL没有被运输，这可能是由于其更高的疏水性。结果表明，C6-HSL处理在植物各部位均产生了响应，而C10-HSL在根中积累会导致毒性作用。除了在植物中运输外，Ortí-Castro等（2008）表明拟南芥通过脂肪酸酰胺水解酶代谢AHLs的可能性，而更早的研究则是在莲花中提出了不同酰基链长度的AHLs（Delalande et al., 2005）。

von Rad等（2008）研究表明，拟南芥C6-HSL处理后并没有真正诱导植物防御，但一些与生长激素相关的基因显著上调或下调，且根明显伸长。生长素与细胞分裂素比值发生显著的变化。Schenk和Schikora（2015）也发现，激素比例的改变对拟南芥的生长有积极作用。C6-HSL处理后，茎部生长最为显著，而较长的酰基链AHLs对植物生长水平有所下降，并对根系生长也产生了相似的影响，表明较短的AHLs对植物的影响更为深远。这与对植物免疫系统的影响形成了鲜明的对比。在植物免疫中，较长的AHL会导致更高的抗性。然而，Ortíz-Castro等（2008）研究表明，与较短或较长的AHL相比，C10-HSL对拟南芥根结构的影响更为显著，但未诱导经典的生长素信号通路发生改变。

Liu等（2012）用C6-或C8-HSLs处理根系后，证实了拟南芥根生长情况。在他们的研究中，gcr1 突变体对这些处理不敏感，这表明 GPCR1 在 AHL 信号转导中发挥了作用。此外，他们还发现，在 AHL 处理后诱导 *GPCR1* 基因的表达，证实 *GPCR1* 在基于 AHL 的根伸长中的作用。Jin 等（2012）的另一项研究也表明，推测 GPCRs 在拟南芥中具有两种作用，并进一步证明这些受体在根伸长反应中对 AHLs 的作用。在绿豆中，AHL 处理诱导了不定根的形成，但仅限于 C3 取代的 3-oxo-C10-HSL。这是由植物中过氧化氢和一氧化氮依赖的 cGMP 信号控制下所介导的生长素依赖信号（Bai et al, 2012）。随后的研究也进一步证实了 AHL 在大麦和豆薯中 ROS 动态中的作用。研究发现，使用不同酰基链的 AHL 处理后，抗氧化酶活性增加（Götz-Rösch et al, 2015）。这些研究不仅揭示了 AHL 是植物检测和信号通路的第一个参与者，还指出由于 AHLs 结构的高度差异和不同植物对这些信号的反应不同，这些相互作用具有复杂性。

4 群体淬灭在植物病理中的作用

4.1 植物群体感应抑制剂

虽然来自植物相关细菌 QS 化合物对植物健康和生长具有积极的作用，但众所周知，植物可产生 QS 抑制剂，这对病原微生物尤其重要。由于 QS 是微生物之间的一种交流方式，干扰这些信号对细菌没有直接的毒性作用，也不会导致耐药菌株发展的选择压力。这种 QS 抑制机制也被称为群体淬灭（QQ；Rasmussen and Givskov, 2006）。植物可以在不同水平上靶向 QS 系统，包括抑制 QS 化合物的生物合成、抑制受体结合以及 QS 化合物的降解（Asfour, 2017）。

抵御 AHLs 的一般策略可以是碱化，从而导致内酯环的打开。这种策略已在植物与胡萝卜软腐果胶杆菌相互作用的过程中被观察到。靶向氧化的 AHLs，如 3-oxo-C14-HSL，掌状海带可分泌氧化卤素化合物，干扰氧化的 AHLs。对几种豆科植物的提取物也证实了 AHL 的降解，尽管仅观察到短酰基链 AHLs 的降解（Delalande et al, 2005；Rasmussen and Givskov, 2006）。

在红藻中进行 QQ 受体水平的关键研究，使用卤代呋喃酮干扰基于 AHL 的 QS（Manefield et al., 1999）。近年来，在多种植物如胡萝卜、番茄和大蒜提取物中发现了针对细菌信号受体的群体感应抑制剂，尽管这些化合物的鉴定往往仍然未知（Truchado et al, 2015）。在大蒜提取物中，至少发现了三种不同的抑制剂，包括二甲基二硫醚、ρ-香豆酸和阿焦烯。ρ-香豆酸在植物中普遍存在，大蒜中 QS 抑制作用的发现，表明在其他植物物种中也有类似的作用，尽管早期的研究指出，该化合物对其他细菌物种有刺激作用（Kalia, 2013 和其中的参考文献）。

从那时起，许多研究调查了不同植物系统中的 QQ 潜力，包括种子、幼苗和植物提取物。在水稻中，种子提取物中存在缺乏内酯环的 AHL 模拟物。使用紫色色素杆菌作为 QS 模式系统来研究 QQ 效应，同时也检测了其他细菌种类。一些细菌菌株中由于这些 AHL 模拟物，QS 生物膜的形成被特异性抑制。在甘蓝中，异硫氰酸酯萝卜硫素和甘油三酸酯通过 QQ 机制显著抑制了铜绿假单胞菌的致病力（Kalia, 2013）。

在植物提取物中，几种植物次生代谢物已被鉴定为 QS 抑制剂。哥伦比亚植物中的单萜二烯、普氏烯、柠檬烯和倍半萜 α-姜柏林烯在大肠杆菌短链 AHL QS 系统中起抑制作用。

柠檬酸萜类化合物柠檬醛对恶臭假单胞菌的长链 QS 系统也有抑制作用。在牛至油中单萜类香芹醛抑制紫色色素杆菌中 QS 调控的生物膜形成。酚类化合物姜黄素和 ε-长春花素分别参与了姜黄和苔草对铜绿假单胞菌的 QQ 作用。从柑橘属提取物中，几种黄酮类化合物如柚皮素抑制哈维弧菌、大肠杆菌和小肠结肠炎耶尔森菌（Y. enterocolitica）的 QS 系统（Helman and Chernin, 2015）。有趣的是，柚皮素似乎抑制 QS 通路的多个水平，包括 AHL 生物合成和 AHL 识别水平（Asfour, 2017；Truchado et al., 2015）。

研究表明，根瘤农杆菌 QS 被植物 GABA（γ-氨基丁酸）和 SA 抑制。GABA 在损伤后积累，SA 在植物防御信号中起重要作用，可诱导 AHL 降解 γ-丁内酯酶的表达。有趣的是，这一机制也有利于根瘤农杆菌，因为高水平的 QS 化合物可能会在植物中诱导强烈的防御反应。此外，这种 AHL 降解酶可降解其他植物化合物，从而使病原菌能够进一步代谢这些化合物，并导致对其他细菌的竞争优势，而其他细菌的 AHLs 也可被该酶降解（Subramoni et al., 2014）。乙酰丁香酮是一种由植物产生的酚类化合物，也被认为可以诱导 *TraM* 的表达，从而抑制 TraR，提供另一个层面的 QS 微调（Lang and Faure, 2014）。

近年来，研究人员也发现几种植物对蜂蜜中的 QS 具有抑制作用。蜂蜜中主要的酚类化合物可降低解淀粉欧文菌和小肠结肠炎耶尔森菌中 AHL 浓度。研究测试了 29 种不同的单花蜂蜜，发现大多数 QS 抑制剂的活性，主要在板栗和椴树蜂蜜。此外，许多精油显示出 QQ 活性，以及药用植物、水果和蔬菜提取物也具有类似的效果（Asfour, 2017），在植物中，QQ 系统无处不在地控制着根际和叶际的细菌种群，同时也为 QS 抑制剂提供了来源，这些抑制剂可用于降低基于植物病原菌带来的作物损失。

4.2 抗植物致病细菌的细菌群体感应抑制剂

无论是致病性还是非致病性，许多细菌可抑制其他物种（包括致病物种）的 QS 系统。考虑到不同细菌之间对营养物质和空间的激烈竞争，它们很自然地发展出各种策略来抑制彼此的信号系统。几种群体感应抑制剂（QSIs）已在 5 个细菌门中被发现：放线菌门（Actinobacteria）、拟杆菌门、蓝藻菌门、厚壁菌门（Firmicutes）和变形菌门（Proteobacteria），这些门均包含大量的物种（Romero et al., 2015）。大多数关于 QSIs 的研究都集中在 AHL 依赖的 QS 系统上。AHL-QSIs 是一种以不同方式降解 AHL 的酶，可分为三组：AHL 内酯酶，它切割高丝氨酸内酯环上的酯键（Torres et al., 2013）；AHL 酰化酶，它水解脂肪酸和高丝氨酸内酯环之间的酰胺键（Terwagne et al., 2013）；AHL 氧化还原酶，它氧化酰基链，从而改变信号分子的结构（Schipper et al., 2009）。在植物相关细菌中，根瘤农杆菌（Carlier et al., 2003）、芽孢杆菌属（Dong et al., 2001）、百脉根中慢生根瘤菌（*M. loti*）（Funami et al., 2005）、砖红色微杆菌（*Microbacterium testaceum*）（Wanget al., 2010）和根瘤菌属（Krysciak et al., 2011）已被报道产生 AHL 内酯酶，而丁香假单胞菌具有两种 AHL 酰化酶（Shepherd and Lindow, 2009）。来自巨大芽孢杆菌（*B. megaterium*）（Chowdhary et al., 2007）和白假丝酵母菌（Ramage et al., 2002；Murataliev et al., 2004）可氧化 AHLs。利用细菌 QSI 酶对抗植物病原菌似乎是一种有效的策略，具有巨大的农学应用潜力（Romero et al., 2015）。

5 植物-微生物相互作用中糖和糖的信号转导

植物与微生物组之间的相互作用是极为重要的,并受到双方的严格调控。在参与者中,糖作为一类重要的生物分子,其作用愈发明显。在第一阶段,即相互作用的建立阶段,植物通过根系分泌物吸引微生物。这些分泌物的组成非常多样化,表明植物在碳方面付出了巨大的成本。在这些化合物中,糖和多糖含量尤为丰富(Chagas et al., 2018)。与有益微生物的相互作用取决于糖从宿主到细菌。例如,在根瘤菌中,糖转运体的表达介导了糖向根瘤菌的有效外排。在致病细菌的存在下,宿主对糖的释放必须严格控制,从而阻止病原菌的生长。在这个意义上,提出了两个假设:当外质糖水平很低时,植物会限制病原菌对糖的可用性。然而,在侵染区通过糖转运体泄漏的糖已被描述为存在。据推测,这些糖可能被用作燃料来增强植物的防御。基于糖信号传导和免疫原理,改变小分子代谢中糖的水平可以激活防御通路(Versluys et al., 2018;Bezrutczyk et al., 2018)。

虽然糖在植物-微生物相互作用中的作用非常明确,但其在 QS 机制中的具体作用尚不清楚。QS 通常调节 EPS 的产生,最终控制生物膜的形成。特别有趣的是,葡聚糖和果聚糖,基于葡聚糖和果聚糖的多聚糖,分别由蔗糖产生(Gangoiti et al., 2018;Versluys et al., 2018)。在苜蓿中华根瘤菌(*Sinorhizobium meliloti*)中,参与 β-葡聚糖合成的 *bgsBA* 操纵子,通过 c-di-GMP 依赖的机制,受到 QS 系统的调控(Pérez-Mendoza et al., 2015)。之前在铜绿假单胞菌中报道,该信使的生物合成是由一个基于 AHL 的 QS 调控(Ueda and Wood, 2009)。在革兰氏阴性细菌中,发现了一个由脂多糖(LPS)组成的外膜,包括一个亲水性脂质 A 和一个以七糖残基为核心的亲水糖链,这些成分来源于一种独特的 ADP-庚糖生物合成途径。一种中间产物,庚糖-1,7-二磷酸(HBP)可引起动物的免疫反应(Gaudet and Gray-Owen, 2016)。HBP 是否在植物中也作为 MAMP 作用尚不清楚。早期的研究报道,通过 AI-2 依赖的 QS 途径控制 LPS 的生物合成,表明 AI-2 信号可能影响 HBP(De Araujo et al., 2010)。有趣的是,AI-2 是一种硼酸盐呋喃糖基,因此,它含有一种类似糖的成分。

来自糖代谢途径的小代谢糖和酶也与 QS 有关,因为细菌利用其代谢状态去调节 QS。最近的两项研究表明,葡萄糖-6-磷酸脱氢酶和异构酶在水稻黄单胞菌致病变种(*X. oryzae* pv. oryzicola)中的作用。在水稻中用这些酶突变的菌株降低致病性,从而降低 EPS 产量。此外,这些突变导致了参与 DSF 信号传导的几个基因的转录变化,从而表明糖相关代谢如何影响该物种的 QS(Guo et al., 2015, 2017)。

然而,在植物相互作用中糖对 QS 的直接信号作用尚不清楚,尽管在文献中可能存在一些迹象。植物纤维二糖影响疥疮链霉菌中抑菌素的产生。这些都是被细菌用作毒素的次生代谢物。Lerat 等(2010)研究表明,纤维二糖和软木脂对促甲状腺素 A 的产生都有积极影响。虽然纤维二糖激活的具体机制尚不清楚,但纤维二糖可能影响 QS,且 QS 调节链霉菌次生代谢物的产生(Du et al., 2011)。在动物病原菌大肠杆菌 O157:H7 中,Lee 等(2011)提出,金合欢蜂蜜中提取的葡萄糖和果糖可抑制生物膜的形成和致病性。有趣的是,即使这些糖的浓度非常低,但这些抑制作用依然可以下调 AI-2 的表达和输入。然而,在共生大肠杆菌 K-12 中,这些浓度对生物膜的形成没有负面影响,表明这种 QQ 机制是如何特异性地针对有害细菌菌株的。这项研究将糖信号与 QS 联系起来,迄今为止,仅在动物病原菌中得到证实,但对植物病原菌可能也有类似的作用仍需进一步研究。

6　总结和展望

病原微生物间基于种群密度依赖的种间和种内信号事件构成植物病理发生的驱动力。病原微生物的 QS 系统机制多种多样，这些病原菌不仅利用它们的 QS 系统对宿主发起协调攻击，而且还通过不同的方式逃避植物的识别。植物致病菌所使用的 QS 信号分子的特征相对明确，如革兰氏阴性细菌 AHL 或革兰氏阳性细菌的寡肽。然而，关于真菌 QS 系统的认识仍然有限，目前的研究主要集中在少数物种上，如白假丝酵母菌或玉米黑粉病菌，主要关注信号分子诱导的形态变化。考虑到植物病原真菌（和卵菌）造成了全球 70%~80% 的作物损失，了解其 QS 系统在侵染过程中对植物的影响至关重要。针对 QS 信号通路阻止病原微生物之间的沟通，即群体淬灭，是一个有非常有前景的替代化学害虫防治剂，然而，这需要对植物微生物及参与植物致病的 QS 系统有广泛的了解，从而研发长期有效的控制方法。

术语表

生物膜　一种多功能的复杂聚合物，由多种胞外聚合物组成，如多糖、蛋白质和 DNA，这些聚合物由其所包含的微生物群落产生的。

生物发光　生物体通过各种生化事件发光的能力。

损伤相关分子模式（DAMP）　由植物损伤而释放的一系列宿主生物分子，其被特定的植物受体识别，从而启动免疫反应。

内稳态　生物体在适应环境变化时保持稳定的自动调节能力。

病原相关分子模式（PAMP）　一系列属于病原体的常见且保守的生物分子，如 flg22 或几丁质，被植物受体感知，从而启动免疫反应。

叶际　植物的地上部分，作为微生物的栖息地。

植物病理发生　生物体感染植物的寄生能力，进而导致对植物的损害。

多态真菌　在其生命周期的不同阶段表现出不同形态形式的真菌。

群体淬灭（QQ）　通过抑制相关化学信号或相关受体的作用来破坏生物体的群体感应系统。

群体感应（QS）　一种微生物种群密度依赖的现象，根据环境中特定阈值浓度下多个信号分子的存在，导致不同基因的差异表达，从而导致不同表型（即毒力、生物发光）的显著变化。

根际　植物根部周围的土壤区域，作为微生物的栖息地。

甜免疫　植物免疫过程中与碳水化合物动态之间的联系。

缩写词

AHL	酰化高丝氨酸内酯
AI	自诱导物
CPS	荚膜多糖
cyclic-di-GMP	双（3′,5′）环二聚鸟苷单磷酸

DAMP	与损伤相关的分子模式
DSF	扩散信号因子
EF-Tu	延伸因子 Tu
EPS	胞外多糖
ETI	效应触发免疫
GABA	γ-氨基丁酸
HBP	庚糖-1,7-磷酸盐
HSL	高丝氨酸内酯
ISR	诱导系统性抗性
JA	茉莉酸
LPS	脂多糖
MAMP	微生物相关的分子模式
MAPK	胞外信号调节激酶
PAMP	病原相关的分子模式
PR protein	发病机制相关蛋白
PRR	模式识别受体
PTI	模式触发免疫
QQ	群体淬灭
QS	群体感应
QSI	群体感应抑制剂
ROS	活性氧
SA	水杨酸
SAR	系统获得抗性
T3SS	Ⅲ型分泌系统
T6SS	Ⅵ型分泌系统
20-DNA	转移的脱氧核糖核酸
Ti	肿瘤诱导

参考文献

ALTHANI A A, MAREI H E, HAMDI W S, et al., 2016. Human microbiome and its association with health and diseases. J. Cell. Physiol., 231 (8): 1688-1694.

ASFOUR H Z, 2017. Antiquorum sensing natural compounds. J. Microsc. Ultrastruc., 6: 2-10.

BAI X, TODD C D, DESIKAN R, et al., 2012. N-3-oxo-decanoyl-L-homoserine-lactone activates auxin-induced adventitious root formation via hydrogen peroxide-and nitric oxide-dependent cyclic GMP signaling in mung bean. Plant Physiol., 158 (2): 725-736.

BALTRUS D A, MCCANN H C, GUTTMAN D S, 2017. Evolution, genomics and epidemiology of Pseudomonas syringae. Mol. Plant Pathol., 18 (1): 152-168.

BEZRUTCZYK M, YANG J, EOM J, et al., 2018. Sugar flux and signaling in plant-microbe interactions. Plant J., 93 (4): 675-685.

BIGEARD J, COLCOMBET J, HIRT H, 2015. Signaling mechanisms in pattern-triggered immunity

(PTI). Mol. Plant, 8 (4): 521-539.

CARLIER A, UROZ S, SMADJA B, et al., 2003. The Ti plasmid of Agrobacterium tumefaciens harbors an attM-paralogous gene, aiiB, also encoding N-acyl homoserine lactonase activity. Appl. Environ. Microbiol., 69 (8): 4989-4993.

CARLIER A, PESSI G, EBERL L, 2014. Microbial biofilms and quorum sensing. In: Lugtenberg B (Ed), Principles of Plant-Microbe Interactions. Springer, Cham, pp: 45-52.

CHA C, GAO P, CHEN Y, et al., 1998. Production of acyl-homoserine lactone quorum-sensing signals by gram-negative plant-associated bacteria. Mol. Plant Microbe Interact., 11 (11): 1119-1129.

CHAGAS F O, PESSOTTI R C, CARABELLO-RODRÍGUEZ A M, et al., 2018. Chemical signaling involved in plantmicrobe interactions. Chem. Soc. Rev., 47 (5): 1652-1704.

CHOI W, MILLER G, WALLACE I, et al., 2017. Orchestrating rapid long-distance signaling in plants with Ca^{2+}, ROS and electrical signals. Plant J., 90 (4): 698-707.

CHOWDHARY P K, KESHAVAN N, NGUYEN H Q, et al., 2007. *Bacillus megaterium* CYP102A1 oxidation of acyl homoserine lactones and acyl homoserines. Biochemistry, 46 (50): 14429-14437.

COMPANT S, MITTER B, COLLI-MULL J G, et al., 2011. Endophytes of grapevine flowers, berries, and seeds: Identification of cultivable bacteria, comparison with other plant parts, and visualization of niches of colonization. Microb. Ecol., 62 (1): 188-197.

DE ARAUJO C, BALESTRINO D, ROTH L, et al., 2010. Quorum sensing affects biofilm formation through lipopolysaccharide synthesis in Klebsiella pneumoniae. Res. Microbiol., 161 (7): 595-603.

DELALANDE L, FAURE F, RAFFOUX A, et al., 2005. N-hexanoyl-L-homoserine lactone, a mediator of bacterial quorum-sensing regulation, exhibits plant-dependent stability and may be inactivated by germinating Lotus corniculatus seedlings. FEMS Microbiol. Ecol., 52 (1): 13-20.

DONG Y H, WANG L H, XU J L, et al., 2001. Quenching quorum-sensingdependent bacterial infection by an N-acyl homoserine lactonase. Nature, 411 (6839): 813-817.

DOW J M, 2017. Diffusible signal factor-dependent quorum sensing in pathogenic bacteria and its exploitation for disease control. J. Appl. Microbiol., 122 (1): 2-11.

DU Y, SHEN X, YU P, et al., 2011. Gamma-butyrolactone regulatory system of *Streptomyces chattanoogensis* links nutrient utilization, metabolism, and development. Appl. Environ. Microbiol., 77 (23): 8415-8426.

DUONG D A, STEVENS A M, 2017. Integrated downstream regulation by the quorum-sensing controlled transcription factors LrhA and RcsA impacts phenotypic outputs associated with virulence in the phytopathogen Pantoea stewartii subsp. stewartii. PeerJ., 5e4145.

FERGUSON B A, DREISBACH T A, PARKS C G, et al., 2003. Coarse-scale population structure of pathogenic Armillaria species in a mixed-conifer forest in the Blue Mountains of northeast Oregon. Can. J. For. Res., 33 (4): 612-623.

FUNAMI J, YOSHIKANE Y, KOBAYASHI H, et al., 2005. 4-Pyridoxolactonase from a symbiotic nitrogen-fixing bacterium Mesorhizobium loti: cloning, expression, and characterization. Biochim. Biophys. Acta, 1753 (2): 234-239.

GALIANA E, FOURRE S, ENGLER G, 2008. Phytophthora parasitica biofilm formation: installation and organization of microcolonies on the surface of a host plant. Environ. Microbiol., 10 (8): 2164-2171.

GANGOITI J, PIJNING T, DIJKHUIZEN L, 2018. Biotechnological potential of novel glycoside hydrolase family 70 enzymes synthesizing α-glucans from starch and sucrose. Biotechnol. Adv., 36 (1): 196-207.

GAO Q, ZHU S, KACHROO P, et al., 2015. Signal regulators of systemic acquired resistance. Front. Plant Sci., 6: 228.

GAUDET R G, GRAY-OWEN S D, 2016. Heptose sounds the alarm: innate sensing of a bacterial sugar stimulates immunity. PLoS Pathog., 12 (9): e1005807.

GIMENEZ-IBANEZ S, SOLANO R, 2013. Nuclear jasmonate and salicylate signaling and crosstalk in defense against pathogens. Front. Plant Sci., 4: 72.

GOLD S, DUNCAN G, BARRETT K, et al., 1994. cAMP regulates morphogenesis in the fungal pathogen Ustilago maydis. Genes Dev., 8 (23): 2805-2816.

GÖTZ C, FEKETE A, GEBEFUEGI I, et al., 2007. Uptake, degradation and chiral discrimination of N-acyl-D/L-homoserine lactones by barley (Hordeum vulgare) and yam bean (Pachyrhizus erosus) plants. Anal. Bioanal. Chem. 389 (5): 1447-1457.

GÖTZ-RÖSCH C, SIEPER T, FEKETE A, et al., 2015. Influence of bacterial N-acyl-homoserine lactones on growth parameters, pigments, antioxidative capacities and the xenobiotic phase II detoxification enzymes in barley and yam bean. Front. Plant Sci., 6: 205.

GUO W, ZOU L, CAI L, et al., 2015. Glucose-6-phosphate dehydrogenase is required for extracellular polysaccharide production, cell motility and the full virulence of Xanthomonas oryzae pv. oryzicola. Microb. Pathog., 78: 87-94.

GUO W, ZOU L, JI Z, et al., 2017. Glucose 6-phosphate isomerase (Pgi) is required for extracellular polysaccharide biosynthesis, DSF signals production and full virulence of Xanthomonas oryzae pv. oryzicola in rice. Physiol. Mol. Plant Pathol., 100: 209-219.

HARTMANN A, SCHIKORA A, 2012. Quorum sensing of bacteria and trans-kingdom interactions of N-acyl homoserine lactones with eukaryotes. J. Chem. Ecol., 38 (6): 704-713.

HELMAN Y, CHERNIN L, 2015. Silencing the mob: disrupting quorum sensing as a means to fight plant disease. Mol. Plant Pathol., 16 (3): 316-329.

HOGAN D A, 2006. Talking to themselves: autoregulation and quorum sensing in fungi. Eukaryot. Cell, 5 (4): 613-619.

JIN G, LIU F, MA H, et al., 2012. Two G-protein-coupled-receptor candidates, Cand2 and Cand7, are involved in Arabidopsis root growth mediated by the bacterial quorumsensing signals N-acyl-homoserine lactones. Biochem. Biophys. Res. Commun., 417 (3): 991-995.

JONES J D G, DANGL J L, 2006. The plant immune system. Nature, 444 (7117): 323-329.

KALIA V C, 2013. Quorum sensing inhibitors: an overview. Biotechnol. Adv., 31 (2): 224-245.

KAMBER T, POTHIER J F, PELLUDAT C, et al., 2017. Role of the type VI secretion systems during disease interactions of Erwinia amylovora with its plant host. BMC Genomics, 18 (1): 628.

KERS J A, CAMERON K D, JOSHI M V, et al., 2005. A large, mobile pathogenicity island confers plant pathogenicity on Streptomyces species. Mol. Microbiol., 55 (4): 1025-1033.

KOCZAN J M, MCGRATH M J, ZHAO Y, et al., 2009. Contribution of Erwinia amylovora exopolysaccharides amylovoran and levan to biofilm formation: implications in pathogenicity. Phytopathology, 99 (11): 1237-1244.

KRYSCIAK D, SCHMEISSER C, PREUSS S, et al., 2011. Involvement of multiple loci in quorum quenching of autoinducer I molecules in the nitrogen-fixing symbiont Rhizobium (Sinorhizobium) sp. strain NGR234. Appl. Environ. Microbiol., 77 (15): 5089-5099.

LANG J, FAURE D, 2014. Functions and regulation of quorum sensing in Agrobacterium tumefaciens. Front. Plant Sci., 5: 14.

LAREEN A, BURTON F, SCHÄFER P, 2016. Plant root-microbe communication in shaping root microbiomes. Plant Mol. Biol., 90 (6): 575-587.

LEE J, PARK J, KIM J, et al., 2011. Low concentrations of honey reduce biofilm formation, quorum sensing, and virulence in *Escherichia coli* O157: H7. Biofouling, 27 (10): 1095-1104.

LEHTIJÄRVI A, DOĞMUş-LEHTIJÄRVI H T, ADAY KAYA A G, et al., 2017. *Armillaria ostoyae* in managed coniferous forests in Kastamonu in Turkey. Forest Pathol., 47 (6): e12364.

LERAT S, SIMAO-BEAUNOIR A, WU R, et al., 2010. Involvement of the plant polymer suberin and the disaccharide cellobiose in triggering thaxtomin A biosynthesis, a phytotoxin produced by the pathogenic agent *Streptomyces scabies*. Phytopathology, 100 (1): 91-96.

LIU F, BIAN Z, JIA Z, et al., 2012. The GCR1 and GPA1 participate in promotion of Arabidopsis primary root elongation induced by N-acyl-homoserine lactones, the bacterial quorum-sensing signals. Mol. Plant Microbe Interact., 25 (5): 677-683.

MANEFIELD M, DE NYS R, KUMAR N, et al., 1999. Evidence that halogenated furanones from *Delisea pulchra* inhibit acylated homoserine lactone (AHL) -mediated gene expression by displacing the AHL signal from its receptor protein. Microbiology 145 (2): 283-291.

MARTIN B D, SCHWAB E, 2013. Current usage of symbiosis and associated terminology. Int. J. Biol., 5 (1): 32-45.

MARTIN-URDIROZ M, OSES-RUIZ M, RYDER L S, et al., 2016. Investigating the biology of plant infection by the rice blast fungus *Magnaporthe oryzae*. Fungal Genet. Biol., 90: 61-68.

MATHESIUS U, MULDERS S, GAO M, et al., 2003. Extensive and specific responses of a eukaryote to bacterial quorum-sensing signals. PNAS, 100 (3): 1444-1449.

MCNALLY R R, ZHAO Y, SUNDIN G W, 2015. Towards understanding fire blight: virulence mechanisms and their regulation in *Erwinia amylovora*. In: Murillo J, Vinatzer B A, Jackson R W, Arnold D L (Eds.), Bacteria Plant Interactions. Caister Academic Press, pp: 61-82.

MENDES R, GARBEVA P, RAAIJMAKERS J M, 2013. The rhizosphere microbiome: significance of plant beneficial, plant pathogenic, and human pathogenic microorganisms. FEMS Microbiol. Rev., 37 (5): 634-663.

MENG X, ZHANG S, 2013. MAPK cascades in plant disease resistance signaling. Annu. Rev. Phytopathol., 51: 245-266.

MOLELEKI L N, PRETORIUS R G, TANUI C K, et al., 2017. A quorum sensing-defective mutant of *Pectobacterium carotovorum* ssp. brasiliense 1692 is attenuated in virulence and unable to occlude xylem tissue of susceptible potato plant stems. Mol. Plant Pathol., 18 (1): 32-44.

MONNET V, GARDAN R, 2015. Quorum - sensing regulators in Gram - positivebacteria: 'cherchez le peptide. Mol. Microbiol., 97 (2): 181-184.

MOORE D, ROBSON G D, TRINCI A P, 2011. 21st Century Guidebook to Fungi with CD. Cambridge, UK: Cambridge University Press.

MURATALIEV M B, TRINH L N, MOSER L V, et al., 2004. Chimeragenesis of the fatty acid binding site of cytochrome P450BM3. Replacement of residues 73-84 with the homologous residues from the insect cytochrome P450 CYP4C7. Biochemistry, 43 (7): 1771-1780.

NEALSON K H, HASTINGS J W, 1979. Bacterial bioluminescence: its control and ecological significance. Microbiol. Rev., 43 (4): 496-518.

NIMTZ M, MORT A, WRAY V, et al., 1996. Structure of stewartan, the capsular exopolysaccharide from the corn pathogen *Erwinia stewartii*. Carbohydr. Res., 288: 189-201.

OERKE E C, 2006. Crop losses to pests. J. Agric. Sci., 144 (1): 31-43.

OLIVEIRA-GARCIA E, VALENT B, 2015. How eukaryotic filamentous pathogens evade plant recognition. Curr. Opin. Microbiol., 26: 92-101.

ORTIÍZ-CASTRO R, MARTÍNEZ-TRUJILLO M, LÒPEZ-BUCIO J, 2008. N-acyl-L-homoserine lactones:

a class of bacterial quorum-sensing signals alter post-embryonic root development in Arabidopsis thaliana. Plant Cell Environ., 31 (10): 1497-1509.

PANG Y, LIU X, MA Y, et al., 2009. Induction of systemic resistance, root colonisation and biocontrol activities of the rhizospheric strain of *Serratia plymuthica* are dependent on *N*-acyl homoserine lactones. Eur. J. Plant Pathol., 124 (2): 261-268.

PARK C, RONALD P C, 2012. Cleavage and nuclear localization of the rice XA21 immune receptor. Nat. Commun., 3: 920.

PÉREZ-MENDOZA D, RODRÍGUEZ-CARVAJAL M Á, ROMERO-JIMÉNEZ L, et al., 2015. Novel mixed-linkage β-glucan activated by c-di-GMP in *Sinorhizobium meliloti*. PNAS, E757-E765.

PIMENTEL D (ED), 2011. Biological Invasions: Economic and Environmental Costs of Alien Plant, Animal, and Microbe Species. CRC Press.

PIQUÉ N, MIÑANA-GALBIS D, MERINO S, et al., 2015. Virulence factors of *Erwinia amylovora*: a review. Int. J. Mol. Sci., 16 (12): 12836-12854.

PIRHONEN M, FLEGO D, HEIKINHEIMO R, et al., 1993. A small diffusible signal molecule is responsible for the global control of virulence and exoenzyme production in the plant pathogen *Erwinia carotovora*. EMBO J., 12 (6): 2467-2476.

QUIÑONES B, DULLA G, LINDOW S E, 2005. Quorum sensing regulates exopolysaccharide production, motility, and virulence in *Pseudomonas syringae*. Mol. Plant Microbe Interact., 18 (7): 682-693.

RAJPUT A, GUPTA A K, KUMAR M, 2015. Prediction and analysis of quorum sensing peptides based on sequence features. PLoS One, 10 (3): e0120066.

RAMAGE G, SAVILLE S P, WICKES B L, et al., 2002. Inhibition of *Candida albicans* biofilm formation by farnesol, a quorum-sensing molecule. Appl. Environ. Microbiol., 68 (11): 5459-5463.

RASMUSSEN T B, GIVSKOV M, 2006. Quorum sensing inhibitors: a bargain of effects. Microbiology, 152 (4): 895-904.

ROMERO M, MAYER C, MURAS A, et al., 2015. Silencing bacterial communication through enzymatic quorumsensing inhibition. In: Kalia V C (Ed), Quorum Sensing vs Quorum Quenching: A Battle with No End in Sight. Springer, India, pp: 219-236.

SANCHEZ-CONTRERAS M, BAUER W D, GAO M, et al., 2007. Quorum-sensing regulation in rhizobia and its role in symbiotic interactions with legumes. Philos. Trans. R. Soc. B, 362: 1149-1163.

SCHENK S T, SCHIKORA A, 2015. AHL-priming functions via oxylipin and salicylic acid. Front. Plant Sci., 5: 784.

SCHENK S T, STEIN E, KOGEL K, et al., 2012. Arabidopsis growth and defense are modulated by bacterial quorum sensing molecules. Plant Signal. Behav., 7 (2): 178-181.

SCHIKORA A, SCHENK S T, STEIN E, et al., 2011. *N*-acyl-homoserine lactone confers resistance toward biotrophic and hemibiotropic pathogens via altered activation of AtMPK6. Plant Physiol., 157 (3): 1407-1418.

SCHIPPER C, HORNUNG C, BIJTENHOORN P, et al., 2009. Metagenome-derived clones encoding two novel lactonase family proteins involved in biofilm inhibition in *Pseudomonas aeruginosa*. Appl. Environ. Microbiol., 75 (1): 224-233.

SCHUHEGGER R, IHRING A, GANTNER S, et al., 2006. Induction of systemic resistance in tomato by *N*-acyl-*L*-homoserine lactone-producing rhizosphere bacteria. Plant Cell Environ., 29 (5): 909-918.

SHAH J, ZEIER J, 2013. Long-distance communication and signal amplification in systemic acquired resistance. Front. Plant Sci., 4: 30.

SHEPHERD R W, LINDOW S E, 2009. Two dissimilar N-acyl-homoserine lactone acylases of Pseudomonas syringae influence colony and biofilm morphology. Appl. Environ. Microbiol., 75 (1): 45-53.

SPELLIG T, BÖLKER M, LOTTSPEICH F, et al., 1994. Pheromones trigger filamentous growth in Ustilago maydis. EMBO J., 13 (7): 1620-1627.

SUBRAMONI S, NATHOO N, KLIMOV E, et al., 2014. Agrobacterium tumefaciens responses to plant-derived signalling molecules. Front. Plant Sci., 5: 322.

TAIZ L, ZEIGER E, 2012. Chapter 13: Secondary metabolites and plant defense. In: Taiz L, Zeiger E (Eds), Plant Physiology. fifth ed. Sinauer Associated Inc, Sunderland, MA, pp: 369-400

TALBOT N J, 2003. On the trail of a cereal killer: exploring the biology of Magnaporthe grisea. Annu. Rev. Microbiol., 57 (1): 177-202.

TERWAGNE M, MIRABELLA A, LEMAIRE J, et al., 2013. Quorum sensing and self-quorum quenching in the intracellular pathogen Brucella melitensis. PLoS One., 8 (12): e82514.

TIAN Y, ZHAO Y, SHI L, et al., 2017. Type VI secretion systems of Erwinia amylovora contribute to bacterial competition, virulence, and exopolysaccharide production. Phytopathology, 107 (6): 654-661.

TORRES M, ROMERO M, PRADO S, et al., 2013. N-acylhomoserine lactonedegrading bacteria isolated from hatchery bivalve larval cultures. Microbiol. Res., 168 (9): 547-554.

TRUCHADO P, LARROSA M, CASTRO-IBÁÑEZ I, et al., 2015. Plant food extracts and phytochemicals: their role as quorum sensing inhibitors. Trends Food Sci. Technol., 43 (2): 189-204.

UDINE C, BRACKMAN G, BAZZINI S, et al., 2013. Phenotypic and genotypic characterisation of Burkholderia cenocepacia J2315 mutants affected in homoserine lactone and diffusible signal factor-based quorum sensing systems suggests interplay between both types of systems. PLoS One., 8 (1): e55112.

UEDA A, WOOD T K, 2009. Connecting quorum sensing, c-di-GMP, Pel polysaccharide, and biofilm formation in Pseudomonas aeruginosa through tyrosine phosphatase TpbA (PA3885). PLoS Pathog., 5 (6): e1000483.

VERMA S C, MIYASHIRO T, 2013. Quorum sensing in the squid-Vibrio symbiosis. Int. J. Mol. Sci., 14 (8): 16386-16401.

VERSLUYS M, TARKOWSKIŁŁ P, VAN DEN ENDE W, 2016. Fructans as DAMPs or MAMPs: evolutionary prospects, cross-tolerance, and multistress resistance potential. Front. Plant Sci., 7: 2061.

VERSLUYS M, KIRTEL O, ÖNER E T, et al., 2018. The Fructan syndrome: evolutionary aspects and common themes among plants and microbes. Plant Cell Environ., 41 (1): 16-38.

VON BODMAN S B, FARRAND S K, 1995. Capsular polysaccharide biosynthesis and pathogenicity in Erwinia stewartii require induction by an N-acylhomoserine lactone autoinducer. J. Bacteriol., 177 (17): 5000-5008.

VON BODMAN S B, BAUER W D, COPLIN D L, 2003. Quorum sensing in plant-pathogenic bacteria. Annu. Rev. Phytopathol., 41 (1): 455-482.

VON RAD U, KLEIN I, DOBREV P I, et al., 2008. Response of Arabidopsis thaliana to N-hexanoyl-DL-homoserine-lactone, a bacterial quorum sensing molecule produced in the rhizosphere. Planta, 229 (1): 73-85.

VORHOLT J A, 2012. Microbial life in the phyllosphere. Nat. Rev. Microbiol., 10 (12): 828.

WANG W Z, MOROHOSHI T, IKENOYA M, et al., 2010. AiiM, a novel class of N-acylhomoserine lactonase from the leaf-associated bacterium Microbacterium testaceum. Appl. Environ. Microbiol., 76 (8): 2524-2530.

YOU Y, MARELLA H, ZENTELLA R, et al., 2006. Use of bacterial quorumsensing components to regulate gene expression in plants. Plant Physiol., 140 (4): 1205-1212.

拓展阅读

RONALD P C, 2011. Small protein-mediated quorum sensing in a gram-negative bacterium: novel targets for control of infectious disease. Discov. Med., 12 (67): 461-470.

第6章　群体感应和肠道微生物组

Angel G. Jimenez*, Vanessa Sperandio†

*Department of Microbiology, University of Texas Southwestern Medical Center, Dallas, TX, United States, †Department of Biochemistry, University of Texas Southwestern Medical Center, Dallas, TX, United States

1 前言

人类与大量微生物共存，微生物主要分布在各种器官中，如皮肤、口腔黏膜、生殖器官和肠道。这个复杂的群落被称为微生物群落，主要由细菌组成，其中许多细菌与宿主有密切的联系，以促进宿主的健康状态。这些生物通常被称为共生微生物，与人类宿主有着丰富而悠久的历史，并进化出了交流和衡量宿主生理的机制，以有效地完成重要的功能，如为宿主提供营养，发展免疫系统，防止病原物的定殖（Gordon and Klaenhammer, 2011）。这些功能在身体的胃肠道（GI）中尤其明显，包含了最丰富和最密集的群体。健康的肠道功能依赖于微生物群落的适当结构和平衡，而群落的破坏或失衡与大量疾病的发生有关，如神经系统疾病、炎症改变和癌症进展（Grenham et al., 2011）。适当的微生物群落结构依赖于细胞间化学信息的交换。研究最多的细菌细胞间通信形式被称为群体感应。当一小部分种群承担群体感应时，细菌种群能有效地同步细菌行为，包括致病因子的发展、毒素的产生和分泌、分泌系统和生物膜的形成，这些过程通过群落作用变得更加有效。群体感应通过被称为自诱导物的小分子来触发整体基因表达的变化。这些分子会随着细菌复制和数量的增加而增加。

群体感应在简单的系统中，如良好的摇晃和稳定含氧的单一培养物中，已得到了广泛的研究。然而，这些系统在多物种如肠道的复杂性上很难进行解释。在这种环境中，细菌必须对一个充满化学信号的环境做出反应，并将这些信号有效地整合成一个可靠的信使。在这种情况下，自诱导物和其他小分子不仅可以与它们的物种或近亲细胞进行交流，还可以与宿主等不同界的细胞进行交流。这拓宽了群体感应的定义，包括多向通信途径如物种间通信和跨界信号。本章将重点讨论肠道中重要的小分子以及宿主、微生物群落和移入的病原菌对这些分子的反应。

2 自诱导物

肠道拥有多种多样的微生物类群，是一个复杂的环境。这些微生物产生大量的化学物质，并依赖于这种复杂的化学物质来调节其基因的表达。一类重要的信号分子被称为自诱导物，它是群体感应的主要信号分子。最常见的一类自诱导物是酰基高丝氨酸内酯（AHLs）。这些分子有一个 N-酰化的高丝氨酸-内酯环和一个 4~18 个碳酰基链（Galloway et al., 2011）。酰基链可能包含在第三个碳上具有修饰，是某些微生物及其亲近微生物所特有的。这些修饰可以改变分子的稳定性，并为它们的传感器提供特异性（von Bodman et al.，2008）。

AHL 是通过 LuxI 酶的活性产生的，LuxI 酶利用 S-腺苷甲硫氨酸（SAM）和脂肪酸代谢的中间体合成 AHLs 的前体物（Case et al.，2008）。AHLs 通过其同源受体 LuxR 被感知，LuxR 是一种与 DNA 结合的转录调控因子（Zhang et al.，2002）。LuxI/LuxR 通常成对发挥作用，共同进化以提高对 AHLs 的敏感性和特异性。LuxR 类型的受体在细胞质中，能检测来自 LuxI 合成酶的同源 AHL。该受体具有 N 端 AHL 结合域和 C 端 DNA 结合域。在没有 AHL 的情况下，LuxR 型受体通常是不稳定的，因为其不能正确折叠并迅速降解。与同源 AHL 结合后，LuxR 型转录因子二聚体与靶基因上游的称为 lux 盒子的短序列结合（Engebrecht et al.，1983；Engebrecht and Silverman，1984；Stevens et al.，1994；Zhang et al.，2002；Zhu and Winans，1999）。

关于与 AHL 感知相关的信号转导机制以及这些小分子在宿主-微生物相互作用中的重要性已有大量研究。研究最多的系统之一是费氏弧菌与其宿主短尾鱿鱼之间的相互作用，涉及的群体感应信号分子为 AHLs。费氏弧菌和夏威夷短尾乌贼有一种天然的共生关系，即细菌在高细胞密度下产生和积累 AHL，从而引起生物发光（Engebrecht et al.，1983；Engebrecht and Silverman，1984）。这种生物发光阻止了夜间对鱿鱼的捕食。这种共生关系证明细菌利用群体感应机制进行宿主-微生物相互作用的有效性。

在哺乳动物中，肠道是体内微生物定殖率最高的器官，因此，肠道作为宿主-微生物相互作用潜力最高的区域。γ变形菌的成员，如肠道沙门菌、大肠杆菌、肺炎克雷伯菌（*Klebsiella pneumoniae*）和肠杆菌（*Enterobacter*），编码 AHL 同源物传感器 LuxR，称为 SdiA，但它们不编码 LuxI 同源物，也不合成 AHLs（Hudaiberdiev et al.，2015）。然而，研究表明，这些细菌实际上有能力感知并响应多种 AHLs（Dyszel et al.，2010；Michael et al.，2001；Nguyen et al.，2015；Sheng et al.，2013；Smith and Ahmer，2003；Smith et al.，2008；Sperandio，2010）。研究表明，这些细菌可能对其他微生物群成员产生的 AHL 有反应。然而，由共同的微生物群成员，如拟杆菌（*Bacteroides*）形成的系统能够产生 AHL 尚缺乏证据。此外，尽管尝试从哺乳动物的肠道中提取 AHL，但未能检测到这些化合物（Hughes et al.，2010）。遗传学上检测 AHL 也很难做到（Swearingen et al.，2013）。以 SdiA 为传感器构建了沙门菌报告系统，并用于接种广泛的宿主。然而，该系统未能在哺乳动物的肠道中检测到任何 AHL，但在海龟或感染小肠结肠炎耶尔森菌的小鼠胃肠道中检测到低水平的 AHL，这些生物已知为 AHL 的生产者（Dyszel et al.，2010；Smith et al.，2008）。然而，在牛瘤胃中已经检测到 AHL，尽管在这种环境中群体感应分子的来源尚不明确（Hughes et al.，2010；Sheng et al.，2013）。在这种情况下，研究表明，肠出血性大肠杆菌（EHEC）O157: H7 通过 SdiA 感知 AHLs，引导其通过牛瘤胃环境中生存，并激活抗酸相关基因，这些基因对于细菌在牛酸性胃中生存至关重要（Hughes et al.，2010）。研究表明，缺乏 SdiA 的沙门菌菌株通过增加粪便排出和转移到全身部位以增强其毒性（Volf et al.，2002）。总之，以上研究表明，在 EHEC 和肠道沙门菌中，SdiA 是一种毒性调节因子。除 AHL 外，SdiA 也被证明可以感知和响应其他分子。最近的研究表明，SdiA 结合和感知 1-辛烷酰-甘油，这种分子在生物体内被广泛用作信号分子（Nguyen et al.，2015）。

尽管 AHL 可以促进种间的通信，但它们也参与种内的相互作用。此外，还有其他群体感应系统，允许物种间的交流，其中研究最多的是细菌的信号分子，称为自动诱导物-2（AI-2）。AI-2 分子是由呋喃酮衍生而来，主要来源于 4,5-二羟基-2,3-戊二酮（DPD），其合成主要与 SAM 代谢相关（Schauder et al., 2001）。*luxS* 基因编码一种 *S*-核糖体同型半胱氨酸裂解酶，是 AI-2 合成所必需的，同时在革兰氏阳性和阴性细菌中都是保守的（Pereira et al., 2013）。

AI-2 被证明存在于人类的胃肠道中（Sperandio et al., 2003）。在肠道中，大部分 AI-2 是由胃肠道中的两个主要门：拟杆菌门和厚壁菌门产生的（Thompson et al., 2015）。这些门，特别是厚壁菌门中的梭状芽孢杆菌（*Clostridia*），在定殖抗性的过程中起到保护作用，防止病原菌的入侵（Itoh and Freter, 1989）。抗生素治疗导致了微生物群落组成的巨大变化。链霉素治疗引起的变化主要属于拟杆菌门，因为它耗尽形成孢子的细菌（梭状芽孢杆菌类）（Sekirov et al., 2008）。最近的研究表明，AI-2 促进了厚壁菌门的再扩张，表明 AI-2 和群体感应在微生物群落组成中发挥作用（Thompson et al., 2015）。研究者设计了大肠杆菌菌株，既可以消耗也可增加肠道中 AI-2 的水平。研究发现，被增加 AI-2 的大肠杆菌菌株在小鼠体内的厚壁菌门水平显著增加，表明 AI-2 可逆转抗生素诱导的生态失调。此外，研究表明，上皮细胞通过诱导白细胞介素-8 等炎症细胞因子的产生来响应 AI-2（Zargar et al., 2015）。最近的一项研究表明，微生物群落产生的 AI-2 可以防止霍乱弧菌侵染（Hsiao et al., 2014）。霍乱弧菌可引起急性和大量的水样腹泻。为此，霍乱弧菌使用群体感应机制来调节其定殖能力和控制其毒力因子的表达。使用宏基因组学分析，Hsiao 等（2014）比较了霍乱弧菌感染患者与健康成人的肠道菌群组成，发现其定殖抗性与肥胖瘤胃球菌（*Ruminococcus obeum*）存在相关。霍乱弧菌毒力基因的表达受群体感应的负调控。因此，通过采用宏转录组学的方法来阐明肥胖瘤胃球菌对霍乱弧菌提供的保护机制。研究发现，在肥胖瘤胃球菌中，*luxS* 同源物的表达水平随着对霍乱弧菌反应的增强而增加。通过将肥胖瘤胃球菌中 *luxS* 基因克隆到一个不能产生 AI-2 诱导启动子的载体中，发现使用这种大肠杆菌菌株定殖的小鼠成功地限制了霍乱弧菌的定殖，表明来自肥胖大肠杆菌的 *luxS* 足以提供霍乱弧菌的保护（Hsiao et al., 2014），如图 6-1 所示。

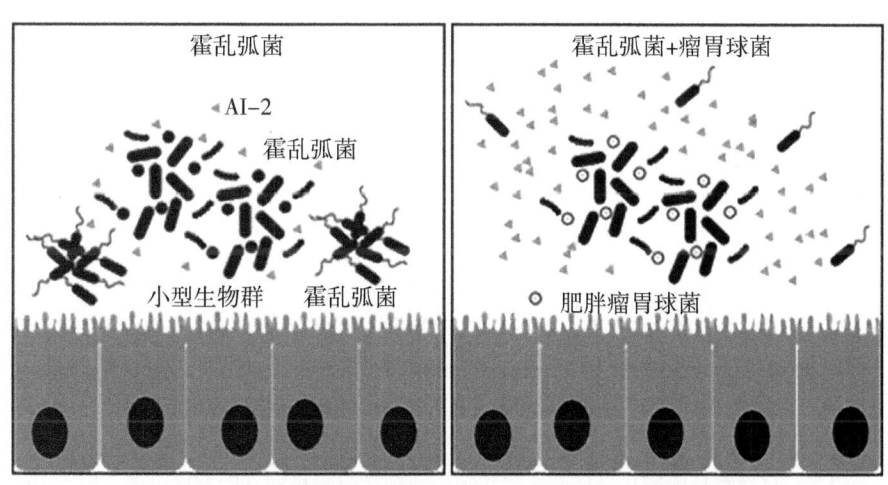

图 6-1 肥胖瘤胃球菌利用群体感应来限制霍乱弧菌的定殖

注：肥胖瘤胃球菌在体内抑制霍乱弧菌的毒力。霍乱弧菌利用群体感应调控其毒力基因的表达。当群体感应分子 AI-2 浓度较低时，霍乱弧菌启动其毒力基因的表达。肥胖瘤胃球菌可产生 AI-2 下调毒力基因的表达，来限制霍乱弧菌的定殖。

另一种较少被描述的信号分子是自诱导物-3（AI-3）。AI-3 是一种甲醇可溶性但尚未确定其特征的信号分子，已被证明是由人类肠道微生物群产生的（Sperandio et al.，2003）。其化学结构或合成途径仍有待阐明。AI-3 被认为是来自酪氨酸，并证明其可以被细菌的儿茶酚胺传感器 QseC 感知（Clarke et al.，2006；Sperandio et al.，2003），其的合成被认为是 LuxS 依赖的。然而，后来的研究表明，LuxS 功能取决于 *LuxS* 突变体中代谢的变化。

3 细菌中的儿茶酚胺信号传导

宿主来源的儿茶酚胺控制着哺乳动物的应激反应。人体中最重要的儿茶酚胺包括多巴胺、肾上腺素（E）和去甲肾上腺素（NE）（Molina and Molina，2006）。这些宿主信号分子在肠道中被发现，能够促进肠道运动、钾和氯的分泌、上皮屏障功能以及炎症等功能。肠道中肾上腺素和去甲肾上腺素主要来源为肾上腺髓质和肠道中的交感肾上腺素能神经元（Asano et al.，2012；Eldrup and Richter，2000；Hörger et al.，1998）。

肠道沙门菌和大肠杆菌对儿茶酚胺的反应通过双组分系统在细胞膜上传递信息。双组分系统通常由组氨酸传感器激酶及其同源反应调节因子组成，后者通常是一种转录因子，在组氨酸激酶磷酸化作用下，这些因子会诱导基因表达的变化（Jung et al.，2012）。使用双组分系统 QseC/B 和 QseE/F 感知儿茶酚胺（Clarke and Sperandio，2005a；Clarke et al.，2006；Hughes et al.，2009；Sperandio et al.，2003；Reading et al.，2009）。QseC 和 QseE 是一种膜结合的组氨酸激酶，它们能够感知肾上腺素和去甲肾上腺素，并通过磷酸化事件将信息分别传递给它们的同源反应调节因子 QseB 和 QseF（图 6-2）。

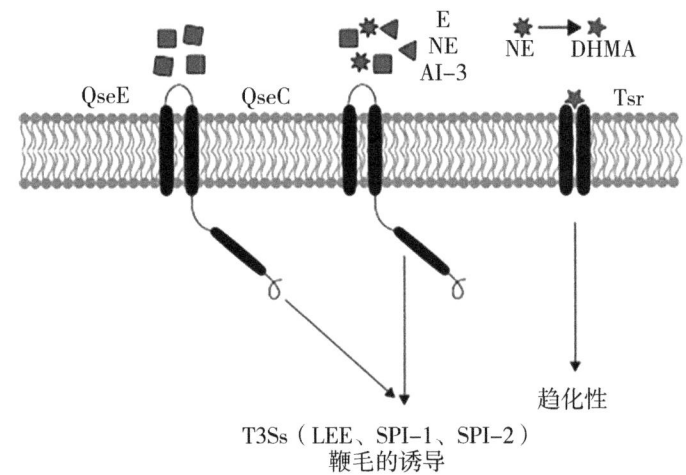

图 6-2　EHEC 和鼠伤寒沙门菌中的 Epi/NE/AI-3 的群体感应

注：肠道中的儿茶酚胺通过 QseC 和 QseE 组氨酸激酶被 EHEC 和鼠伤寒沙门菌感知。感受器激酶 QseE 感知肾上腺素；这导致自磷酸化事件的发生和随后磷酸盐被转移到其同源反应调节因子 QseF 上。QseC 在肾上腺素、去甲肾上腺素和 AI-3 的作用下发生自磷酸化。QseC 可将其磷酸转移至其同源的反应调节器 QseB；这导致了与运动性相关基因的诱导，如鞭毛基因的诱导。QseC 还可以磷酸化其非同源反应调节因子 QseF 和 KdpE，进而诱导毒力相关基因的表达。去甲肾上腺素可以通过两步法转化为 DHMA。当 DHMA 被丝氨酸趋化因子 Tsr 感知时，DHMA 可以作为一种趋化剂。

这些反应调节因子继续诱导基因表达的变化，从而控制鞭毛运动和病原体如沙门菌和EHEC的毒性（Clarke and Sperandio, 2005a, b; Clarke et al., 2006; Hughes et al., 2009; Moreira et al., 2010; Moreira and Sperandio, 2012）。多种动植物病原体已被证明编码一个功能性的 qseC 基因，而 qseC 在这些病原体的毒力基因调控中发挥着重要作用（Kendall and Sperandio, 2016）。在与 E/NE 结合后，QseC 会自身磷酸化，并将磷酸盐转移到 QseB 中。此外，QseC 还可以磷酸化非同源反应调节因子 QseF 和 KdpE。而 QseE 只能将其磷酸盐转移至其同源反应调节因子 QseF 上。QseC 已被证明对体内定殖至关重要，因为缺失 QseC 的 EHEC、啮齿枸橼酸杆菌（*Citrobacter rodentium*）和肠道沙门菌血清型鼠伤寒沙门菌在牛、兔和小鼠感染模型中表现出被减弱的致病性（Clarke et al., 2006; Hughes et al., 2009; Moreira et al., 2010, 2016; Rasko et al., 2008; Sharma and Casey, 2014）。

3,4-二羟基甘露聚糖酸（DHMA）是微生物，特别是肠杆菌科（Enterobacteriaceae）中NE代谢的产物，已被证明以 QseC 依赖的方式诱导 EHEC 的毒力基因表达。当 NE 通过 QseCE 感知时，它诱导酪胺氧化酶（*tynA*）和芳香醛脱氢酶（*feaB*）的表达，将 NE 转化为 DHMA。研究还表明，DHMA 在 EHEC 中作为趋化分子，通过 EHEC 丝氨酸化学受体可能帮助 EHEC 实现组织定殖（图 6-2; Pasupuleti et al., 2014）。

4 肠道中的营养信号

微生物群和宿主衍生的代谢物对肠道病原菌和疾病发展具有重要影响。然而，共生代谢物和宿主代谢物如何影响病原菌的致病性还不是很清楚。微生物群落被认为是抵御肠道病原菌的屏障，这一过程被称为定殖抗性（Bohnhoff et al., 1954），其原因被认为是对有限营养供应的激烈竞争。肠道病原菌进化出机制以绕过这一屏障，从而战胜常驻微生物群。许多营养物质，无论是宿主、饮食，还是微生物，都为进入的病原体提供化学线索，并用于适当地测量微生物群组成、宿主生理状态以及在肠道内的位置，以便在适当的时候发挥其毒力（Bäumler and Sperandio, 2016）。

EHEC 可以感知其营养环境，并协调其毒力产生。研究表明，在营养缺乏（糖异生）和富营养（糖酵解）条件下，EHEC 优先表达其毒力基因（Njoroge et al., 2012, 2013）。EHEC 通过激活两个转录因子 Cra 和 KdpE 的表达来进行控制。Cra 是碳代谢调节器，负责测量环境中碳源水平的波动并控制靶基因的表达。在糖酵解或营养丰富的条件下，细菌会积累果糖-1-磷酸和果糖-1,6-二磷酸。这些代谢中间体的积累会抑制 Cra 的活性，因为这些代谢物与 Cra 结合，降低其与靶基因的结合力。KdpE 是 KdpDE 双组分系统的响应调节器（Heermann et al., 2003, 2009, 2014; Heermann and Jung, 2010; Jung et al., 2000; Kraxenberger et al., 2012）。在营养不良的条件下，这些转录因子通过与 *ler* 启动子相互作用，诱导毒力基因表达，其中 *ler* 是肠上皮细胞消退（LEE）致病岛的主调控因子（Carlson-Banning and Sperandio, 2016; Hughes et al., 2009; Njoroge et al., 2012, 2013）。

胃肠道内的黏液层为微生物与上皮细胞之间提供物理分离，从而发挥防御微生物的作用。黏液层的主要成分是黏蛋白，其中80%的总重量由 *O*-链聚糖高度修饰黏蛋白组成（Johansson et al., 2013）。在黏液层中发现的主要糖包括 *N*-乙酰半乳糖胺（GalNAc）、*N*-乙酰神经氨酸（NANA）、*N*-乙酰氨基葡萄糖（GlcNAc）、甘露糖、半乳糖，同时还含有海藻糖和唾液酸（Marcobal et al., 2013）。微生物群的一些成员能编码水解酶，这些酶可以从黏蛋白中释放半乳糖和唾液酸，从而进入黏蛋白衍生的其他聚糖中。在许多黏液层中发现了多形拟杆菌（*B. thetaiotaomicron*），其编码多种参与多糖分解代谢的位点（Koropatkin et al., 2012；Porter and Martens, 2017）。当饮食中缺乏复杂碳水化合物，特别是纤维时，多形拟杆菌在降解黏蛋白衍生糖方面表现出特别的有效性。研究表明，黏液层的降解会释放出葡聚糖，这些葡聚糖可以被其他微生物或侵入的病原菌作为碳源（Desai et al., 2016）利用。

一项研究利用转录组学分析，比较了无菌小鼠和转基因小鼠中沙门菌基因表达的变化，探讨了黏蛋白衍生糖作为病原菌碳源的重要性（Ng et al., 2013）。研究表明，涉及利用唾液酸和海藻糖的基因发生了显著变化。沙门菌缺乏从黏液层释放糖所需的水解酶，然而，多形拟杆菌来源的水解酶可释放这些糖。Ng 等（2013）发现，在沙门菌中产生了 *nanA* 和 *fucI* 基因代谢的突变体，其分别参与了唾液酸和海藻糖的利用。唾液酸和海藻糖代谢的缺失，降低了使用多形拟杆菌菌株重组小鼠中野生型亲本菌株的竞争能力。研究还表明，与沙门菌类似，艰难梭菌（*C. difficile*）也利用微生物来源的唾液酸，从而在肠道中扩张。链霉素治疗增加了肠道中唾液酸的水平，这可能解释了这些病原菌在抗生素治疗后扩展的机制（图 6-3）。

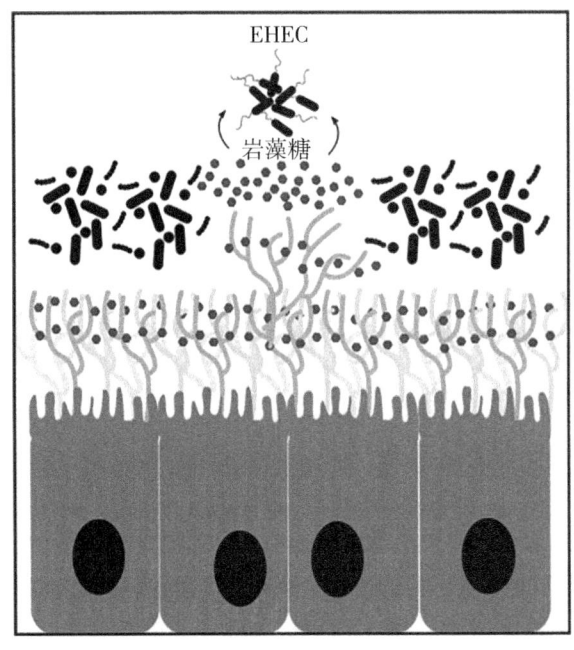
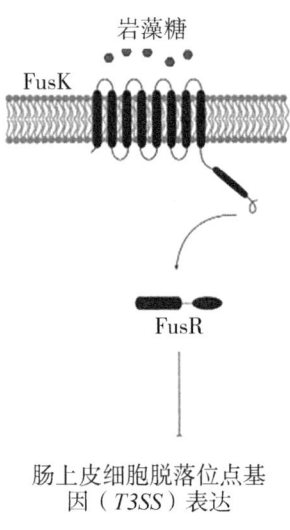

图 6-3　微生物来源的琥珀酸盐和病原体定殖

注：琥珀酸是拟杆菌属分泌的最常见产物之一。然而，在没有炎症刺激的情况下，它被微生物群中的次级发酵产物进一步代谢，在没有炎症刺激的情况下，很少在肠道中积累。当炎症发生时，主要的琥珀酸消费者被耗尽，从而导致琥珀酸的积累。病原体如 EHEC、啮齿枸橼酸杆菌和艰难梭菌利用这种微生物衍生的代谢物在肠道环境中扩展。艰难梭菌结合利用膳食衍生碳水化合物，如山梨醇与琥珀酸转化为丁酸，以扩大和超过常驻微生物群。EHEC 可以利用碳代谢中的 Cra 传感器感知糖异生代谢物琥珀酸，从而驱动毒力相关基因的表达。

拟杆菌门的成员因其强大和多样的酶学能力而被大量研究。这些生物体可降解各种复杂的碳水化合物。缺乏降解这些碳水化合物能力的生物体则可利用微生物群在分解过程中释放出的糖（Desai et al., 2016）。海藻糖是一种微生物来源糖，研究表明，在黏液层的岩藻糖基化是微生物群所依赖的但在无菌小鼠中不存在（Bry et al., 1996; Hooper et al., 1999）。黏液层糖的释放也依赖于微生物群编码的岩藻糖苷酶的活性。EHEC 编码一种名为 FusKR 的双组分系统，该系统可以感知游离海藻糖，从而抑制 EHEC 的毒性（图 6-4；Pacheco et al., 2012）。FusR 是该双组分系统的同源反应调节因子，在被 FusK 磷酸化后，继续抑制 EHEC 中 LEE 致病性岛基因的表达（Pacheco et al., 2012）。

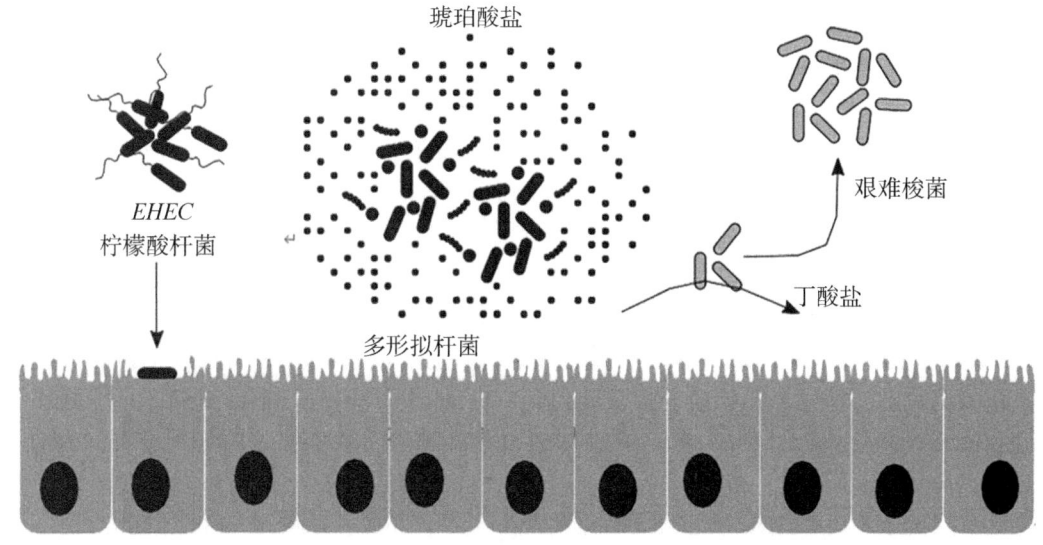

图 6-4　黏蛋白来源的碳水化合物对 EHEC 毒力相关基因的调控
注：在食物来源的复杂碳水化合物缺乏的情况下，多形拟杆菌可以转向代谢其黏蛋白来源的糖。它能够从黏液层中释放岩藻糖，并利用肠道中的其他细菌。EHEC 通过 Fus KR 调控系统感知岩藻糖，从而抑制编码 T3SS 基因的表达。

肠道微生物群落研究中最常关注的功能是强大的代谢能力，这一功能受到广泛的研究。当利用复杂碳水化合物时，微生物可以产生多种化学物质。不可消化的碳水化合物，如纤维素、果胶或宿主黏蛋白糖蛋白，经过代谢后会生成短链脂肪酸（SCFA）。在肠道中，复杂的碳水化合物的终产物通过厌氧发酵形成。肠道内的代谢物可能存在显著差异，并可能受到微生物群结构变化的影响，如饮食、抗生素治疗或感染性结肠炎。因此，肠道中 SCFA 的存在取决于微生物群落组成和可用的碳水化合物。然而，一般来说，肠道中最常见的 SCFA 是醋酸、丙酸、丁酸和甲酸盐（Macfarlane and Macfarlane, 2003）。微生物群落对可用碳源的发酵可以产生发酵中间产物，如乳酸和琥珀酸。由抗生素治疗或病原菌艰难梭菌感染引起的肠道炎症已被证明会增加肠道中琥珀酸的水平（Lawley et al., 2012）。研究表明，由微生物群落产生的琥珀酸可以帮助增强肠道病原菌中毒力基因的表达和定殖（Curtis et al., 2014; Ferreyra et al., 2014）。微生物来源的琥珀酸上调了 EHEC 中毒力基因的表达，EHEC 通过代谢调节剂 Cra 感知微生物群产生的琥珀酸。在营养丰富的条件下，如当葡萄糖和其他单糖充足时，Cra 与 ler 启动子（ler 是所有 LEE 基因的主激活因子）相互作用的能力就会降低。当葡萄糖浓度受限时，大肠杆菌可以转向糖异生代谢，从而有效利用替代碳源如琥珀酸，并以 Cra-依赖的方式增加毒力相关基因的表达。感染了啮齿枸橼酸杆菌的小鼠通过抗生素处理会耗尽其微生物群落，而使用多形拟杆菌重组的小鼠则会增加致病性。代谢组学分析显示，用多形拟杆菌重组的小鼠在肠道中显示出更高的琥珀酸水平（Curtis et al., 2014）。与 EHEC

类似，啮齿枸橼酸杆菌能利用微生物群落来源的琥珀酸来促进其侵染（图6-3）。Ferreyra等（2014）将无菌小鼠进行了试验，用单独多形拟杆菌或艰难梭菌，或二者共同处理，研究艰难梭菌感染对微生物群落的影响。这些小鼠被喂食富含多糖或缺乏多糖的食物。当艰难梭菌与多形拟杆菌共定殖时，该病原菌的细菌负荷也有所增加。这种扩张是由于艰难梭菌能够利用琥珀酸作为碳源，通过琥珀酸到丁酸的发酵途径进行代谢。与亲本菌株相比，不能运输琥珀酸的艰难梭菌突变体处于竞争劣势。这表明依赖于微生物来源的琥珀酸的艰难梭菌在肠道内经过抗生素处理后能够实现扩增（图6-3）。

肠道中最丰富的SCFAs是醋酸盐、丙酸盐、丁酸盐和甲酸盐，它们在结肠中浓度最高，而在小肠中浓度较低（Louis et al., 2014; Macfarlane and Macfarlane, 2003）。一些病原菌无论是在结肠还是在小肠中均可以感知SCFAs，这些物质为病原体识别提供了线索。醋酸和丙酸分布在小肠和大肠中的浓度大致相同，但它们在胃肠道中的分布却有所不同，SCFA甲酸和丁酸被认为为细菌提供了特定的信号。甲酸盐在小肠，特别是回肠中含量较高，而丁酸盐在结肠中浓度较高（Louis et al., 2014）。丁酸盐已被证明可诱导EHEC中毒力相关基因的表达，如T3SS的表达增加以及细菌对Caco-2细胞的附着增强（Nakanishi et al., 2009; Takao et al., 2014）。在EHEC中，丁酸盐通过依赖亮氨酸响应蛋白（Lrp）的机制，能增加鞭毛基因的表达，从而增强细菌的运动能力（Tobe et al., 2011）。研究表明，丁酸盐诱导编码黏附素 *iha* 基因的表达（Herold et al., 2009）。EHEC对丁酸盐的感知依赖于Lrp的转录调控因子（Nakanishi et al., 2009; Tobe et al., 2011）。相比之下，丁酸和丙酸则降低了沙门菌毒力基因的表达（Gantois et al., 2006; Hung et al., 2013; Lawhon et al., 2002）。

丁酸盐是胃肠道中发现的主要短链脂肪酸之一，其产生依赖于微生物群，尤其是梭状芽孢杆菌对复杂碳水化合物的代谢。梭状芽孢杆菌被认为是肠道微生物群的重要成员，能有效地防止病原菌的定殖。然而，链霉素治疗会显著减少梭状芽孢杆菌的种群，从而为共同致病性菌株如大肠杆菌和沙门菌的扩增创造条件。由于中性粒细胞的活性，在肠道沙门菌血清型鼠伤寒杆菌感染期间，梭状芽孢杆菌菌群的消耗也独立于抗生素治疗（Gill et al., 2012）。最近研究表明，由于致病因子的消耗，梭状芽孢杆菌的消耗依赖于炎症反应，这一菌群的消耗伴随着丁酸盐水平的显著下降（Rivera-Chávez et al., 2016）。丁酸盐是结肠细胞的主要能源之一，结肠细胞利用β氧化途径将丁酸盐氧化为二氧化碳，并利用氧气作为末端电子受体，从而消耗大量的氧气（图6-5; Colgan and Taylor, 2010）。当丁酸盐因抗生素治疗或感染而缺失或减少时，结肠细胞会转向一种无氧发酵代谢。由于氧气泄漏到管腔内，使病原菌如鼠伤寒沙门菌和啮齿枸橼酸杆菌能利用有氧呼吸来抵抗微生物群的抑制（Lopez et al., 2016; Rivera-Chavez et al., 2016）。丁酸盐可作为丁酸盐传感器过氧化物酶体增殖物激活受体γ（PPAR-γ）的激动剂。丁酸盐激活PPAR-γ信号通路，促进结肠细胞参与β氧化。因此，PPAR-γ在维持结肠缺氧环境具有重要意义（Byndloss et al., 2017）。后续研究表明，梭状芽孢杆菌的消耗以及随后的丁酸盐减少，诱导了肠道中乳酸的积累。乳酸的增加是由于结肠细胞代谢的变化，β-氧化转向乳酸发酵，乳酸脱氢酶抑制剂足以降低乳酸水平，这表明乳酸是宿主来源的。通过抗生素治疗或感染导致丁酸盐的消耗，会导致乳酸的积累和腔内氧浓度的增加。鼠伤寒沙门菌可以利用这种环境，通过有氧呼吸积累的乳酸在管腔内扩增（图6-5, Gillis et al., 2018）。

某些宿主来源的营养物质也可以引起微生物群结构的改变，并有助于肠道病原菌的扩增。上皮细胞的周转和脱落为宿主提供了营养物质的来源，如乙醇胺。乙醇胺是原核生物和

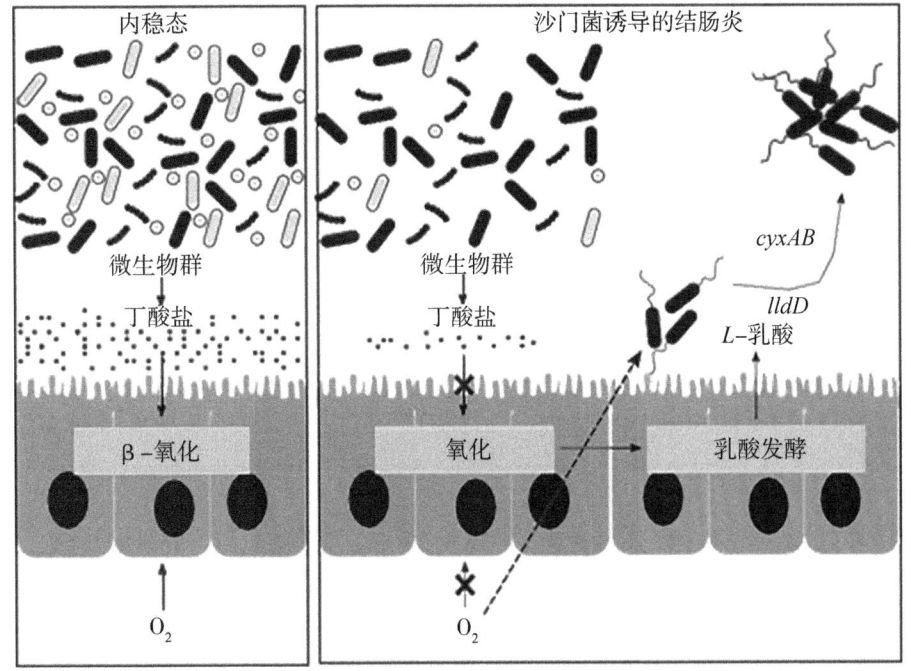

图 6-5 沙门菌（*Salmonella*）诱导的结肠炎触发了肠腔内氧气和 *L*-乳酸的积累

注：鼠伤寒沙门菌利用两个 T3SS 诱导肠道炎症。炎症反应的启动触发了上皮细胞释放的活性氧和活性氮。炎症的诱导导致共生梭状芽孢杆菌的消耗，这是肠道中主要的丁酸盐生产者。丁酸盐通过 β-氧化成为结肠细胞的主要能量来源。β-氧化利用氧气，防止氧气泄漏到肠腔内。当丁酸盐浓度下降时，结肠细胞转向乳酸发酵，这导致氧和 *L*-乳酸在腔内的积累。鼠伤寒沙门菌可以利用这种营养物质，有氧刺激其在腔内扩张。

真核生物细胞膜的主要成分之一（Gibellini and Smith，2010）。它主要是磷脂酰乙醇胺，动植物细胞中的磷脂类物质属于第二丰度。尽管由于肠细胞的周转，在胃肠道中大量发现了乙醇胺，但其常驻微生物群并不使用乙醇胺（Bertin et al.，2011）。通过基因组挖掘分析，试图找到肠道感染的决定因素，发现乙醇胺的利用途径对单核增生李斯特菌（*Listeria monocytogenes*）、产气荚膜梭菌（*C. perfringens*）以及肠道沙门菌等肠道病原菌至关重要。研究表明，EHEC、粪肠球菌（*Enterococcus faecalis*）和沙门菌等病原菌利用现有的乙醇胺，将其作为非竞争性氮源，以扩张和竞争微生物群（Bertin et al.，2011；Maadani et al.，2007；Thiennimitr et al.，2011）。由于存在乙醇胺利用（*eut*）位点，因此，乙醇胺在这些生物体中的分解代谢是可能存在的，这个操纵子编码的基因对乙醇胺的感知、运输和分解代谢中起着重要作用（Garsin，2010）。

此外，乙醇胺还可作为诱导沙门菌和 EHEC 毒力基因的信号。对乙醇胺的感知主要是通过 EutR 转录调控来实现的（Kendall et al.，2012；Luzader et al.，2013）。EutR 可对乙醇胺进行自动调节，并调节 *eut* 操纵子的表达。除调控 *eut* 操纵子外，EutR 还直接调控 EHEC 中 LEE 的表达和沙门菌中 SPI-2 致病性岛的表达（Anderson et al.，2015）。乙醇胺以 EutR 依赖的方式诱导毒力基因的表达。然而，*EutR* 的缺失未使 EHEC 或沙门菌对乙醇胺完全无反应，这表明可能存在其他传感器的参与。此外，在沙门菌小鼠模型中，缺失对乙醇胺利用至

关重要的 *EutR* 和 *EutB* 基因会导致竞争劣势（Anderson et al.，2015；Kendall et al.，2012）。对乙醇胺的感知和利用对该病原菌的建立和传播具有重要意义。最近的研究表明，与 EHEC 和沙门菌相比，另一种病原菌艰难梭菌对乙醇胺的感知和利用因毒力相关基因的下调，而降低了病理学发病率。

5　结论

在这里，我们讨论了共生菌和致病菌如何感知胃肠道中不同的环境信号，以促进其在宿主中的定殖。对于宿主、微生物群和进入的肠道病原菌之间的复杂关系的理解才刚刚开始。随着组学技术如基因组学、转录组学和代谢组学等组学技术的不断进步，将更好地了解这些复杂种群之间的潜在的相互作用关系。在胃肠道中的寄主与细菌之间的相互作用是互惠灵活的。微生物群的组成可以随着饮食、宿主遗传和免疫过程的变化而变化。然而，宿主的生理功能也可因代谢变化导致微生物群的改变而受到影响。这些肠道代谢环境的变化可能导致对外来病原菌易感性的改变。本书中展示了微生物群组成的变化导致代谢物的波动，这些代谢物可被病原菌感知，以调节其毒性。一些病原菌能利用宿主和微生物来源的代谢物更好地在肠道中定殖并促进疾病发生，而其他病原菌则可能因毒性降低而受到阻碍，或由于营养物质的限制而被常驻微生物群击败。

宿主、微生物群和肠道病原菌之间的代谢物交换为预防或治疗肠道感染带来了令人兴奋的可能性。一个有吸引力的思路是追求益生元治疗，以促进宿主与微生物群之间相互作用的变化，从而限制肠道病原菌的进入。理解这些相互作用也为专注于通过抗病毒方法来改善感染治疗的发展奠定基础（Rasko et al.，2008）。

术语表

儿茶酚胺　由色氨酸衍生的单胺类化合物主要由肾上腺产生，并作为参与战斗或逃跑反应的激素。肾上腺素、去甲肾上腺素和多巴胺是典型的儿茶酚胺。

定殖抗性　栖息地中的微生物通过多种机制抑制新的或有害微生物的定殖。

生态失调　表示微生物群落结构从"正常"状态开始发生的变化。

无菌　表示没有常驻微生物群，无菌或清洁的动物。

跨界信号　是指不同界的生物体之间的化学相互作用，如细菌和哺乳动物之间的化学相互作用。

种间信号传导　不同物种但属于同一生物体之间的化学信号。

种内信号传导　同一物种成员之间的化学信号。

微生物群落　指在特定地点包括细菌、古细菌、原虫、真菌和病毒在内的微生物群落。

缩写词

AHL	酰基高丝氨酸内酯
AI	自诱导物
DHMA	3,4-二羟基杏仁酸

DPD	4,5-二羟基-2,3-戊二酮
E/Epi	肾上腺素
EHEC	肠出血性大肠杆菌
GalNAc	N-乙酰半乳糖胺
GI	胃肠的
GlcNAc	N-乙酰葡糖胺
LEE	肠细胞流出位点
NANA	N-乙酰神经氨酸
NE	去甲肾上腺素
PPAR-γ	过氧化物酶体增殖物激活受体 γ
SAM	S-腺苷甲硫氨酸
SCFA	短链脂肪酸
T3SS	三型分泌系统

参考文献

ANDERSON C J, CLARK D E, ADLI M, et al., 2015. Ethanolamine signaling promotes *Salmonella* niche recognition and adaptation during infection. PLoS Pathog., 11: e1005278.

ASANO Y, HIRAMOTO T, NISHINO R, et al., 2012. Critical role of gut microbiota in the production of biologically active, free catecholamines in the gut lumen of mice. Am. J. Physiol. Gastrointest. Liver Physiol., 303: G1288-G1295.

BÄUMLER A J, SPERANDIO V, 2016. Interactions between the microbiota and pathogenic bacteria in the gut. Nature, 535: 85-93.

BERTIN Y, GIRARDEAU J P, CHAUCHEYRAS-DURAND F, et al., 2011. Enterohaemorrhagic *Escherichia coli* gains a competitive advantage by using ethanolamine as a nitrogen source in the bovine intestinal content. Environ. Microbiol., 13: 365-377.

BOHNHOFF M, DRAKE B L, MILLER C P, 1954. Effect of streptomycin on susceptibility of intestinal tract to experimental Salmonella infection. Proc. Soc. Exp. Biol. Med., 86: 132-137.

BRY L, FALK P C, MIDTVEDT T, et al., 1996. A model of host-microbial interactions in an open mammalian ecosystem. Science, 273: 1380-1383.

BYNDLOSS M X, OLSAN E E, RIVERA-CHÁVEZ F, et al., 2017. Microbiota-activated PPAR-γ signaling inhibits dysbiotic Enterobacteriaceae expansion. Science, 357: 570-575.

CARLSON-BANNING K M, SPERANDIO V, 2016. Catabolite and oxygen regulation of Enterohemorrhagic *Escherichia coli* virulence. MBio, 7: e01852-16.

CASE R J, LABBATE M, KJELLEBERg S, 2008. AHL-driven quorum-sensing circuits: Their frequency and function among the Proteobacteria. ISME J., 2: 345-349.

CLARKE M B, SPERANDIO V, 2005a. Transcriptional autoregulation by quorum sensing Escherichia coli regulators B and C (QseBC) in enterohaemorrhagic *E. coli* (EHEC). Mol. Microbiol., 58: 441-455.

CLARKE M B, SPERANDIO V, 2005b. Transcriptional regulation of flhDC by QseBC and sigma (FliA) in enterohaemorrhagic *Escherichia coli*. Mol. Microbiol., 57: 1734-1749.

CLARKE M B, HUGHES D T, ZHU C, et al., 2006. The QseC sensor kinase: a bacterial adrenergic receptor. Proc. Natl. Acad. Sci. U. S. A., 103: 10420-10425.

COLGAN S P, TAYLOR C T, 2010. Hypoxia: an alarm signal during intestinal inflammation. Nat. Rev. Gastroenterol. Hepatol., 7: 281-287.

CURTIS M M, HU Z, KLIMKO C, et al., 2014. The gut commensal Bacteroides thetaiotaomicron exacerbates enteric infection through modification of the metabolic landscape. Cell Host Microbe, 16: 759-769.

DESAI M S, SEEKATZ A M, KOROPATKIN N M, et al., 2016. A dietary fiber-deprived gut microbiota degrades the colonic mucus barrier and enhances pathogen susceptibility. Cell, 167: 1339-1353.

DYSZEL J L, SMITH J N, LUCAS D E, et al., 2010. *Salmonella enterica serovar Typhimurium* can detect acyl homoserine lactone production by *Yersinia enterocolitica* in mice. J. Bacteriol., 192: 29-37.

ELDRUP E, RICHTER E A, 2000. DOPA, dopamine, and DOPAC concentrations in the rat gastrointestinal tract decrease during fasting. Am. J. Physiol. Endocrinol. Metab., 279: E815-E822.

ENGEBRECHT J, SILVERMAN M, 1984. Identification of genes and gene products necessary for bacterial bioluminescence. Proc. Natl. Acad. Sci. U. S. A., 81: 4154-4158.

ENGEBRECHT J, NEALSON K, SILVERMAN M, 1983. Bacterial bioluminescence: Isolation and genetic analysis of functions from *Vibrio fischeri*. Cell, 32: 773-781.

FERREYRA J A, WU K J, HRYCKOWIAN A J, et al., 2014. Gut microbiotaproduced succinate promotes *C. difficile* infection after antibiotic treatment or motility disturbance. Cell Host Microbe, 16: 770-777.

GALLOWAY W R, HODGKINSON J T, BOWDEN S D, et al., 2011. Quorum sensing in Gramnegative bacteria: small-molecule modulation of AHL and AI-2 quorum sensing pathways. Chem. Rev., 111: 28-67.

GANTOIS I, DUCATELLE R, PASMANS F, et al., 2006. Butyrate specifically down-regulates salmonella pathogenicity island 1 gene expression. Appl. Environ. Microbiol., 72: 946-949.

GARSIN D A, 2010. Ethanolamine utilization in bacterial pathogens: roles and regulation. Nat. Rev. Microbiol., 8: 290-295.

GIBELLINI F, SMITH T K, 2010. The Kennedy pathway-De novo synthesis of phosphatidylethanolamine and phosphatidylcholine. IUBMB Life, 62: 414-428.

GILL N, FERREIRA R B, ANTUNES L C, et al., 2012. Neutrophil elastase alters the murine gut microbiota resulting in enhanced Salmonella colonization. PLoS One, 7: e49646.

GILLIS C C, HUGHES E R, SPIGA L, et al., 2018. Dysbiosis-associated change in host metabolism generates lactate to support Salmonella growth. Cell Host Microbe, 23: 54-64.

GORDON J I, KLAENHAMMER T R, 2011. A rendezvous with our microbes. Proc. Natl. Acad. Sci. U. S. A., 108: 4513-4515.

GRENHAM S, CLARKE G, CRYAN J F, et al., 2011. Brain-gut-microbe communication in health and disease. Front. Physiol., 2: 94.

HEERMANN R, JUNG K, 2010. The complexity of the 'simple' two-component system KdpD/KdpE in *Escherichia coli*. FEMS Microbiol. Lett., 304: 97-106.

HEERMANN R, ALTENDORF K, JUNG K, 2003. The N-terminal input domain of the sensor kinase KdpD of *Escherichia coli* stabilizes the interaction between the cognate response regulator KdpE and the corresponding DNA-binding site. J. Biol. Chem., 278: 51277-51284.

HEERMANN R, WEBER A, MAYER B, et al., 2009. The universal stress protein UspC scaffolds the KdpD/KdpE signaling cascade of *Escherichia coli* under salt stress. J. Mol. Biol., 386: 134-148.

HEERMANN R, ZIGANN K, GAYER S, et al., 2014. Dynamics of an interactive network composed of a bacterial two-component system, a transporter and K^+ as mediator. PLoS One, 9: e89671.

HEROLD S, PATON J C, SRIMANOTE P, et al., 2009. Differential effects of short-chain fatty acids and iron on expression of iha in Shiga-toxigenic *Escherichia coli*. Microbiology, 155: 3554-3563.

HOOPER L V, XU J, FALK P G, et al., 1999. A molecular sensor that allows a gut commensal to control its nutrient foundation in a competitive ecosystem. Proc. Natl. Acad. Sci. U. S. A., 96: 9833-9838.

HÖRGER S, SCHULTHEISS G, DIENER M, 1998. Segment-specific effects of epinephrine on ion transport in the colon of the rat. Am. J. Physiol., 275: G1367-G1376.

HSIAO A, AHMED A M, SUBRAMANIAN S, et al., 2014. Members of the human gut microbiota involved in recovery from *Vibrio cholerae* infection. Nature, 515: 423-426.

HUDAIBERDIEV S, CHOUDHARY K S, VERA ALVAREZ R, et al., 2015. Census of solo LuxR genes in prokaryotic genomes. Front. Cell. Infect. Microbiol., 5: 20.

HUGHES D T, CLARKE M B, YAMAMOTO K, et al., 2009. The QseC adrenergic signaling cascade in Enterohemorrhagic *E. coli* (EHEC). PLoS Pathog., 5: e1000553.

HUGHES D T, TEREKHOVA D A, LIOU L, et al., 2010. Chemical sensing in mammalian host-bacterial commensal associations. Proc. Natl. Acad. Sci. U. S. A., 107: 9831-9836.

HUNG C C, GARNER C D, SLAUCH J M, et al., 2013. The intestinal fatty acid propionate inhibits Salmonella invasion through the post-translational control of HilD. Mol. Microbiol., 87: 1045-1060.

ITOH K, FRETER R, 1989. Control of *Escherichia coli* populations by a combination of indigenous clostridia and lactobacilli in gnotobiotic mice and continuous-flow cultures. Infect. Immun., 57: 559-565.

JOHANSSON M E, SJÖVALL H, HANSSON G C, 2013. The gastrointestinal mucus system in health and disease. Nat. Rev. Gastroenterol. Hepatol., 10: 352-361.

JUNG K, VEEN M, ALTENDORF K, 2000. K^+ and ionic strength directly influence the autophosphorylation activity of the putative turgor sensor KdpD of *Escherichia coli*. J. Biol. Chem., 275: 40142-40147.

JUNG K, FRIED L, BEHR S, et al., 2012. Histidine kinases and response regulators in networks. Curr. Opin. Microbiol., 15: 118-124.

KENDALL M M, SPERANDIO V, 2016. What a dinner party! Mechanisms and functions of interkingdom signaling in host-pathogen associations. MBio, 7: e01748.

KENDALL M M, GRUBER C C, PARKER C T, et al., 2012. Ethanolamine controls expression of genes encoding components involved in interkingdom signaling and virulence in enterohemorrhagic *Escherichia coli* O157: H7. MBio, 3: e00050-12.

KOROPATKIN N M, CAMERON E A, MARTENS E C, 2012. How glycan metabolism shapes the human gut microbiota. Nat. Rev. Microbiol., 10: 323-335.

KRAXENBERGER T, FRIED L, BEHR S, et al., 2012. First insights into the unexplored two-component system YehU/YehT in *Escherichia coli*. J. Bacteriol., 194: 4272-4284.

LAWHON S D, MAURER R, SUYEMOTO M, et al., 2002. Intestinal short-chain fatty acids alter *Salmonella typhimurium* invasion gene expression and virulence through BarA/SirA. Mol. Microbiol., 46: 1451-1464.

LAWLEY T D, CLARE S, WALKER A W, et al., 2012. Targeted restoration of the intestinal microbiota with a simple, defined bacteriotherapy resolves relapsing *Clostridium difficile* disease in mice. PLoS Pathog., 8e1002995.

LOPEZ C A, MILLER B M, RIVERA-CHÁVEZ F, et al., 2016. Virulence factors enhance *Citrobacter rodentium* expansion through aerobic respiration. Science, 353: 1249-1253.

LOUIS P, HOLD G L, FLINT H J, 2014. The gut microbiota, bacterial metabolites and colorectal cancer. Nat. Rev. Microbiol., 12: 661-672.

LUZADER D H, CLARK D E, GONYAR L A, et al., 2013. EutR is a direct regulator of genes that contribute to metabolism and virulence in enterohemorrhagic *Escherichia coli* O157: H7. J. Bacteriol., 195: 4947-4953.

MAADANI A, FOX K A, MYLONAKIS E, et al., 2007. *Enterococcus faecalis* mutations affecting virulence in the Caenorhabditis elegans model host. Infect. Immun., 75: 2634-2637.

MACFARLANE S, MACFARLANE G T, 2003. Regulation of short-chain fatty acid production. Proc. Nutr. Soc., 62: 67-72.

MARCOBAL A, SOUTHWICK A M, EARLE K A, et al., 2013. A refined palate: Bacterial consumption of host glycans in the gut. Glycobiology, 23: 1038-1046.

MICHAEL B, SMITH J N, SWIFT S, et al., 2001. SdiA of *Salmonella enterica* is a LuxR homolog that detects mixed microbial communities. J. Bacteriol., 183: 5733-5742.

MOLINA P E, MOLINA P E, 2006. Endocrine Physiology. Lange Medical Books/McGraw-Hill, New York.

MOREIRA C G, SPERANDIO V, 2012. Interplay between the QseC and QseE bacterial adrenergic sensor kinases in *Salmonella enterica serovar Typhimurium* pathogenesis. Infect. Immun., 80: 4344-4353.

MOREIRA C G, WEINSHENKER D, SPERANDIO V, 2010. QseC mediates *Salmonella enterica serovar Typhimurium* virulence in vitro and in vivo. Infect. Immun., 78: 914-926.

MOREIRA C G, RUSSELL R, MISHRA A A, et al., 2016. Bacterial adrenergic sensors regulate virulence of enteric pathogens in the gut. MBio, 7: e00826-16.

NAKANISHI N, TASHIRO K, KUHARA S, et al., 2009. Regulation of virulence by butyrate sensing in enterohaemorrhagic *Escherichia coli*. Microbiology, 155: 521-530.

NG K M, FERREYRA J A, HIGGINBOTTOM S K, et al., 2013. Microbiota-liberated host sugars facilitate postantibiotic expansion of enteric pathogens. Nature, 502: 96-99.

NGUYEN Y, NGUYEN N X, ROGERS J L, et al., 2015. Structural and mechanistic roles of novel chemical ligands on the SdiA quorum-sensing transcription regulator. MBio, 6: e02429-14.

NJOROGE J W, NGUYEN Y, CURTIS M M, et al., 2012. Virulence meets metabolism: Cra and KdpE gene regulation in enterohemorrhagic *Escherichia coli*. MBio, 3: e00280-12.

NJOROGE J W, GRUBER C, SPERANDIO V, 2013. The interacting Cra and KdpE regulators are involved in the expression of multiple virulence factors in enterohemorrhagic *Escherichia coli*. J. Bacteriol., 195: 2499-2508.

PACHECO A R, CURTIS M M, RITCHIE J M, et al., 2012. Fucose sensing regulates bacterial intestinal colonization. Nature, 492: 113-117.

PASUPULETI S, SULE N, COHN W B, et al., 2014. Chemotaxis of *Escherichia coli* to norepinephrine (NE) requires conversion of NE to 3, 4-dihydroxymandelic acid. J. Bacteriol., 196: 3992-4000.

PEREIRA C S, THOMPSON J A, XAVIER K B, 2013. AI-2-mediated signalling in bacteria. FEMS Microbiol. Rev., 37: 156-181.

PORTER N T, MARTENS E C, 2017. The critical roles of polysaccharides in gut microbial ecology and physiology. Annu. Rev. Microbiol., 71: 349-369.

RASKO D A, MOREIRA C G, LI DE R, et al., 2008. Targeting QseC signaling and virulence for antibiotic development. Science, 321: 1078-1080.

READING N C, RASKO D A, TORRES A G, et al., 2009. The two-component system QseEF and the membrane protein QseG link adrenergic and stress sensing to bacterial pathogenesis. Proc. Natl. Acad. Sci. U. S. A., 106: 5889-5894.

RIVERA-CHÁVEZ F, ZHANG L F, FABER F, et al., 2016. Depletion of butyrate-producing clostridia from the gut microbiota drives an aerobic luminal expansion of Salmonella. Cell Host Microbe, 19: 443-454.

SCHAUDER S, SHOKAT K, SURETTE M G, et al., 2001. The LuxS family of bacterial autoinducers: biosynthesis of a novel quorum-sensing signal molecule. Mol. Microbiol., 41: 463-476.

SEKIROV I, TAM N M, JOGOVA M, et al., 2008. Antibiotic-induced perturbations of the intestinal microbiota alter host susceptibility to enteric infection. Infect. Immun., 76: 4726-4736.

SHARMA V K, CASEY T A, 2014. *Escherichia coli* O157: H7 lacking the qseBC-encoded quorum-sensing system outcompetes the parental strain in colonization of cattle intestines. Appl. Environ. Microbiol., 80: 1882-1892.

SHENG H, NGUYEN Y N, HOVDE C J, et al., 2013. SdiA aids enterohemorrhagic *Escherichia coli* carriage by cattle fed a forage or grain diet. Infect. Immun., 81: 3472-3478.

SMITH J N, AHMER B M, 2003. Detection of other microbial species by Salmonella: Expression of the SdiA regulon. J. Bacteriol., 185: 1357-1366.

SMITH J N, DYSZEL J L, SOARES J A, et al., 2008. SdiA, an *N*-acylhomoserine lactone receptor, becomes active during the transit of *Salmonella enterica* through the gastrointestinal tract of turtles. PLoS One, 3e2826.

SPERANDIO V, 2010. SdiA sensing of acyl-homoserine lactones by enterohemorrhagic *E. coli* (EHEC) serotype O157: H7 in the bovine rumen. Gut Microbes, 1: 432-435.

SPERANDIO V, TORRES A G, JARVIS B, et al., 2003. Bacteria-host communication: The language of hormones. Proc. Natl. Acad. Sci. U. S. A., 100: 8951-8956.

STEVENS A M, DOLAN K M, GREENBERG E P, 1994. Synergistic binding of the *Vibrio fischeri* LuxR transcriptional activator domain and RNA polymerase to the lux promoter region. Proc. Natl. Acad. Sci. U. S. A., 91: 12619-12623.

SWEARINGEN M C, SABAG-DAIGLE A, AHMER B M, 2013. Are there acyl-homoserine lactones within mammalian intestines? J. Bacteriol., 195: 173-179.

TAKAO M, YEN H, TOBE T, 2014. LeuO enhances butyrate-induced virulence expression through a positive regulatory loop in enterohaemorrhagic *Escherichia coli*. Mol. Microbiol., 93: 1302-1313.

THIENNIMITR P, WINTER S E, WINTER M G, et al., 2011. Intestinal inflammation allows Salmonella to use ethanolamine to compete with the microbiota. Proc. Natl. Acad. Sci. U. S. A., 108: 17480-17485.

THOMPSON J A, OLIVEIRA R A, DJUKOVIC A, et al., 2015. Manipulation of the quorum sensing signal AI-2 affects the antibiotic-treated gut microbiota. Cell Rep., 10: 1861-1871.

TOBE T, NAKANISHI N, SUGIMOTO N, 2011. Activation of motility by sensing short-chain fatty acids via two steps in a flagellar gene regulatory cascade in enterohemorrhagic *Escherichia coli*. Infect. Immun., 79: 1016-1024.

VOLF J, SEVCIK M, HAVLICKOVA H, et al., 2002. Role of SdiA in *Salmonella enterica* serovar Typhimurium physiology and virulence. Arch. Microbiol., 178: 94-101.

VON BODMAN S B, WILLEY J M, DIGGLE S P, 2008. Cell-cell communication in bacteria: united we stand. J. Bacteriol., 190: 4377-4391.

ZARGAR A, QUAN D N, CARTER K K, et al., 2015. Bacterial secretions of nonpathogenic *Escherichia coli* elicit inflammatory pathways: a closer investigation of interkingdom signaling. MBio, 6: e00025.

ZHANG R G, PAPPAS K M, BRACE J L, et al., 2002. Structure of a bacterial quorum-sensing transcription factor complexed with pheromone and DNA. Nature, 417: 971-974.

ZHU J, WINANS S C, 1999. Autoinducer binding by the quorum-sensing regulator TraR increases affinity for target promoters in vitro and decreases TraR turnover rates in whole cells. Proc. Natl. Acad. Sci. U. S. A., 96: 4832-4837.

第3部分
群体感应抑制

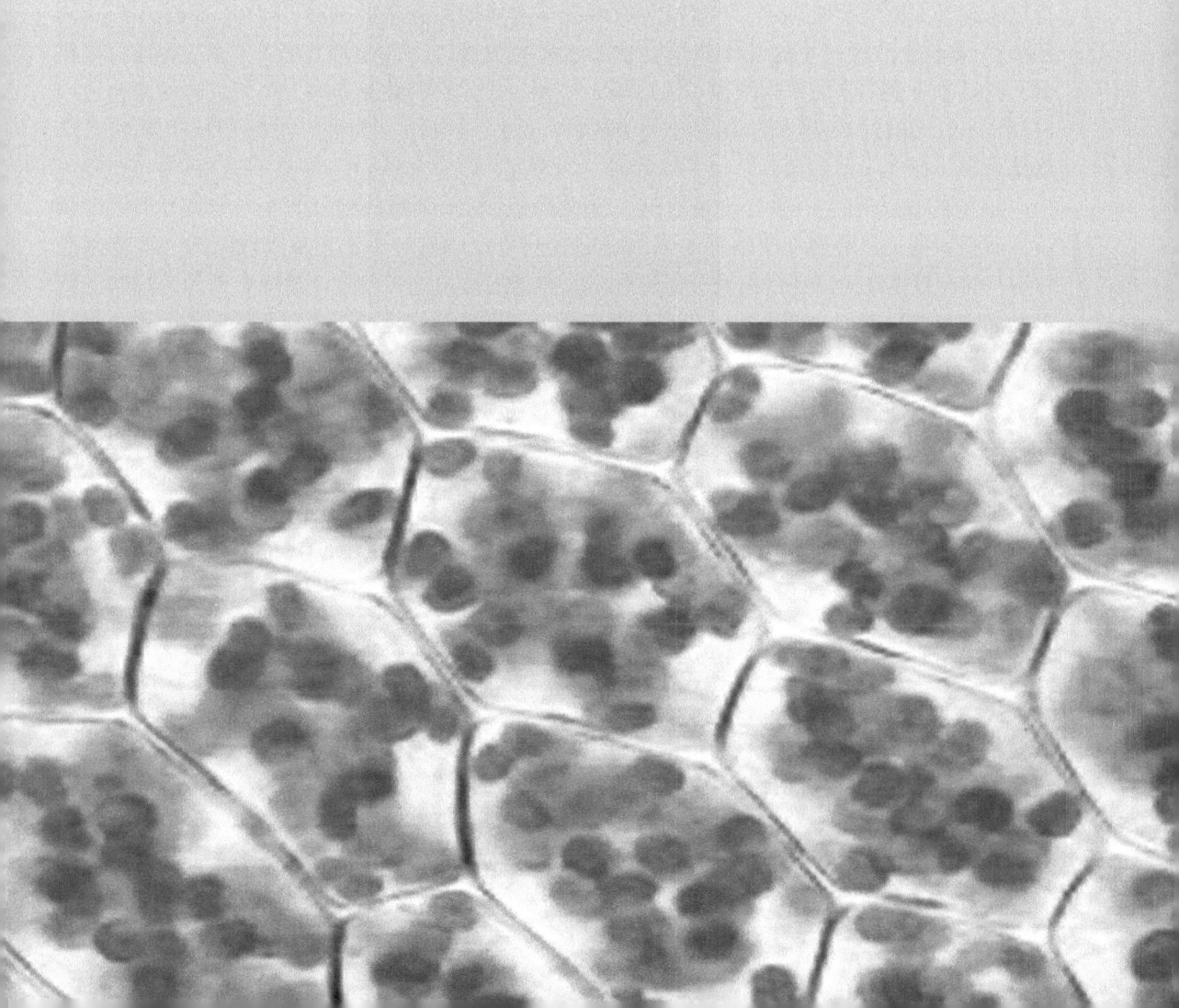

第7章　生物膜中酶促群体淬灭

Jan Vogel，Wim J. Quax

Groningen Research Institute of Pharmacy, Department of Chemical and
Pharmaceutical Biology, University of Groningen, Groningen, The Netherlands

1　前言

　　20世纪初，德国科学家保罗·埃利希试图寻找一种"灵丹妙药"，旨在针对一种特定目标，在不伤害人体的情况下，消除疾病。这项研究不仅可以更好地理解人类的免疫系统，也可以更好地理解相关的病原体。1928年，亚历山大·弗莱明在一个琼脂平板上发现了一种"灵丹妙药"：一种产生青霉素的青霉菌（*Penicillium*）菌株，该菌株能有效地杀死金黄色葡萄球菌。这一事件标志着一场医疗革命的开始。然而，亚历山大·弗莱明在1945年接受诺贝尔生理学或医学奖时表示，"滥用药物可能导致对耐药细菌的选择"（Rosenblatt-Farrell，2009）。事实上，在过去的几十年里，耐药细菌的数量已惊人地增加；并且在过去的40年里，几乎没有任何新的抗生素被引入市场。这种治疗方案的缺乏是对人类最危险的威胁之一 [Gill et al.，2015；美国传染病学会（IDSA）et al.，2011]。因此，对新型抗生素的需求很高。

　　20世纪60年代初，发光细菌费氏弧菌的发现，揭示了这些细菌的发光与其种群密度相关。研究表明，只有当细菌培养的细胞密度达到特定阈值时，发光基因被激活。群体感应（QS）的有效性取决于特定信号分子的分泌。有趣的是，这一现象曾被视为是深海细菌的特异性而被忽略，然而，如今几乎在每个生态位上，细菌群体会通过各种可扩散的分子进行交流，因此利用干扰QS系统进行新型、急需的抗菌治疗开辟了新的可能性。为实现这一目的，理解QS系统至关重要。

　　这篇综述提出，干扰细菌QS系统是一种很有前途的对抗细菌感染的方法。各种酶已被证明能够通过降解QS信号分子，从而有效降低细菌的毒力（Grandclement et al.，2015）。

　　群体淬灭（QQ）酶在各个研究领域中转化为实际应用具有巨大的潜力，从防止海水中结构性的生物污垢到医疗设备的功能化，这些酶能有效地阻碍生物膜的形成（Bzdrenga et al.，2017）。然而，本章重点关注医学视角以及通过群体淬灭酶提供的机会。

2　表层上的生命：细菌生物膜

　　细菌生物膜是细菌对各种外部因素做出协调反应，从而改变其表型的一个典型例子。反过来，这种表型增加了细菌在表面上的附着力和持久性。生物膜被描述为细菌的多细胞群落，在特定环境中为细菌提供了许多好处，因此，具有显著的进化优势。

　　固着细菌在许多方面不同于浮游生物个体。生活在生物膜中的细菌有一个共同特征，是能够产生细胞外基质。这种基质不仅提供结构支持，而且是耐受抗菌化合物的关键因素。该基质由不同的大分子、蛋白质、脂质、细胞外 DNA 和多糖组成（Billings et al., 2013）。生物膜的形成分为五个基本阶段（图 7-1）。

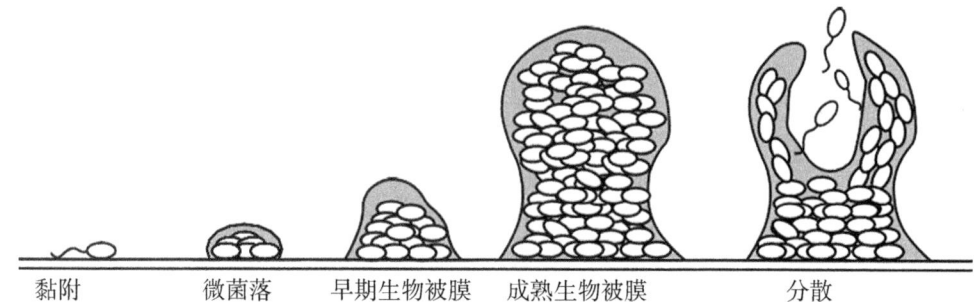

黏附　　　微菌落　　早期生物被膜　成熟生物被膜　　　分散

图 7-1　生物膜形成的示意

注：第一阶段，细菌附着在表面；第二个阶段，细菌形成微菌落，产生深灰色的 eDNA 和 EPS 化合物，菌落从早期的生物被膜发展到成熟的生物被膜，表现出复杂的三维结构；最后一个阶段，生物被膜分散阶段，单个细胞从生物被膜上分离。

首先，细胞附着在表面上，然后通过从环境中聚集和招募浮游的细胞。例如，铜绿假单胞菌在表面上聚集是通过第Ⅳ型鞭毛介导的蹭行运动来进行的。二元细胞分裂对于最初的生物膜聚集和该结构的后续分化是至关重要的（Stoodley et al., 2002）。在后期阶段，生物膜的变化通过单个细菌的表型分化，从附着的微菌落转变为三维结构。例如，在初始附着阶段，铜绿假单胞菌中细胞外聚合物质（EPS）基因的表达在 15 min 内上调。这些成分中最明显的基质成分被称为 PsI 和海藻酸盐（Billings et al., 2013；Chang, 2018）。这标志着从可逆附着到不可逆附着的转变，也被描述为从弱相互作用到与底物永久结合的转变（Stoodley et al., 2002）。微生物在生物膜上的扩散，标志着生物膜的最后阶段，可以主动调节，也可以通过流体流动侵蚀等机械影响结果（Nadell et al., 2016）。新扩散的铜绿假单胞菌细胞被证明比浮游细胞的毒性更强（Chua et al., 2014）。

条件致病菌铜绿假单胞菌是研究细菌生物膜的模式生物。铜绿假单胞菌 PAO1 菌株依赖于 3-oxo-C12-HSL 介导的 *lasI* QS 回路能形成分化的生物膜。LasI/LasR QS 系统由 AHL 合酶 LasI 及其同源反应调节因子 LasR 组成。LasR 与 3-oxo-C12-HSL 结合后作为转录因子，激活包括毒力因子在内的各种基因，并在一个正反馈通路中激活 LasR 的表达。Δ*lasI* 突变体形成的生物膜比铜绿假单胞菌 PAO1 野生型薄得多。通过添加外源 3-oxo-C12-HSL 可以恢复野生型表型（De Kievit and Iglewski, 2000）。然而，随着时间的推移，*lasI* 基因的表达水平随着生物膜厚度的增加而降低，而 *rhlI* 的表达水平保持不变（Stoodley et al., 2002）。

3　生物膜和抗生素耐药性

生物膜的形成是微生物之间高度复杂且协调作用的结果。生物膜形成需要由 QS 介导进行高度沟通来完成（Solano et al., 2014）。生物膜降低了细菌对抗生素治疗的可能性，并增加了细菌在表面持续存在的能力（Flemming et al., 2016）。事实上，细菌构成的生物膜被认为是所有生态位中的主要生活形式（Solano et al., 2014）。因此，对抗细菌性感染意味着理解并最终控制生物膜的形成与功能。

以往研究表明，在医生治疗的所有感染中，有60%与细菌构成的生物膜有关（Fux et al.，2005）。然而，由于各种原因，细菌生物膜的处理成为一个主要问题。大多数关于抗菌剂活性的研究都是在浮游和分裂细菌上进行的。因此，当细菌在生物膜中持续存在时，确定的最小抑制浓度值（MIC）可能无法代表真实的疾病状态（Fux et al.，2005）。这些细菌在生物膜中的可进入性存在差异，可能导致剂量不足和长期使用抗生素的问题。此外，在细菌生物膜内，促进了包括耐药基因在内的水平基因转移，新出现的抗生素耐药性问题变得显著。因此，在细菌构成的生物膜中，对抗生素的敏感性较低，并且能耐受更高浓度的抗生素。在大多数情况下，感染在治疗结束时并没有完全清除。

在这种情况下，生物膜基质在细菌对抗生素的耐受性和保护细菌免受宿主免疫系统的攻击中起着重要作用（Fux et al.，2005）。在一些细菌中，暴露于亚抑制浓度的抗生素甚至会诱导黏液表型。额外的基质成分导致形成更厚的生物膜（Fux et al.，2005）。由铜绿假单胞菌产生的生物膜中，分泌的海藻酸盐对白细胞的吞噬作用具有保护作用，并可作为抗氨基糖苷类抗生素的吸收屏障（Bayer et al.，1991）。

与普遍认为的观点相反，细胞外基质对抗生素扩散提供了严格的屏障，目前，尚无证据证实这一假设。扩散研究表明，抗生素确实能够穿透细菌生物膜，而细胞外基质的保护作用是由于与抗生素的特殊相互作用（如电荷）所致（Tseng et al.，2013）。此外，有时酶被整合到外层基质中，例如，在铜绿假单胞菌生物膜中，β-内酰胺酶被整合到外膜，降解β-内酰胺类抗生素，从而促进抗生素耐药性（Heydari and Eftekhar，2015）。

具有生物膜内层的细菌与其稳定期细菌之间有许多相似之处。与游离态细菌相比，由于不同的代谢状态，稳定期细胞对抗生素的耐受性更强（Spoering and Lewis，2001）。因此，细菌在生物膜中对抗生素耐受性是其表型的结果，而非通过突变介导的内在耐药性，后者可能适用于游离态细胞（Fux et al.，2005）。有趣的是，外排泵的表达被认为是浮游铜绿假单胞菌内在抗性的关键因素，但对成熟生物膜整体的抗生素耐受性影响甚微（Ciofu and Tolker-Nielsen，2010）。然而，最近的研究发现，在铜绿假单胞菌中外排泵 MexAB-OprM 和 MexCD-oprJ 确实参与了生物膜对阿奇霉素的耐受性（Ciofu and Tolker-Nielsen，2010）。

3.1 多物种生物膜

在自然界中，单物种的生物膜往往是个例。囊性纤维化（CF）患者的肺部就是一个典型例子：他们的电解质平衡受到破坏，导致黏稠、脱水黏液为微生物提供了一个理想的栖息地。年轻 CF 患者的肺部通常被金黄色葡萄球菌和流感嗜血杆菌（*Haemophilus influenza*）定殖，在进一步感染过程中，60%~70%的 CF 患者表现出铜绿假单胞菌的定殖，其中50%代表优势物种（O'Brien and Fothergill，2017）。在大多数情况下，会伴随洋葱伯克霍尔德氏菌的二次感染，其涉及多种细菌的慢性感染，使 CF 患者接受终身治疗。此外，主要的病原菌包括真菌和酵母菌，如烟曲霉（*A. fumigatus*）和白假丝酵母菌。研究证实，50%的 CF 患者存在曲霉属和假丝酵母菌属的感染。不用说，肺组织的慢性感染对患者构成巨大的威胁，往往导致患者死亡（O'Brien and Fothergill，2017）。

牙龈卟啉单胞菌（*Porphyromonas gingivalis*）和格氏链球菌（*S. gordonii*）栖息在人类口腔中，它们可以通过自诱导物-2（AI-2）信号形成混合生物膜（McNab et al.，2003）。AI-2 介导革兰氏阴性和革兰氏阳性细菌之间的通信，由 LuxS 型合成酶合成（McNab et al.，

2003；Xavier and Bassler，2003）。研究表明，这两株菌 LuxS 合成酶的缺陷突变体在聚苯乙烯表面上无法形成生物膜（McNab et al.，2003）。此外，在混合群落中观察到一种协同的生活方式：混合群落的生物体积比二者单独培养时更大，并通过试验证实了这一假设（Elias and Banin，2012）。结果表明，与野生型菌株互惠共生的生物膜相比，由内氏放线菌（*Actinomyces naeslundii*）和 LuxS 缺陷型口腔链球菌（*S. oralis*）组成的混合生物膜更容易分散（Rickard et al.，2006）。

3.2 铜绿假单胞菌生物膜的交流

细菌生物膜在囊性纤维化（CF）患者肺部铜绿假单胞菌持续感染中起着至关重要的作用。然而，由于 QS 在生物膜形成中起着重要作用，使用 QQ 酶干扰生物膜形成可能有助于使这些病原体更容易受到治疗的影响，最终也更容易受到宿主免疫系统的攻击。为了强调 QS 系统的破坏对病原体的影响，重要的是阐明几个对毒力和持续性至关重要的 QS 机制。如前所述，海藻盐是生物膜结构元素的一部分，可以作为抗生素的保护层，同时使细菌免受免疫系统白细胞的吞噬（Leid et al.，2005）。EPS 的产生受到 QS 调控；研究表明，QQ 能损害生物膜的形成。然而，众所周知，依赖于 *lasI* 的 QS 信号在成熟的生物膜中会减少（De Kievit et al.，2001）。此外，研究表明，在某些环境条件下，QS 通过增加共同的有利物质为整个细菌群落提供好处。研究人员模拟了一种环境，在这种环境中表达生长所需的蛋白酶（群体感应生长培养基，QSM）。同一研究小组发现，QS 系统的损伤会导致生长速度明显减慢，这表明在某些条件下，QQ 可以减少病原菌的生长（Diggle et al.，2007）。

3.3 群体感应信号

通过细菌间通信来抑制生物膜的形成，对于了解 QS 网络的工作原理及其使用的信号是至关重要的，细菌利用广泛不同的分子进行交流。同时，对于单个细菌来说，识别来自同一物种的信号并区分外来分子是至关重要的。

QS 分子或自诱导物（AIs）包括多肽、*N*-酰基高丝氨酸内酯（AHLs）和喹啉酮。这些分子不仅具有不同的性质，而且在代谢成本上也存在差异。例如，金黄色葡萄球菌的信号肽 AgrD 的产生需要 184 个 ATP 分子，而 C4-HSL 合成仅需要 8 个 ATP 分子（Keller and Surette，2006）。在革兰氏阳性细菌中，信号分子主要是小肽，主动分泌到环境中。它们通过膜结合受体识别，或被内化后在细胞质中被识别（Monnet and Gardan，2015）。

研究最多的信号分子是革兰氏阴性细菌使用的 AHLs。在这里，酰基链的长度决定了整个分子的特异性（Papenfort and Bassler，2016）。不同的细菌在 QS 系统中使用不同的信号分子。

自诱导物-2（AI-2）是信号分子的一部分，这些信号分子均来自 4,5-二羟基-2,3-戊二酮（DPD），由 LuxS 合成酶产生。据报道，这种蛋白质存在于 500 多种细菌物种中，到目前为止，AI-2 仍然是最常见的信号分子（Pereira et al.，2013）。AI-2 被认为是微生物种间的通信分子（Papenfort and Bassler，2016）。有趣的是，铜绿假单胞菌的基因组不编码 *luxS* 基因，尽管该细菌已被报道对 AI-2 有反应（Duan et al.，2003）。

大约 76% 已注释 LuxR 型受体蛋白被归类为 LuxR-单独类受体，表明这些是 AHL 依赖的反应调节因子，没有相应的 LuxI 型合成酶。也许 LuxR 单独类受体最好的例子是来自铜绿假单胞菌的 QscR。在纳摩尔浓度的 C8-HSL、C10-HSL、3-oxo-C10-HSL、C12-HSL、3-oxo-

C12-HSL 和 C14-HSL 存在下，该受体控制其靶基因（Lee et al.，2006）。在这种情况下，很容易得出结论，即非 LuxI 型合酶可以合成各种各样的自诱导物。更有可能的是，这些单独的类受体使细菌能阻止与微生物之间的竞争或合作通信。

3.4 群体淬灭酶

QS 分子的酶降解一直是多年来研究的热点。在此背景下，研究得最好的群体感应淬灭（QQ）酶包括：(i) 内酯酶类，它催化高丝氨酸内酯环的打开；(ii) 酰化酶类，水解同型丝氨酸内酯环与脂肪酸链之间的酰胺键；(iii) 氧化还原酶类，能将 3-oxo-AHLs 转化为 3-羟基 AHLs，这类酶对相应反应调节因子并没有表现出相同的亲和力（图 7-2）。

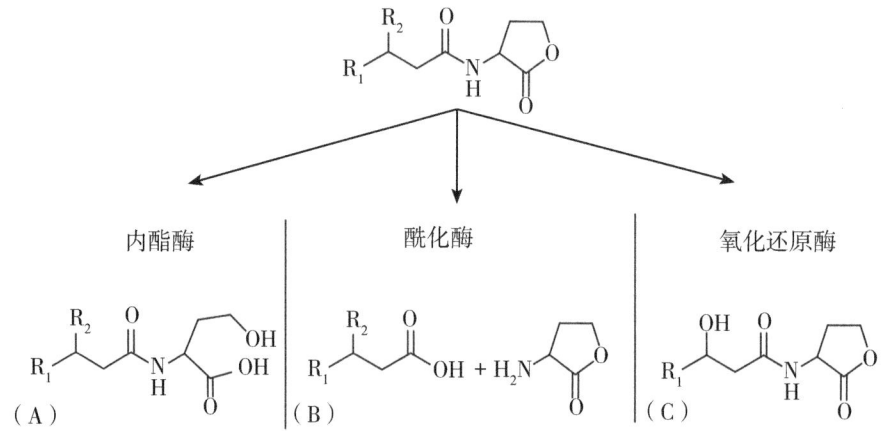

图 7-2 AHL QQ 酶

注：AHL QQ 酸包括三类酶，分别是（A）打开高丝氨酸内酯环的内酯酶；(B) 水解酰基链和高丝氨酸内酯环之间酰胺键的酰化酶；
(C) 通过氧化或还原酰基链而不是降解 AHLs 的氧化还原酶。

3.4.1 内酯酶

AHL 内酯酶对 AHL 的内酯环具有活性。AHL 内酯酶的一个突出例子是来自芽孢杆菌 240B1 的金属 β-内酰胺酶 AiiA。这些酶的一个特征性是 Zn^{2+} 结合 HXHXDH 基序，有趣的是，它们属于同一酶超家族，其介导对 β-内酰胺类抗生素的耐药性（Wang et al.，1999）。研究人员发现，在铜绿假单胞菌中表达蜡样芽孢杆菌（B. cereus）A24 的 aiiA 基因时，毒力因子的产生减少，运动能力也降低（Michel et al.，2002）。

磷酸三酯酶样内酯酶（PLLs）是一组具有广泛底物谱的内酯酶。它们最初被描述为对氧磷酶，但后来发现它们是天然 AHL-内酯酶，对有机磷农药具有混杂的活性（Afriat-Jurnou et al.，2012）。

一个非常有前途的 QQ 候选内酯酶是 SsoPox，来自嗜极端古菌硫黄矿硫化叶菌。它的极端稳定性、广泛温度范围上的活性及 pH 值水平均是值得关注。此外，SsoPox 对有机溶剂、表面活性剂和蛋白酶具有显著的耐受性（Hiblot et al.，2012）。这些特性使其在生物技术的应用中特别有趣（Bzdrenga et al.，2017；Rémy et al.，2016）。蛋白质结晶和蛋白质工程方面已开展研究，以提高催化效率（Bzdrenga et al.，2017）。研究证实了 SsoPox 蛋白工程变体的有效性和可行性。该蛋白显示了一个氨基酸的变化，特别是残基 W263，其定位于蛋白质的活性中心。色氨酸变为异亮氨酸使内酯酶对 3-oxo-C12-HSL 的催化效率更高。在铜绿假单胞菌 PAO1 培养中存在 QQ 内酯酶,研究人员报道,蛋白酶产量减少了 90%,铜绿假单胞菌铁载体的产量几乎完全消失。此外,生物膜的形成也减少了 90%。同样的研究也表明,当酶以戊二醛作为交联化合物固定在聚氨酯涂层中时,SsoPox 的淬灭能力表现出类似的结果 (Guendouze et al., 2017)。

在 QQ 的背景下,内酯酶具有非常有益的特性,因为内酯酶可识别并切割 AHL 的内酯环,而酰基链长度在识别分子过程中起次要的作用。这意味着内酯酶可以降解不同链长的广谱 AHL 分子 (Fetzner, 2015)。这些研究强调了 QQ 酶在抗感染性疾病中的可行性。在这种情况下,需要提到的是,打开的内酯环可以在酸性条件下再循环利用 (Fetzner, 2015; Grandclement et al., 2015)。相反,由酰化酶介导的 QQ 则提供了信号分子的不可逆降解。

例如,在抗细菌感染时,QQ 酶被证明可以降低人类病原菌鲍氏不动杆菌 (*Acinetobacter baumannii*) 的生物膜形成 (Chow et al., 2014)。在过去的几年里,由于鲍氏不动杆菌是一种医院获得的具有高发病率的菌种,已成为医学研究的重点 (Chow et al., 2014)。临床分离的不动杆菌生物膜形成依赖于 QS (Niu et al., 2008)。不幸的是,鲍氏不动杆菌基因的 QS 调控尚未得到充分的描述。据报道,63% 分离鲍氏不动杆菌菌株产生不只一种以上的 AHL (Peleg et al., 2008)。然而,鲍氏不动杆菌 ATCC 17978 的完整基因组序列表明,只有 LuxI 型合酶 AbaI 参与了 AHL 的合成。科学家们在鲍氏不动杆菌 M2 中发现了至少 5 种 AHLs 信号分子,假定的 3-羟基-C12-HSL 附近,而在临床上分离的主要是鲍氏不动杆菌 S1,其主要 QS 分子是 3-羟基-C10-HSL (Chow et al., 2014)。与能产生 AHLs 同一菌株相比,鲍氏不动杆菌 M2 中一个 Δ*abaI* 缺失突变体能显著降低生物膜的形成 (30%~40%)。当外源添加信号分子到鲍氏不动杆菌 M2 培养中,可以恢复生物膜 (Niu et al., 2008)。此外,研究表明,通过工程 QQ 内酯酶的酶降解 QS 信号,可显著减少生物膜的形成 (Chow et al., 2014)。事实是,生物膜比浮游细菌更难治疗,这些结果为鲍氏不动杆菌的成功治疗策略提供了更大的机会。

研究人员对来自卡氏地芽孢杆菌 (*Geobacillus kaustophilus*, GKL) 的耐热 QQ 内酯酶进行了定向进化,并探索了其对鲍氏不动杆菌 S1 生物膜形成能力的淬灭作用 (Chow et al., 2014)。该酶属于酰胺水解酶超家族中的磷酸三酯酶类内酯酶 (PLL) 家族。由于不同的临床分离株鲍氏不动杆菌具有不同长度的 AHLs,因此,强调具有混杂活性的更广谱的 AHL 在对抗生物膜方面具有明显的优势 (Chow et al., 2014)。

3.4.2 群体淬灭酶

细菌的灵活性和适应性极强,因此,一些细菌能使用 AHLs 作为能源和碳源不足为奇。争论贪噬菌 (*Variovorax paradoxus*) 将 AHLs 分解成高丝氨酸内酯环和相应的脂肪酸,并将 AHLs 作为氮源 (Leadbetter and Greenberg, 2000)。条件性致病菌铜绿假单胞菌拥有几种能降解 AHLs 的酰化酶,其中最突出的酶是 PvdQ (Sio et al., 2006)。大多数已鉴定的 AHL 酰化酶属于 Ntn-水解酶超家族 (Utari et al., 2017)。这个超家族非常多样化,其活性中心以一个有趣且非常独特的 αββα 折叠结构脱颖而出。

Ntn-水解酶一直是研究的焦点,其中一些成员被应用于工业。例如,青霉素酰化酶可以切断青霉素分子的酰基链,生成 6-氨基青霉酸 (6-APA) 和相应的有机酸。6-APA 是许多半合成 β-内酰胺类抗生素的重要中间体 (Hewitt et al., 2000)。Ntn-水解酶的反应依赖于一个 *N*-末端亲核基团,这是这个超家族的命名来源。在活性中心内,亲核基团可以是 *N*-末端的丝氨酸、苏氨酸或半胱氨酸，其对 AHL 的酰胺键发起亲核攻击（Utari et al., 2017）。这个家族中一个值得注意的成员是来自克鲁维菌属（*Kluyvera citrophila*）青霉素 G 酰化酶（PGA），最近发现它能降解链长为 6~8 个碳的 AHLs（Fetzner et al., 2015）。

在铜绿假单胞菌基因组中，酰化酶 PvdQ 被鉴定出来，仅此于另外两个酶即 HacB 和 QuiP（Wahjudi et al., 2011）。这些酰化酶的生理功能尚未完全阐明；然而，PvdQ 在铜绿假单胞菌铁载体产生过程中起着关键作用（Drake and Gulick, 2011）。PvdQ 可切割超过 10 个碳的侧链 AHL（Koch et al., 2014）。在秀丽隐杆线虫侵染铜绿假单胞菌的模型中，结果表明，PvdQ 足以降解 3-oxo-C12-HSL 信号，从而显著降低了 QS 依赖的致病因子。在这个试验中，提高了秀丽隐杆线虫的存活率（Papaioannou et al., 2009）。此外，PvdQ 底物特异性从切割长链 AHLs 转变为接受 C8-HSLs 作为底物。在洋葱伯霍尔德菌中两个单点突变（Lα146W；Fβ24Y）涉及 AHL 的活性。在随后的试验中发现，在洋葱伯霍尔德菌侵染的情况下这种工程酶可以保护大蜡螟幼虫（Koch et al., 2014）。

3.4.3 氧化还原酶类

QQ 氧化还原酶类通过氧化或还原第三个碳的羰基团或氧化 ω-1、ω-2 和 ω-3 碳来靶向 AHL 分子的侧链。这类酶不会破坏信号而是对其进行修饰，从而有效地调控同源受体对信号分子的识别（Chen et al., 2013）。例如，红斑红球菌菌株 W2 除具有内酯酶和酰化酶外，还具有氧化还原酶，能够将长链为 8~12 个碳的 3-氧代-HSL 还原为其 3-羟基-HSL 的形式（Phane Uroz et al., 2005）。BpiB09 是一种氧化还原酶，催化 3-氧代-HSL 还原为 3-羟基-HSL，当在铜绿假单胞菌中表达时，其对 QS 系统产生影响，导致铜绿假单胞菌素的产生减少，以及运动性和生物膜形成的变化。此外，在秀丽隐杆线虫侵染铜绿假单胞菌 PAO1 的模型中观察到其毒力减弱（Bijtenhoorn et al., 2011）。

3.4.4 抗 AI-2 的群体淬灭酶

LuxS 不仅是 AI-2 信号分子的合成酶，也是活化甲基循环（AMC）的一部分，因此，导致了认为 AI-2 分子可能是 AMC 的副产物。值得注意的是，AHL 类的自诱导物和 CAI1 是以 *S*-4 腺苷蛋氨酸（SAM）分子作为底物合成的。一个有趣的发现是，不同的细菌可以感知 AI-2 而不自己合成。例如，铜绿假单胞菌菌株通过 AI-2 上调 *rhlA*（鼠李糖脂生物合成）、*lasB*（弹性蛋白酶）、*exoT*（外毒素）、*phzA1* 和 *phzA2*（非乃嗪合成）以及 *fliC*（鞭毛成分）基因（Pereira et al., 2013）。在大肠杆菌细胞存在的情况下，AI-2 信号通过膜转运体被细菌细胞主动吸收，并被受体激酶 LsrK 磷酸化为磷酸化-AI-2。磷酸化的自诱导因子可以与转录抑制因子 LsrR 结合，导致靶基因的激活（Pereira et al., 2013）。有趣的是，激酶 LsrK 一旦外源添加到哈维弧菌、鼠伤寒沙门菌和大肠杆菌的培养物中，降低了其对群体感应的反应。LsrK 的 QQ 活性背后的机制被认为是由于细胞外 AI-2 的磷酸化，禁止其运输进入细胞。此外，AI-2 的修饰使分子变得不稳定。

3.5 淬灭喹诺酮依赖的 QS 信号

在铜绿假单胞菌中，PQS 信号被证明在毒力因子产生和生物膜形成中起重要作用（Pustelny et al., 2009）。除铜绿假单胞菌外，只有伯克霍尔德菌属能利用 2-烷基-4(1H)-喹啉酮（AQ）信号进行群体感应（Fetzner et al., 2015）。酶降解这种信号的可能性表明 QQ 酶以非常选择性的方式被使用，特别针对致病菌。到目前为止，只有 QQ 酶能降解 AQ，是双加氧酶 Hod（1H-3-羟基-4-氧代喹啉 2,4-二氧化酶）。QQ 酶催化杂环 PQS 的断裂生成 N-十八烷基苯甲酸和一氧化碳。Hod 的外源添加到铜绿假单胞菌培养物中导致各种毒力因子的表达减少（Pustelny et al., 2009）。该酶是从土壤细菌节杆菌属 Rue61a 中分离得到。在土壤环境中，AQs 是由植物等高等生物作为次生代谢产物产生的。Fetzner 和同事（2015）假设共享这种生态位的细菌可能进化出 AQ 降解酶。一个尚未鉴定的土壤分离物具有 PQS 降解活性的这一发现支持了这个假设。

3.6 群体淬灭酶的应用

在 QS 和 QQ 特征方面，最为典型的生物之一是革兰氏阴性细菌铜绿假单胞菌。2017 年，该细菌被列为世界卫生组织研究和开发新型抗菌药物的优先病原体名单中的第二位。在美国的医疗环境中收集的数据显示，所有一般医疗保健相关感染的 7.5% 与铜绿假单胞菌有关（Sievert et al., 2013）。铜绿假单胞菌具有两个分层有序的 QS 通路，通过 N-酰基高丝氨酸内酯（AHLs）信号分子进行控制。Las 系统依赖于 3-oxo-C12-HSL，而第二个系统，即 Rlh 调控子，利用 N-丁酰基高丝氨酸内酯（C4 HSL）。这两种信号都可以被 QQ 酶降解，这对毒力因子的产生，包括生物膜形成有重要影响（Papaioannou et al., 2009）。此外，QQ 酰化酶可以区分不同 AHLs 链长，这为更具体地靶向 QS 通路提供了可能性。

3.7 表面的功能化

在工业应用中，表面上的生物膜经常出现问题，如膜生物反应器膜。包括 QQ 酶在内的策略已经被测试用于抵消这方面生物膜的形成（Kim et al., 2011）。此外，在医疗设备上的生物膜形成是多种严重疾病的主要原因。在临床上，为对抗表面上的细菌，人们付出了巨大的努力。任何外来物质进入人体都会对健康构成巨大威胁，并使身体变得易感染。这种情况下，医疗植入物是一个重要例子。骨科植入物成为许多细菌生长的表面，这在抗生素抗性日益增加的背景下成为一个日益严重的问题。这些感染往往很难治疗（Gbejuade et al., 2015），强调了对非抗生素依赖性方法的迫切需求，以清除或破坏表面上的细菌生物膜。通常用于骨科植入物的常见材料有不锈钢、钴铬、聚甲基丙烯酸甲酯（PMMA）和一系列聚合物生物材料（Gbejuade et al., 2015）。这些材料的所有表面都容易被生物膜定殖。感染可能在手术后立即发生，分为早期感染（术后<3 个月）或延迟感染（术后 3~24 个月），以及非常晚期感染（>24 个月）。这些感染的处理对医生而言是一个重大挑战。在这种情况下，标准程序包括长时间的静脉抗菌治疗。如果这些措施没有效果，可能导致严重的后果，即肢体截肢或患者死亡（Osmon et al., 2013）。因此，迫切需要制订预防性对抗感染的策略以防止生物膜形成。最近，几种方法被探索用于酶促涂覆各种表面，以干扰细菌的生物膜形成。Swartjes 等（2013）在 PDMMA 表面上固定了 DNaseI，从而创建了具有抗生物膜特性的涂层。细胞外 DNA 是细胞外聚合物的一个组成部分，这些分子被认为是最长的 EPS 分子，并支持生物膜中稳定的纤维状网络结构的形成（Flemming et al., 2016）。在金黄色葡萄球菌中，eDNA 的释放对于生物膜形成的初始附着至关重要（Swartjes et al., 2013）。PMMA 被多巴胺耦合层涂覆后，浸入 DNaseI 溶液中。使用质粒 DNA 对该方法制备的固定化酶的功能进行检测。科学家们报告称，在 30 min 后，质粒 DNA 被完全消化。这证明了在涂层过程后的 8 h 内没有活性损失。当研究人员观察到细菌最初附着在涂层表面时，他们发现，在孵育 60 min 后，与未涂层的 PMMA 相比，对铜绿假单胞菌的附着减少了 99%，对金黄色葡萄球菌的附着减少了95%。结果表明，酶在生物技术应用中的可行性，以及具有生物膜抑制特性的功能化表面潜力。

可以看到，为赋予其抗生物膜特性，酶的生物膜抑制作用和表面功能化在QQ酶的应用中得到了体现。最近，使用QQ酰化酶对表面进行了涂层处理，显著减少了铜绿假单胞菌生物膜的形成。该酰化酶针对AHL信号分子，并被应用于尿道导管的硅胶表面（Ivanova et al., 2015b）。导尿管通常由硅胶制成，因其疏水性，可以抵消细菌的附着（Ivanova et al., 2015b）。然而，临床上使用导尿管会对患者带来风险，导致尿路感染，进而导致住院时间延长（Jacobsen et al., 2008）。在这项研究中，商业上可获得来自麦曲霉蜂蜜曲霉（A. melleus）的AHL酰化酶，该酶被证明可以降解AHL信号分子。本研究中的硅胶经过十二烷基硫酸钠（SDS）清洗，并用氨丙基三乙氧基硅烷（APTES）溶液进行预处理，以引入氨基团。随后，利用线性聚乙烯亚胺（PEI）作为一种聚合阳离子，对表面进行酰化酶涂层处理。在特定pH值条件，使酶具有多阴离子性质。涂层过程以一层（LbL）方式重复进行，创建多层表面涂层。通过人类成纤维细胞的细胞毒性试验证实，涂层材料不具有毒性效果。值得注意的是，PEI被认为是有毒的，但研究表明，涂层中的化合物不会引起任何不良反应。研究表明，这可能是由于与酰化酶的相互作用，中和了正电荷。此外，通过将涂层硅胶暴露于人工尿液中进行了稳定性测试。经过7 d孵育期后，仍可观察到表面的抗生物膜活性。所有结果都证实了涂层的稳健性，并突显了QQ酶的潜力。此外，在涂层过程中，酶活性的损失仅为20%。固定化酶对几种AHL模型化合物具有活性，包括C4-HSL、C6-HSL、3-oxo-C10-HSL和3-oxo-C12-HSL。与未经处理和涂层硅胶表面相比，24 h后铜绿假单胞菌生物膜的生物量降低了75%。同一小组通过首次整合两种酶抑制方法进一步推动了研究（Ivanova et al., 2015a）。在之前的研究中，利用蜂蜜曲霉中的一种酰化酶作为QQ酶，此外，该涂层方法与解淀粉芽孢杆菌（B. amyloliquefaciens）的α-淀粉酶相结合，这两种酶都是商业可得的。选择淀粉酶作为生物膜抑制酶是基于以下事实：如前所描述，EPS组分包括多糖等物质，而该酶可催化上述多糖中的1,4-α-糖苷键的水解。因此，α-淀粉酶通过降解EPS组分来靶向生物膜的结构成分（Craigen et al., 2011）。总共依次应用10层酰化酶和α-淀粉酶，这导致了两种不同的结构，一种以酰化酶作为末端层，另一种以淀粉酶作为最外层。固定化的α-淀粉酶对释放还原糖量的测定显示，对纯化铜绿假单胞菌EPS组分具有活性。在静态条件下，与未处理的硅胶相比，单独固化的酰化酶和α-淀粉酶使每个处理中铜绿假单胞菌生物膜形成减少30%。与末端酰化酶的混合涂层相比，生物膜形成减少至60%，而末端淀粉酶的涂层则减少了38%。这强调了混合涂层的协同效应，同时也表明其组装结构对生物膜抑制效果的重要作用。使用酰化酶进行QQ已被证明是一种有效的方法；然而，AHLs仅代表一种信使分子，主要在革兰氏阴性QS系统中发挥作用。尽管EPS组分确实有所变化，但1,4-α-糖苷键的多糖仍可被α-淀粉酶水解（Ivanova et al., 2015a）。在动态铜绿假单胞菌生物膜抑制测试中，使用涂层导管在人类膀胱的物理模型中进行试验，其结果可与静态环境下的结果相当。据报道，酰端混合LbL涂层的生物膜抑制率为70%。值得注意的是，以淀粉酶为最外层的涂层仅降低了10%。在随后的动物模型中，研究监测了插管兔的自发性细菌感染，使用了末端酰化酶的混合涂层。研究人员报告了7 d的导管使用情况，导管切片显示，生物膜的形成存在差异。在球囊部分，与对照组硅胶导管相比，生物膜减少70%，而在轴中可减少30%（Ivanova et al., 2015b）。

3.8 膜生物反应器

如今，QQ 酶在废水处理中是一个重要的应用领域，特别是在膜生物反应器（MBR）的维护中。这些设备结合了活性污泥和微观或纳米过滤的技术。MBR 是一项有前途的技术，但各参数仍需要改进以最大化其潜力。在使用过程中，过滤膜会暴露于生物污垢（Lade et al.，2014）。为了确保长期性能并减少因生物污垢而产生的成本，QQ 酶提供了一种有效的解决方案。细菌生物膜的形成在 MBR 膜的生物污染中起着关键作用。为了解决这一问题，各种抗菌化合物和抗生素被测试。用猪肾酰化酶 I 涂覆在 MBR 纳滤（NF）膜上，随后将细胞膜放入流动细胞中孵育 5 d。结果表明，与未处理的 NF 膜相比，铜绿假单胞菌生物膜的相对体积仅为 24%（Kim et al.，2011）。这展示了 QQ 酶在医学领域之外的技术应用潜力。

3.9 吸入式乳酸酶类

Hraiech 等（2014）研究了乳酸酶 SsoPox-I 对由铜绿假单胞菌引起的大鼠肺炎的影响。正如所提到的，许多体外研究证实了酶促 QQ 的潜力；然而，体内试验仍然非常有限。研究表明，乳酸酶能有效降低铜绿假单胞菌的毒力因子和生物膜的产生（Hiblot et al.，2012）。Migiyama 等（2013）的前期试验表明，在急性肺炎小鼠模型中，表达乳酸酶 AiiM 的铜绿假单胞菌 PAO1 肺损伤较轻，并显著地提高急性肺炎小鼠模型的存活率。此外，结果表明，在铜绿假单胞菌中共表达 AiiM 可减少 AHL 介导的毒力因子的产生，并减轻对人肺上皮细胞的细胞毒性。

Hraiech 等（2014）研究在大鼠铜绿假单胞菌 PAO1 肺感染模型中外源服用 QQ 内酯酶 SsoPox-I 的可能性，探讨该酶是否能够减少感染引起的死亡率。本章所述，嗜极端古菌硫黄矿硫化叶菌的生物工程内酯酶 SsoPox 对 3-oxo-C12-HSL 表现出增强的亲和力，是 LasI/LasR QS 通路中的信号分子，在毒力因子表达中起重要作用（Guendouze et al.，2017；Hiblot et al.，2012）。大鼠感染 2.5×10^8 个铜绿假单胞菌 PAO1 菌落形成单位（CFU），总体积为 250 μL。然而，在这个模型中，生物膜本身的影响并未得到进一步阐明。动物被侵染 2 d 后，通过气管插入导管进行 SsoPox-I 给药以及细菌感染。该酶的给药遵循与感染方案相同，通过导管直接在感染后给予 250 μL 的 1mg/mL SsoPox-I 的 PBS 溶液，同时在另一组中延迟 3 h 给予。在这项研究中，值得注意的是，将酶给予动物的肺环境是耐受的。在测试动物中，没有一例对酶表现出不良耐受性，肺部的肉眼检查也未显示损伤。组织学评估也证实了组织未受损。这一发现标志着 QQ 酶作为医学治疗应用研究的重要进展。在前期的体外试验中，已显示向铜绿假单胞菌 PAO1 培养基中添加 SsoPox-I 可降低蛋白酶活性和生物膜形成。感染模型显示，感染后立即接受治疗的大鼠在监测 50 h 内死亡率降低。一个非常重要的观察结果，在经过治疗、未经治疗和延迟治疗的测试动物组中，细菌计数处于相当的水平，但毒力有所不同。

这些例子表明，干扰细菌间通信对细菌的致病性具有重要影响。

4 讨论和展望

本章重点讨论酶在 QQ 中的作用及其潜在应用。各种 QQ 酶的例子展示了其在这个领域的巨大潜力。正如之前提到的，QQ 不仅通过酶降解信号分子来实现，还考虑了抑制受体或合成酶的小分子。

　　细菌生活在混合群落中，暴露于各种不同细菌种类的信号分子中（Bokhove et al., 2010）。越来越多的证据表明，细菌间的串扰不仅限于同一属，甚至能跨越不同界之间的障碍（Tang and Zhang, 2014）。近年来，大量的研究阐明植物根际微生物群落依赖于 QS 界间通信的共生作用（Sanchez-Contreras et al., 2007）。在过去的 15 年里，关于 QQ 酶及其应用潜力的研究已取得显著进展（Fetzner, 2015）。在细胞外环境中，AI 的酶降解提供了许多优势，使这一领域的研究在不同的用途上非常广泛，尤其在医学领域展现出良好的前景（Gill et al., 2015）。其中一些酶可以区分不同的 AHLs 链长，从而选择性地依赖于不同信号分子的 QS 通路（Bokhove et al., 2010）。此外，这些酶还可以固定在表面上，从而在局部定义的区域内使 QQ 效应成为可能（Ivanova et al., 2015a）。人们认为，AIs 的外部降解不太可能导致耐药性的形成，因为它不会对细菌产生直接的选择性压力。一般来说，QQ 不会杀死细胞，因为它以一种非致命的方式干扰基因表达，因此，不会使细胞暴露在强烈的选择性压力下（Bzdrenga et al., 2017）。对 QQ 酶耐药性形成的风险评估是非常困难，因为细菌间通信的干扰对生物体本身并没有造成直接的进化压力（Maeda et al., 2012）。Diggle 等（2007）表明，当选择细菌培养依赖于 QS 调控的蛋白酶时，与没有 QQ 酶的相同菌株相比，这会导致严重的生长障碍。在该模型中，QQ 对种群的适应性产生了直接影响（Diggle et al., 2007）。另外，这也被视为一个不利因素，因为可能的治疗策略并不会导致清除细菌感染，而是依赖于宿主免疫系统或急性病例的综合治疗。有关 β-内酰胺酶抑制剂与抗生素联合使用的典型研究表明，这些抑制剂已在临床上得到应用（Gill et al., 2015）。将酶作为治疗选择意味着这些化合物必须被注入患者体内，从而使患者暴露于对酶的稳定性和功能性不利的各种因素。例如，铜绿假单胞菌表达蛋白酶 LasA，可能会影响酶的效果。因此，问题是：将酶作为一种医疗治疗是否可行？事实上，在临床上囊性纤维化（CF）管理中，DNase I 已被用作一种喷雾剂（Bakker and Tiddens, 2007）。此外，之前提出的 QQ 酶作为涂层的应用，凸显了固定化酶的稳定性及其在临床应用中选择性（Ivanova et al., 2015b）。

术语表

　　自诱导物（AI）　在文献中，通常被描述为细菌的信号分子，尤其是在 QS 背景下，因为早期发现这些分子能诱导其自身的产生。

　　囊性纤维化　一种遗传性疾病，导致肺部产生大量黏液，有利于致病菌的定殖，从而引发慢性肺部感染。其他症状可能包括鼻窦感染、生长不良、脂肪粪便以及手指和脚趾的杵状畸形。

　　分层涂层　一种在表面上交替沉积相反电荷材料层的技术。

　　膜生物反应器（MBR）　一种用于废水处理的系统，通过过滤步骤的顺序来去除颗粒物。其主要优点是高容积负荷，仅受膜污染的限制。

　　群体感应淬灭（QQ）　细菌群体感应系统的干扰。

缩写词

　　AHL　　　　　　　　　　N-酰基高丝氨酸内酯

AI	自诱导物
AI-2	自诱导物-2
AQ	2-烷基-4(1H)-喹诺酮
CAI-1	霍乱病自诱导物-1
CF	囊性纤维化
CFU	群落形成单位
C4-HSL	N-丁酰高丝氨酸内酯
C6-HSL	N-己酰高丝氨酸内酯
C8-HSL	N-辛酰-1-同型丝氨酸内酯
3-oxo-C6-HSL	N-(3-氧-己酰)-L-高丝氨酸内酯
3-oxo-C8-HSL	N-(3-氧-辛酰)-L-高丝氨酸内酯
3-oxo-C12-HSL	N-(3-氧-十二酰)-L-高丝氨酸内酯
DSF	扩散信号因子
LbL	逐层
MIC	最小抑菌浓度
PQS	假单胞菌喹诺酮信号
QS	群体感应
QQ	群体淬灭

致谢

感谢 Ykelien L. Boersma 博士对手稿的批判性阅读,同时感谢 Robbert H. Cool 博士对手稿的有益探讨。

参考文献

AFRIAT-JURNOU L, JACKSON C J, TAWFIK D S, 2012. Reconstructing a missing link in the evolution of a recently diverged phosphotriesterase by active-site loop remodeling. Biochemistry, 51: 6047-6055.

BAKKER E, TIDDENS H, 2007. Pharmacology, clinical efficacy and safety of recombinant human DNase in cystic fibrosis. Expert Rev. Respir. Med., 1 (3): 317-329.

BAYER A S, SPEERT D P, PARK S, et al., 1991. Functional role of mucoid exopolysaccharide (alginate) in antibiotic-induced and polymorphonuclear leukocyte-mediated killing of *Pseudomonas aeruginosa*. Infect. Immun., 59: 302-308.

BIJTENHOORN P, MAYERHOFER H, MÜLLER-DIECKMANN J, et al., 2011. A novel metagenomic short-chain dehydrogenase/reductase attenuates *Pseudomonas aeruginosa* biofilm formation and virulence on *Caenorhabditis elegans*. PLoS One, 6: e26278.

BILLINGS N, RAMIREZ MILLAN M, CALDARA M, et al., 2013. The extracellular matrix component Psl provides fast-acting antibiotic defense in *Pseudomonas aeruginosa* biofilms. PLoS Pathog., 9e1003526.

BOKHOVE M, NADAL JIMENEZ P, QUAX W J, et al., 2010. The quorum-quenching N-acyl homoserine lactone acylase PvdQ is an Ntn-hydrolase with an unusual substrate-binding pocket. Proc. Natl. Acad. Sci. U. S. A., 107: 686-691.

BZDRENGA J, DAUDE D, REMY B, et al., 2017. Biotechnological applications of quorum quenching enzymes. Chem. Biol. Interact., 267: 104-115.

CHANG C Y, 2018. Surface sensing for biofilm formation in *Pseudomonas aeruginosa*. Front. Microbiol., 8: 2671.

CHEN F, GAO Y, CHEN X, et al., 2013. Quorum quenching enzymes and their application in degrading signal molecules to block quorum sensing-dependent infection. Int. J. Mol. Sci., 14: 17477-17500.

CHOW J Y, YANG Y, TAY S B, et al., 2014. Disruption of biofilm formation by the human pathogen *Acinetobacter baumannii* using engineered quorum-quenching lactonases. Antimicrob. Agents Chemother., 58: 1802-1805.

CHUA S L, LIU Y, YAM J K H, et al., 2014. Dispersed cells represent a distinct stage in the transition from bacterial biofilm to planktonic lifestyles. Nat. Commun., 5: 4462.

CIOFU O, TOLKER-NIELSEN T, 2010. Antibiotic tolerance and resistance in biofilms. In: Biofilm Infections. Springer New York, New York, NY, pp: 215-229.

CRAIGEN B, DASHIFF A, KADOURI D E, 2011. The use of commercially available alpha-amylase compounds to inhibit and remove *Staphylococcus aureus* biofilms. Open Microbiol. J., 5: 21-31.

DE KIEVIT T R, IGLEWSKI B H, 2000. Bacterial quorum sensing in pathogenic relationships. Infect. Immun., 68: 4839-4849.

DE KIEVIT T R, GILLIS R, MARX S, et al., 2001. Quorum-sensing genes in *Pseudomonas aeruginosa* biofilms: their role and expression patterns. Appl. Environ. Microbiol., 67: 1865-1873.

DIGGLE S P, GRIFFIN A S, CAMPBELL G S, et al., 2007. Cooperation and conflict in quorum-sensing bacterial populations. Nature, 450: 411-414.

DRAKE E J, GULICK A M, 2011. Structural characterization and high-throughput screening of inhibitors of PvdQ, an NTN hydrolase involved in pyoverdine synthesis. ACS Chem. Biol., 6: 1277-1286.

DUAN K, DAMMEL C, STEIN J, et al., 2003. Modulation of *Pseudomonas aeruginosa* gene expression by host microflora through interspecies communication. Mol. Microbiol., 50: 1477-1491.

ELIAS S, BANIN E, 2012. Multi-species biofilms: living with friendly neighbors. FEMS Microbiol. Rev., 36: 990-1004.

FETZNER S, 2015. Quorum quenching enzymes. J. Biotechnol., 201: 2-14.

FLEMMING H C, WINGENDER J, SZEWZYK U, et al., 2016. Biofilms: an emergent form of bacterial life. Nat. Rev. Microbiol., 14: 563-575.

FUX C A, COSTERTON J W, STEWART P S, et al., 2005. Survival strategies of infectious biofilms. Trends Microbiol., 13: 34-40.

GBEJUADE H O, LOVERING A M, WEBB J C, 2015. The role of microbial biofilms in prosthetic joint infections. Acta Orthop., 86: 147-158.

GILL E E, FRANCO O L, HANCOCK R E W, 2015. Antibiotic adjuvants: diverse strategies for controlling drugresistant pathogens. Chem. Biol. Drug Des., 85: 56-78.

GRANDCLEMENT C, TANNIERES M, MORERA S, et al., 2015. Quorum quenching: role in nature and applied developments. FEMS Microbiol. Rev., 40: 86-116.

GUENDOUZE A, PLENER L, BZDRENGA J, et al., 2017. Effect of quorum quenching lactonase in clinical isolates of *Pseudomonas aeruginosa* and comparison with quorum sensing inhibitors. Front. Microbiol., 8: 227.

HEWITT L, KASCHE V, LUMMER K, et al., 2000. Structure of a slow processing precursor penicillin acylase from *Escherichia coli* reveals the linker peptide blocking the active-site cleft. J. Mol. Biol., 302: 887-898.

HEYDARI S, EFTEKHAR F, 2015. Biofilm formation and β-lactamase production in burn isolates of *Pseudomonas aeruginosa*. Jundishapur J. Microbiol., 8: e15514.

HIBLOT J, GOTTHARD G, CHABRIERE E, et al., 2012. Characterisation of the organophosphate hydrolase catalytic activity of SsoPox. Sci. Rep., 2: 779.

HRAIECH S, HIBLOT J, LAFLEUR J, et al., 2014. Inhaled lactonase reduces *Pseudomonas aeruginosa* quorum sensing and mortality in rat pneumonia. PLoS ONE, 9 (10): e107125.

INFECTIOUS DISEASES SOCIETY OF AMERICA (IDSA), SPELLBERG B, BLASER M, et al., 2011. Combating antimicrobial resistance: policy recommendations to save lives. Clin. Infect. Dis., 52 (Suppl. 5): S397-S428.

IVANOVA K, FERNANDES M M, FRANCESKO A, et al., 2015a. Quorum quenching and matrix-degrading enzymes in multilayer coatings synergistically prevent bacterial biofilm formation on urinary catheters. ACS Appl. Mater. Interfaces, 7: 27066-27077.

IVANOVA K, FERNANDES M M, MENDOZA E, et al., 2015b. Enzyme multilayer coatings inhibit *Pseudomonas aeruginosa* biofilm formation on urinary catheters. Appl. Microbiol. Biotechnol., 99: 4373-4385.

JACOBSEN S M, STICKLER D J, MOBLEY H L T, et al., 2008. Complicated catheter-associated urinary tract infections due to *Escherichia coli* and Proteus mirabilis. Clin. Microbiol. Rev., 21: 26-59.

KELLER L, SURETTE M G, 2006. Communication in bacteria: an ecological and evolutionary perspective. Nat. Rev. Microbiol., 4: 249-258.

KIM J H, CHOI D C, YEON K M, et al., 2011. Enzyme-immobilized nanofiltration membrane to mitigate biofouling based on quorum quenching. Environ. Sci. Technol., 45: 1601-1607.

KOCH G, NADAL-JIMENEZ P, REIS C R, et al., 2014. Reducing virulence of the human pathogen *Burkholderia* by altering the substrate specificity of the quorum-quenching acylase PvdQ. Proc. Natl. Acad. Sci. U. S. A., 111: 1568-1573.

LADE H, PAUL D, KWEON J H, 2014. Quorum quenching mediated approaches for control of membrane biofouling. Int. J. Biol. Sci., 10: 550-565.

LEADBETTER J R, GREENBERG E P, 2000. Metabolism of acyl-homoserine lactone quorum-sensing signals by *Variovorax paradoxus*. J. Bacteriol., 182: 6921-6926.

LEE J H, LEQUETTE Y, GREENBERG E P, 2006. Activity of purified QscR, a *Pseudomonas aeruginosa* orphan quorum-sensing transcription factor. Mol. Microbiol., 59: 602-609.

LEID J G, WILLSON C J, SHIRTLIFF M E, et al., 2005. The exopolysaccharide alginate protects *Pseudomonas aeruginosa* biofilm bacteria from IFN-γ-mediated macrophage killing. J. Immunol., 175 (11): 7512-7518.

MAEDA T, GARCÍA-CONTRERAS R, PU M, et al., 2012. Quorum quenching quandary: resistance to antivirulence compounds. ISME J., 6: 493-501.

MCNAB R, FORD S K, EL-SABAENY A, et al., 2003. LuxS-based signaling in Streptococcus gordonii: autoinducer 2 controls carbohydrate metabolism and biofilm formation with *Porphyromonas gingivalis*. J. Bacteriol., 185: 274-284.

MICHEL L, HARMS H, HEURLIER K, et al., 2002. Genetically programmed autoinducer destruction reduces virulence gene expression and swarming motility in *Pseudomonas aeruginosa* PAO1. Microbiology, 148: 923-932.

MIGIYAMA Y, KANEKO Y, YANAGIHARA K, et al., 2013. Efficacy of AiiM, an N-acylhomoserine lactonase, against *Pseudomonas aeruginosa* in a mouse model of acute pneumonia. Antimicrob. Agents Chemother., 57: 3653-3658.

MONNET V, GARDAN R, 2015. Quorum-sensing regulators in Gram-positive bacteria: "cherchez le peptide" Mol. Microbiol., 97: 181-184.

NADELL C D, DRESCHER K, FOSTER K R, 2016. Spatial structure, cooperation and competition in biofilms. Nat. Rev. Microbiol., 14: 589-600.

NIU C, CLEMMER K M, BONOMO R A, et al., 2008. Isolation and characterization of an autoinducer synthase from Acinetobacter baumannii. J. Bacteriol., 190: 3386-3392.

O'BRIEN S, FOTHERGILL J L, 2017. The role of multispecies social interactions in shaping *Pseudomonas aeruginosa* pathogenicity in the cystic fibrosis lung. FEMS Microbiol. Lett., 364 (15): fnx128.

OSMON D R, BERBARI E F, BERENDT A R, et al., 2013. Diagnosis and Management of Prosthetic Joint Infection: Clinical Practice Guidelines by the Infectious Diseases Society of America. Clin. Infect. Dis., 56: e1-e25.

PAPAIOANNOU E, WAHJUDI M, NADAL-JIMENEZ P, et al., 2009. Quorum-quenching acylase reduces the virulence of *Pseudomonas aeruginosa* in a *Caenorhabditis elegans* infection model. Antimicrob. Agents Chemother., 53: 4891-4897.

PAPENFORT K, BASSLER B L, 2016. Quorum sensing signal-response systems in Gram-negative bacteria. Nat. Rev. Microbiol., 14 (9): 576-588.

PELEG A Y, SEIFERT H, PATERSON D L, 2008. *Acinetobacter baumannii*: emergence of a successful pathogen. Clin. Microbiol. Rev., 21: 538-582.

PEREIRA C S, THOMPSON J A, XAVIER K B, 2013. AI-2-mediated signalling in bacteria. FEMS Microbiol. Rev., 37: 156-181.

PHANE UROZ S, CHHABRA S R, CÁMARA M, et al., 2005. N-Acylhomoserine lactone quorum-sensing molecules are modified and degraded by *Rhodococcus erythropolis* W2 by both amidolytic and novel oxidoreductase activities. Microbiology, 151: 3313-3322.

PUSTELNY C, ALBERS A, BÜLDT-KARENTZOPOULOS K, et al., 2009. Dioxygenase-mediated quenching of quinolone-dependent quorum sensing in *Pseudomonas aeruginosa*. Chem. Biol., 16: 1259-1267.

RÉMY B, PLENER L, POIRIER L, et al., 2016. Harnessing hyperthermostable lactonase from *Sulfolobus solfataricus* for biotechnological applications. Sci. Rep., 6: 37780.

RICKARD A H, PALMER R J, BLEHERT D S, et al., 2006. Autoinducer 2: a concentration-dependent signal for mutualistic bacterial biofilm growth. Mol. Microbiol., 60: 1446-1456.

ROSENBLATT-FARRELL N, 2009. The landscape of antibiotic resistance. Environ. Health Perspect., 117: A244-A250.

SANCHEZ-CONTRERAS M, BAUER W D, GAO M, et al., 2007. Quorum-sensing regulation in rhizobia and its role in symbiotic interactions with legumes. Philos. Trans. R. Soc. Lond. Ser. B Biol. Sci., 362: 1149-1163.

SIEVERT D M, RICKS P, EDWARDS J R, et al., 2013. Antimicrobial resistant pathogens associated with healthcare-associated infections summary of data reported to the National Healthcare Safety Network at the Centers for Disease Control and Prevention: 2009-2010. Infect. Control Hosp. Epidemiol., 34: 1-14.

SIO C F, OTTEN L G, COOL R H, et al., 2006. Quorum quenching by an N-acyl-homoserine lactone acylase from *Pseudomonas aeruginosa* PAO1. Infect. Immun., 74: 1673-1682.

SOLANO C, ECHEVERZ M, LASA I, 2014. Biofilm dispersion and quorum sensing. Curr. Opin. Microbiol., 18: 96-104.

SPOERING A L, LEWIS K, 2001. Biofilms and planktonic cells of *Pseudomonas aeruginosa* have similar resistance to killing by antimicrobials. J. Bacteriol., 183: 6746-6751.

STOODLEY P, SAUER K, DAVIES D G, et al., 2002. Biofilms as complex differentiated communities. Annu. Rev. Microbiol., 56: 187-209.

SWARTJES J J T M, DAS T, SHARIFI S, et al., 2013. A functional DNase I coating to prevent adhesion of bacteria and the formation of biofilm. Adv. Funct. Mater., 23: 2843-2849.

TANG K, ZHANG X H, 2014. Quorum quenching agents: resources for antivirulence therapy. Mar. Drugs, 12: 3245-3282.

TSENG B S, ZHANG W, HARRISON J J, et al., 2013. The extracellular matrix protects *Pseudomonas aeruginosa* biofilms by limiting the penetration of tobramycin. Environ. Microbiol., 15: 2865-2878.

UTARI P D, VOGEL J, QUAX W J, 2017. Deciphering physiological functions of AHL quorum quenching acylases. Front. Microbiol., 8: 1123.

WAHJUDI M, PAPAIOANNOU E, HENDRAWATI O, et al., 2011. PA0305 of *Pseudomonas aeruginosa* is a quorum quenching acylhomoserine lactone acylase belonging to the Ntn hydrolase superfamily. Microbiology, 157: 2042-2055.

WANG Z, FAST W, VALENTINE A M, et al., 1999. Metallo-β-lactamase: structure and mechanism. Curr. Opin. Chem. Biol., 3: 614-622.

XAVIER K B, BASSLER B L, 2003. LuxS quorum sensing: more than just a numbers game. Curr. Opin. Microbiol., 6: 191-197.

第8章 多酚对微生物细胞间通信的影响

Filomena Nazzaro[*], Florinda Fratianni[*], Antonio d'Acierno[*],
Vincenzo De Feo[†], Fernando Jesus Ayala-Zavala[‡], Adriano Gomes-Cruz[§],
Daniel Granato[¶], Raffaele Coppola[‖]

[*]Institute of Food Science, ISA-CNR, Avellino, Italy, [†]Department of Pharmacy, University of Salerno, Salerno, Italy, [‡]Center for Research in Nutrition and Development, A.C (CIAD AC), Hermosillo, Mexico, [§]Federal Institute of Education, Science and Technology of Rio de Janeiro (IFRJ), Department of Food, Rio de Janeiro, Brazil, [¶]State University of Ponta Grossa (UEPG), Department of Food Engineering, Ponta Grossa, Brazil, k DiAAA, Department of Agricultural, Environmental and Food Sciences, University of Molise, Campobasso, Italy

1 什么是群体感应？

微生物通过细胞间通信系统来协调微生物之间的相互作用以及与高等生物之间的联系。该通信系统即所谓的群体感应（QS）系统，仅当微生物达到特定的细胞密度时，才会发生这种现象。在单个生物体中执行此行为是无效的，但当一群微生物（细菌-细菌，细菌-真菌，真菌-真菌）同时存在时，该行为便会变得有效。QS系统能够调控多种生物活动，包括孢子形成、生物发光、毒力因子表达、生物膜形成和交配。此外，一些QS系统本身可以作为毒力因子，因为它们对宿主细胞具有毒性和/或可以调节宿主免疫。该系统通过微生物在生长的特定阶段产生化学性质的信号分子，也称为自诱导物，当微生物的浓度超过某一阈值时，这些信号分子的积累将导致特定基因的激活或抑制。只有当存在特定且足够数量的细胞（称为群体）时，这些分子才能有效积累（Bassler，2002）。总体而言，QS系统是基于以下关键元素：（a）自诱导物，（b）信号合成酶，（c）信号受体，（d）信号响应调节器，（e）受调控基因（形成所谓的QS调节子；Nazzaro et al., 2013）。细菌使用的QS系统可以分为三种主要类型：主要用于革兰氏阴性细菌的LuxR/I型系统；主要用于革兰氏阳性细菌的肽信号系统；以及用于物种间通信的*lux*S/AI-2系统。此外，AI-3/肾上腺素/去甲肾上腺素被作为一种跨界信号系统（Reading and Sperandio，2006）。革兰氏阴性细菌通常产生酰化的高丝氨酸内酯（AHLs）作为自诱导物，这些物质由LuxI型酶（信号合成酶）合成，且由*lux*操纵子的第一个基因编码。在低水平细菌细胞密度条件下，*lux*操纵子转录水平较低，无法激活LuxR，只有当细胞密度增加并且信号水平达到特定阈值时，LuxR才能被激活。因此，通过lux操纵子激活LuxR/3-oxo-C6-HSL复合物转录，引起其他基因的表达，包括费氏弧菌编码荧光素酶的*lux*AB基因和荧光素酶底物酶的*lux*CDE基因，从而产生生物发光。这些分子通过细菌膜被动扩散，并根据细胞密度成比例地在细胞内和细胞外积累。在超过25种革兰氏阴性细菌中已鉴定出QS通路，有效地通过QS机制将基因表达与细胞种群密度的波动耦合起来（图 8-1A）。

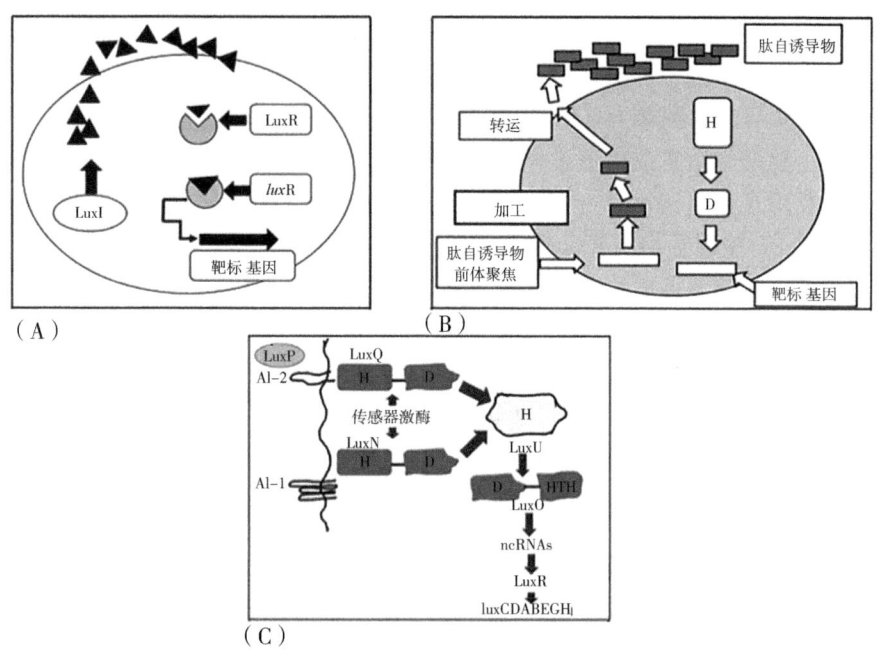

图 8-1　群体感应机制在革兰氏阴性细菌（A）、革兰氏阳性细菌（B），以及革兰氏阳性和革兰氏阴性混合体系（例如，展示了海洋细菌哈维弧菌中的 Lux S/AI-2 信号系统（C）

注：（A）革兰氏阴性细菌群体感应系统。（B）革兰氏阳性细菌中的群体感应系统，H 和 D 分别磷酸化了信号蛋白组氨酸和天冬氨酸残基。（C）哈维弧菌 QS 系统代表了革兰氏阳性和革兰氏阴性成分混合的典型系统。（A 和 B）由 Nazzaro F Fratianni F Coppola R 在 2013 年修改，Quorum sensing and phytochemicals. Int. J. Mol. Sci., 14, 12607-12619；(C) 由 Reading, N C, Sperandio, V 于 2006 年修改，Quorum sensing: the many languages of bacteria. FEMS Microbiol. Lett., 254: 1-11.

费氏弧菌、铜绿假单胞菌、荧光假单胞菌、绿针假单胞菌（*P. chlororaphis*），胡萝卜软腐欧文菌和根瘤农杆菌系统可以被认为是最容易理解的系统。此外，紫色色素杆菌的群体感应机制通常用于评估几种天然分子的 QS 活性和 QS 抑制活性。在这类微生物中，对 AHLs 的表型响应涉及多种因子的产生，包括抗生素、氰化氢、蛋白酶、几丁质酶，尤其是紫色素，这是一种具有抗菌活性且不溶于水的紫色色素（Chernin et al., 1998）。革兰氏阳性细菌通过利用前体加工产生的分泌肽作为 QS 自诱导物来调节对细胞种群密度增加的各种过程（图 8-1B）。这些信号动态地携带到细胞外，与膜结合受体蛋白的外部结构域相互作用。由磷酸化级联反应产生的信号转导最终导致 DNA 结合蛋白的激活：这一事件影响特定基因的转录，因此，每个受体蛋白对特定的肽信号具有极强的选择性。在革兰氏阳性细菌中，肽信号前体位点存在于前体蛋白的序列中。随后，这个前体蛋白被剪切，加工后产生的肽自诱导物信号通常通过 ATP 结合盒（ABC）转运体运出细胞。与革兰氏阴性细菌类似，革兰氏阳性细菌也可以采用多个自诱导物和受体。然而，一些肽可以专门在细菌细胞外起作用，引起特定基因的表达，从而转化为激活一系列不同行为的变化。对这些和其他类型的化学信号的组合和响应，可能会导致细菌对其自身种群以及周围其他物种的种群密度进行普查。对每个信号的明确响应，或基于对各种信号进行组合的响应，使细菌能够根据群体内存在的物种不断调节其行为。在乳酸菌（LAB）中，QS 涉及由膜上的组氨酸激酶直接感知多肽，然后将信号传递给细胞内的反应调节因子。该调控因子激活靶基因的转录，靶基因通常包含诱导分子的结构基因。双组分信号转导机制对于在 LAB 中发现的几种自诱物的转录激活和产生至关重要，这些自诱导物主要是细菌素或细菌素样肽，可作为反应调节剂并激活靶基因的转录，例如可导致它们进入黏附过程（Sturme et al., 2005）。在海洋细菌费氏弧菌中，存在 LuxS/AI-2 信号系统，其通过 QS 控制生物发光。费氏弧菌中的 QS 系统是革兰氏阳性和革兰氏阴性系统混合的一个典型。该细菌显示出两个 QS 系统：系统 1 中，自诱导物-1（AI-1）是 AHL，主要参与种内信号传导；系统 2 中，自诱导物-2（AI-2）是呋喃硼酸二酯，参与种间信号传导（图 8-1C）。哈维弧菌使用 LuxN 和 LuxQ 两种传感器激酶分别识别 AI-1 和 AI-2。LuxQ 识别 AI-2 及其复合的胞质受体 LuxP。在感知到这些信号后，激酶通过 LuxU 和 LuxO 这个复杂的磷酸化系统去磷酸化后转化为磷酸酶。这一过程不再激活非编码（nc）RNA 的转录，因此，LuxR mRNA 不被降解，LuxR 激活荧光素酶操纵子的转录（Reading and Sperandio, 2006）。

一般来说，几种微生物的 QS 被报道参与不同发酵食品，如发酵面团、乳制品、发酵蔬菜和葡萄酒，这表明 QS 在这些食品的发酵过程中起着重要的作用（Lebeer et al., 2007）。这个系统有前途的应用是开发自溶性启动 LAB，在奶酪成熟的早期阶段会溶解，从而降低细胞内酶的释放，有助于风味形成（Kuipers et al., 1998）。对哺乳动物肠道微生物组的了解，揭示了 QS 如何影响微生物组物种组成。对 QS 的深入理解可能导致对控制传染病如霍乱等新方法的研究，在宿主生物体中，不同机制如何进化以影响细菌 QS 从而塑造它们的微生物组。例如，由许多细菌产生的 AI-2 分子能促进厚壁菌门相对于拟杆菌门在肠道中定殖（Thompson et al., 2015）。此外，肠道共生菌产生的这种分子可以限制霍乱弧菌感染（Bäumler and Sperandio, 2016）。尽管哺乳动物肠道中细菌的相互作用要复杂得多，但现在已确定 QS 参与这些微生物群落，并提供了干预所谓的肠道菌群失调机会（Whiteley et al., 2017）。在真核生物中，包括酵母和霉菌，QS 是一个相对较新的研究和应用领域。事实上，直到在致病真菌白假丝酵母菌中发现 E-法尼醇作为 QS 分子（QSM）之前，QS 几乎是未知的（Hornby et al., 2001）。这种外源分子在早期黏附时可以抑制生物膜的形成。法尼醇也会在成熟生物膜的上清液中积累，其中刺激酵母细胞的产生，同时作为菌丝形成的抑制剂也可刺激它们的扩散（Peleg et al., 2010）。此外，芳香醇、十二醇、酪醇和 γ-丁内酯等其他分子被鉴定为真核生物（如丝状真菌曲霉和青霉菌）中 QS 过程的介质。

2 多酚

2.1 结构与分布

多酚是植物的次生代谢产物，因其独特的结构化学特性，对于保护植物免受紫外线辐射、病原体和捕食者的侵害，且对植物的生长和生存至关重要。多酚可以根据环的数量和结构基团将一个环与另一个环连接起来，分为以下几类：酚酸类（具有单环结构）、苯乙烯类、类黄酮类和木酚类（显示多环结构）（图 8-2）。

单个化合物之间的结构差异也影响着多酚化合物在生物利用度和生物特性方面的差异（Manach et al., 2009）。在自然界中，多酚化合物通常与糖和有机酸结合。它们可以分为两大类：黄酮类和非黄酮酚类，广泛分布于水果、蔬菜、草药和香料中。

黄酮类酚类化合物具有基本结构，基于两个苯环（A 和 B）通过一个杂环吡喃酮 C 环相连。非黄酮类酚类化合物是一组更为杂合的化合物，范围从最简单的类别，如羟基肉桂酸和苯甲酸（Selma et al., 2009），到更复杂的化合物，包括白藜芦醇、木脂素水解单宁，以及没食子酸和鞣花单宁，其主要成分为没食子酸和六羟基二苯甲酸，经水解后形成鞣花酸。在这一过程中，从酚类物质中脱去糖苷或葡萄糖醛酸酯基团，即早期解离，引起苷元（Rechner et al., 2004）的形成。黄酮醇是一类具有平面结构的黄酮类化合物，因其酚羟基位置不同而不同，主要存在于十字花科植物、茶、红酒和苹果中（Hollman and Katan, 1999）。它们在肠道微生物的作用下被活跃地降解，并在黄酮类 C 环断裂后被 A 和 B 环衍生的更简单、更小的酚类化合物所取代（Aura, 2008；Quirós Sauceda et al., 2017）。在肠道中，黄酮醇首先可能通过 C 环裂解，然后通过脱羟基化反应转化。糖基化的类型也会影响黄酮类化合物在肠道中的稳定性。黄烷酮主要存在于柑橘属植物中，其结构为 2,3-二氢-2-苯基色素-4-1，具有非平面的黄酮吡喃环。它们的生物利用度比黄酮醇或黄烷更高（Selma et al., 2009）。这可能与结肠微生物对其他黄酮类化合物的降解较少有关；因此，它们更适合在肠道远端部位被吸收。黄烷酮以糖苷形式存在，通常为芦丁糖苷和新橘皮糖苷。在这种形式下，它们也能在一些中草药中发现，如薄荷。黄烷酮糖苷，如柚皮苷，首先经过脱糖基化反应，生成无糖基的柚皮素；柚皮素然后分解，产生间苯三酚和 3-（4-羟基苯基）丙酸。这些产物可以进一步脱羟基化，生成 3-苯基丙酸（Rechner et al., 2004）。黄酮-3-醇和原花青素是一类非常复杂的多酚类化合物，由简单的黄酮-3-醇（如儿茶素、表儿茶素、没食子儿茶素、表没食子儿茶素和相应的没食子酸酯）生成，代谢后形成聚合的原花青素或缩合单宁。它们是膳食酚类摄入的主要成分，主要来自水果、茶和葡萄酒。在小肠消化以及从小肠到肝脏的转移过程中，黄酮-3-醇被迅速代谢为不同的 O-硫酸盐、O-葡萄糖醛酸酯和 O-甲基化形式（Kuhnle et al., 2000）。

细菌和人类代谢的共同作用降解了儿茶素和表儿茶素（Meselhy et al., 1997；Rechner et al., 2004）。花青素是一类重要的黄酮类化合物，广泛分布于有色水果和花朵中，负责紫色、紫罗兰色、蓝色和红色的植物色素。可食用植物中花青素的主要来源是葡萄科（如葡萄）和蔷薇科（如覆盆子、草莓、樱桃、黑莓、李子、苹果、桃子等），在较小程度上，花青素还存在于茄科（如茄子和番茄）、虎耳草科（如红色和黑色醋栗）、十字花科（如红卷心菜）和杜鹃花科（如蓝莓和蔓越莓）中。花青素通常以糖苷形式存在于水果和蔬菜中（Granato et al., 2015；Pascual-Teresa and Sanchez-Ballesta, 2007）。在肠道中，花青素相对于各自的葡萄糖苷要更稳定，可能是因为它们在小肠中受到葡萄糖苷酶的攻击。对于 p-香豆酸（Nazzaro et al., 2015）的保护作用或将分子固定在食物和安全级聚合物（如藻酸盐）中，例如，海藻酸盐可以稳定分子，从而在肠道运输过程中保护它们（Fratianni et al., 2010）。常见的与肠道微生物相关的花青素代谢物包括丁香酸（3,5-二甲氧基-4-羟基苯甲酸）、香草酸（3-甲氧基-4-羟基苯甲酸）、邻苯三酚醛（2,4,6-三羟基苯甲醛）、邻苯三酚酸（2,4,6-三羟基苯甲酸）、没食子酸（3,4,5-三羟基苯甲酸）和 3-O-甲基没食子酸（Keppler and Humpf, 2005）。异黄酮是一类非类固醇雌激素，其化学结构与雌激素相似，主要存在于大豆坚果、豆豉和红三叶草中。由于其较高的分子量和亲水性，大部分异黄酮无法

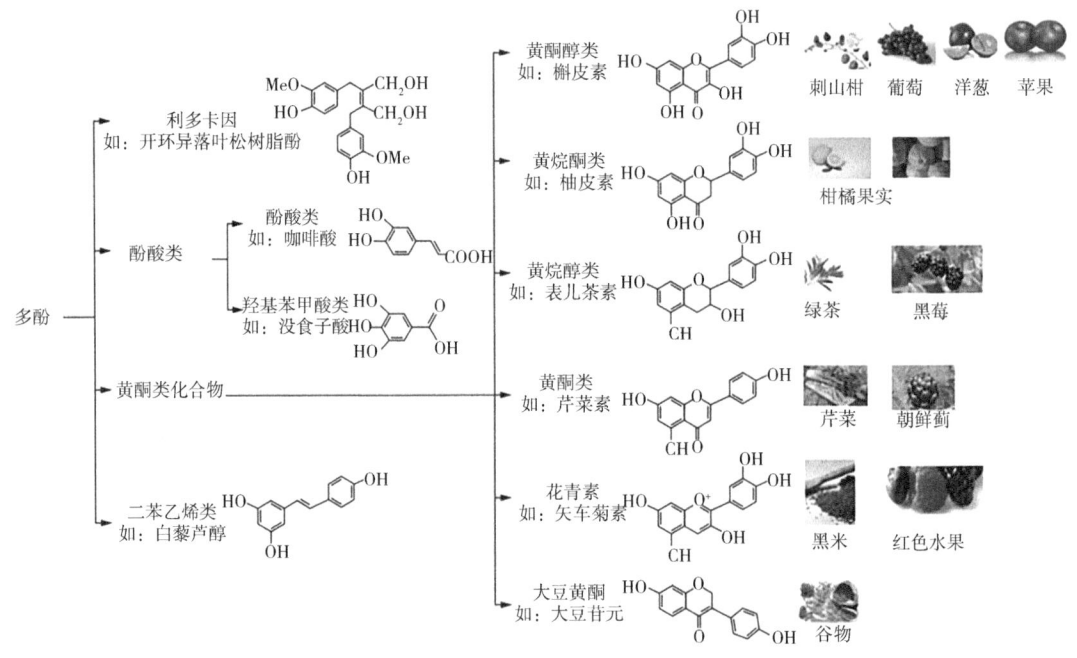

图 8-2 多酚类物质在水果、蔬菜和中草药中的分类和存在形式

直接被肠细胞吸收。异黄酮的活性主要是通过肠道微生物中的肠道 β-葡萄糖苷酶将葡萄糖苷转化为主要的生物活性糖苷（如黄豆异黄酮、大豆苷、根黄酮和大豆异黄酮）而产生的 (Setchell et al., 2002)。非黄酮类酚类化合物包括可水解单宁，如没食子单宁和鞣花单宁，在水解过程中分别产生没食子酸和鞣花酸，从而生成尿脂素 A 和尿脂素 B (Cerdá et al., 2004)。没食子单宁主要存在于浆果中，主要负责独特的涩味。而鞣花单宁通常存在于一些浆果、坚果、少数水果和橡木陈酿葡萄酒中。这些分子均可被微生物转化。木脂素被认为是植物雌激素，包括亚麻木脂素、马泰木酚、松脂醇、落叶松树脂醇、异落叶松脂素和丁香脂素，它们主要存在于水果和蔬菜中，同时在茶、谷物、咖啡和酒精饮料中也含有少量 (Scalbert and Williamson, 2000)。它们的代谢涉及哺乳动物和肠道微生物酶活性 (Possemiers et al., 2007)。木脂素转化为植物雌激素的生物活化主要由微生物组完成 (Borriello et al., 1985)。羟基肉桂酸通常存在于许多植物源食品中，具体包括 p-香豆酸（4-羟基肉桂酸）、咖啡酸（3,4-二羟基肉桂酸）、阿魏酸（4-羟基-3-甲氧基肉桂酸）和芥酸（4-羟基-3,5-二甲氧基肉桂酸），这些化合物在奎宁酸和酒石酸（绿原酸、咖啡酸等）的酯类中含量最丰富。阿魏酸和 p-香豆酸存在于菠菜和谷物麸皮中的阿拉伯木聚糖中。绿原酸在咖啡中非常丰富，但它也是形成其他次生代谢物的多个代谢途径的关键环之一（http://www.genome.jp/kegg/pathway.html）。羟基肉桂酸是非常重要的植物成分，具有多种生物学特性，包括化学预防作用 (Karlsson et al., 2005)，其在人体肠道中的吸收和微生物转化后表现出这些特性 (Gonthier et al., 2006; Kroon and Williams, 1999)。细菌酯酶释放出游离酸，然后这些酸再被代谢成其他化合物，如苯丙酸，进一步转化为苯乙酸。二苯基乙烯是存在于木材或软植物组织中的化合物，对健康有益处。除将反式白藜芦醇加氢转化为二氢白藜芦醇外，这些化合物不会被微生物转化 (Walle et al., 2004)。苯甲酸，苯甲酸盐和苯甲酸酯在水果中很常见，尤其是浆果（蔓越莓）。这些分子代表了最常见的代谢物，源

于微生物对不同代谢物的降解，包括类黄酮和非类黄酮酚类化合物。属于这一类物质的主要分子有没食子酸、对羟基苯甲酸、香草酸、丁香酸和原儿茶酸；它们甚至可以被微生物转化为更简单的分子，如邻苯三酚、邻苯二酚和邻甲基邻苯二酚，这些分子更容易被肠道吸收。

2.2 多酚的抗菌活性

植物化学物质的结构差异导致其生物利用度存在显著差异（Manach et al., 2009）。微生物在多酚类物质代谢中的核心作用是产生更简单的代谢产物，这些代谢产物通常会影响其生物活性（Van Duynhoven et al., 2011），如酚酸和短链脂肪酸，其中一些酚酸可以通过肠黏膜被吸收。同时，多酚及其代谢产物能影响肠道生态调节的微生物群落（Selma et al., 2009）。许多酚类化合物作为细菌感染的抑制剂在体外已表现出潜在的抗菌活性，表明一些酚类化合物可作为抗人类感染的抗菌剂（Fratianni et al., 2013；Selma et al., 2009）。因此，假设微生物群落的组成与植物化学物质之间存在着严格的联系，这对人类健康产生重大影响，随后，这些分子在微生物细胞间通信机制中可能发挥潜在作用。几种多酚化合物可作为不同病原菌的抑制剂。柚皮素和槲皮素对大肠杆菌的生长表现出总体和剂量依赖的抑制作用。相较于糖苷化合物，它们表现出更高的抑制效果。这种效果可以通过这些分子中糖基的存在/缺失来解释。除芦丁外，检测纯多酚化合物，主要是槲皮素（黄酮醇）和柚皮素（黄烷酮），在不同浓度下均能降低大肠杆菌、金黄色葡萄球菌和鼠伤寒沙门氏菌的生长（Parkar et al., 2008）。一些酚类糖苷，例如表儿茶素、儿茶素、3-O-甲基没食子酸、没食子酸和咖啡酸，能抑制不同病原菌的生长（Lee et al., 2006），如艰难梭菌、产气荚膜梭菌和一些拟杆菌属。咖啡酸对沙门菌、假单胞菌、大肠杆菌、梭状芽孢杆菌及拟杆菌具有强力的抑制作用。原花青素（聚合单宁）也会影响不同细菌的生长，主要肠杆菌、拟杆菌、普氏菌（*Prevotella*）和卟啉单胞菌等（Smith and Mackie, 2004）。非黄酮类化合物，如白藜芦醇，对几种病原体具有体外抗菌活性。鞣花丹宁在体内水解释放鞣花酸，并肠道微生物转化为尿石素（Espin et al., 2013），已知尿石素对某些病原体，如梭状芽孢杆菌和金黄色葡萄球菌，具有强烈的抑制作用（Bialonska et al., 2009）。

2.3 多酚作为群体淬灭剂

食品行业的消费者通常需要天然产品，因为这些产品在传统应用中具有治疗价值。近年来，许多研究集中于 QS 在食品行业微生物菌株中的作用；根据相关报告，在肉类、蔬菜和乳制品等食品产品中已检测到一些不同浓度的信号分子。食品加工的时间以及储存条件可能会影响 QSMs 的分泌。因此，食品由于其含有丰富的营养成分而成为微生物生长和生物膜形成的可用培养介质。同时，食物基质中微生物的生长也可能受到不同参数的影响，包括食物中不同微生物菌株之间的通信，以及随后信号分子产生的能力。因此，对生物基质中存在的微生物群落和相关 QS 机制的深入了解，有助于制定一种新的方法，不仅可以预防病原体和腐败微生物的生长，还可以在食品中检测到其生物膜基质。在这方面，有助于加深对植物治疗特性的研究，以及群体淬灭（QQ）机制在生物技术应用中的研究。近年来，强调植物食品提取物和植物化学物质具有 QS 抑制剂（QSIs）的潜力（Koh et al., 2013）。它们成功的一个因素可归因于其与理想 QSIs 相似，涉及化学稳定性和高效性，对人体健康无害。尽管分子机制仍不完全清楚，但仍认为一些植物化学物质可能会干扰 QS 机制中的关键成分（Dong et al., 2007）。如前所述，富含植物来源食物和蔬菜的饮食可能会影响肠道微生物群落，防止病原体在肠道系统中的定殖和生物膜形成（El-Hamid，2016；Yang et al.，2018）。QQ 作用机制是阻止小信号分子到达相关信号受体，和/或控制微生物代谢中的基因表达，从而使累积行为无法受到信号机制的调控。基本上，植物提取物中干扰 QS 系统的能力可能提供一种抵抗细菌入侵的防御能力。植物中天然存在的生物活性成分为探索和开发新型药物和抗菌药物打开了新的大门，并可作为天然防腐剂对食品工业产生积极影响。许多研究强调了植物提取物和植物化学物质干预种内和种间 QS 通信系统的能力（Al-Refi，2016；Yang et al.，2018）。几种水果和蔬菜中存在的植物化学物质显示出灭活 QS 机制的能力（Korukluoglu and Gulgor，2017）。因此，QQ 机制的基础是植物保护自身免受病原体攻击（Bacha et al.，2016）。最初对植物提取物在细菌中抗 QS 活性的研究，旨在阐明它们如何抑制已建立的特定 QS 诱导基因的表达（Schuster et al.，2013），从而降低与细菌毒力相关的因素，并预防细菌的不良反应（Suga and Smith，2003）。由于信号分子被抑制，微生物活性可以通过 QQ 机制而不是 QS 机制来调控（Chen et al.，2013）。总体而言，生物组分的目标是微生物 QS 系统，化合物的作用是通过信号分子的失活、降解或修饰来实现。DNA 芯片研究和实时荧光定量 PCR（qRT-PCR）数据表明，植物提取物对各种细菌物种的 QS 相关基因下调的作用，被认为是降低其毒力的主要动机之一（Tolmacheva et al.，2014）。植物成分可阻止由 *luxI* 编码的 AHL 合成酶合成的信号分子，降解信号分子和/或靶向 luxR 信号受体。植物食品提取物和植物化学物质最常见的作用机制与它们在化学结构上与 QS 信号（AHL）的相似性，以及它们破坏信号受体（LuxR/LasR；Rasmussen and Givskov，2006）的能力相关。最后，QS 基因表达和信号分子水平的降低影响毒力因子的产生，这为过去使用植物提取物以及它们未来可能以何种方式对抗细菌感染提供了更多见解（Vandeputte et al.，2010）。众所周知，从植物中分离出的 QS 拮抗剂成分是黄酮类化合物，几乎存在于所有植物中。黄酮类化合物赋予花朵和水果特有的颜色，同时在保护植物免受害虫侵害方面也起着重要作用。此外，它们还可以作为一种重要抵抗微生物之间通信的 QQ 机制（Bacha et al.，2016；Koh et al.，2013）。穿心莲内酯是从穿心莲中分离出的半日花烷型双环二萜，而姜黄素是一种由生姜科植物产生的黄色化学成分，可作为抑制剂影响铜绿假单胞菌 QS 调控的毒力。其他植物源活性成分，如鞣花酸和漆黄素，对停乳链球菌（*S. dysgalactiae*）生物膜形成具有一定的抑制作用。水杨酸可抑制铜绿假单胞菌中 AHL 的产生；另外，丹宁酸和反-肉桂醛也可能像水杨酸一样抑制 AHL 合酶（Chang et al.，2014）。蔓越莓、树莓、牛至、迷迭香、罗勒、甘蓝、姜黄和生姜的提取物显示对不同微生物（如铜绿假单胞菌和大肠杆菌 O157:H7，是已知最重要的食物源病原菌）的 QQ 机制（Vattem et al.，2007）。犬蔷薇、黑檀香、核桃、欧洲栗、迷迭香提取物对金黄色葡萄球菌具有 QQ 抑制能力，且柑橘提取物可阻止小肠结肠炎耶尔森菌菌株间的通信（Korukluoglu and Gulgor，2017）。一些富含单宁的植物具有抑制群体聚集以及在病原细菌菌株中抗毒力因子的能力。未来，具有 QQ 活性的植物提取物可能会取代医学中的药物，并在食品工业中替代所有（或大量）合成分子作为防腐剂和/或添加剂。这些药物的使用可能会促进新的生物防治策略的发展（Koh et al.，2013；Mahmoudi et al.，2014；Shukla and Bhathena，2016）。大多数拮抗剂仅对特定病原体具有窄谱活性。这一特性可能在多菌环境中对抗特定类型的病原菌时提供一种有趣的策略，但这种狭窄的拮抗作用在临床应用中的价值有限。因此，包括抗生素和 QS 抑制剂在内的联合疗法可能提供协同效应（Rutherford and Bassler，2012）。意大利腊菊提取物可抑制变形链球菌细胞在玻璃表面上的黏附；而蔓越莓汁则能抑制远缘链球菌（*S. sobrinus*）和血链球菌

(S. sanguinis)生物膜的形成（Abachi et al., 2016）。金丝桃属植物的乙醇和乙酸乙酯提取物对紫色色素杆菌具有QQ活性，限制其紫色素产生（Fratianni et al., 2013）。几种植物的乙醇提取物能降低不同病原体生物膜的形成。具体而言，黄樟、凡夫兰、柽柳、茴香、花椒、月桂、鼠尾草、小麦、苦艾、芙蓉、百里香、石榴、龙舌兰、长叶薄荷和马齿苋的提取物对铜绿假单胞菌生物膜形成的抑制作用在20%~80%（Korukluoglu and Gulgor, 2017）。据报道，香叶天竺葵、茴香、彩叶草、迷迭香、大荨麻、德国洋甘菊、鸡冠刺桐和苦瓜使用乙醇提取物可抑制或降低生物膜的形成（Al-Refi, 2016）。橡树皮富含水解单宁、邻苯三酚单宁和缩合单宁-原花青素，作为QQ分子起作用（Deryabin and Tolmacheva, 2015）。钮子树含有栎木鞣花素和栗木鞣花素具有QQ活性且可水解单宁（Adonizio et al., 2008; Vandeputte et al., 2010）。大蒜提取物含有至少3种不同的QS抑制剂，其活性取决于其浓度。几种植物化学物质也可作为真菌QS机制的抑制剂：植物抗菌素，白藜芦醇（3,5,4′-三羟基二苯乙烯），天然存在于葡萄和其他蔬菜中，是一种抗真菌剂；芫荽提取物富含α-蒎烯、β-红没药烯、p-花香烃、己醛和芳樟醇具有抗真菌的QQ活性（Savoia, 2012）。柳叶马齿苋的果实、叶子和茎具有QQ活性，这是由于一些多酚类物质的存在和协同作用，如没食子酸、咖啡酸和绿原酸，对金黄色葡萄球菌生物膜形成和紫色色素杆菌产生的紫罗兰素具有特别强的抑制作用（Noumi et al., 2017）。一些植物的分泌物类似细菌AHLs，并随后影响这些植物相关细菌中的QS调节行为（Brackman et al., 2009），紫花苜蓿能调控不同生物中的AhyR、CviR和LuxR报告活性（Gao et al., 2003）。姜黄通过产生姜黄素，可以阻断铜绿假单胞菌PA01中毒力基因的表达（Rudrappa and Bais, 2008）。一些苹果品种的提取物和成分表现出QSI活性，主要是由于不同多酚类化合物的存在，如芦丁，表儿茶素和羟基肉桂酸，能以协同的方式对紫色色素杆菌表现出QQ行为（Fratianni et al., 2011）。月桂、苦菜、迷迭香、硬骨凌霄、茉莉花、银白杨和黑松的提取物中也观察到了QQ活性，所有这些提取物能降解紫色素的含量（Al-Hussaini and Mahasneh, 2009）。黄酮类化合物能够干扰QS系统，从而影响相关的生理过程（Truchado et al., 2015）。这些分子能显著减少铜绿假单胞菌中吡氰蓝和弹性蛋白酶的产生，而不影响细菌的生长。类黄酮类化合物，如槲皮素、柚皮素、芹菜素和山奈酚，抑制哈维弧菌BB886和MM32中HAI-1或AI-2介导的生物发光。柚皮素和槲皮素降低铜绿假单胞菌PAO1中几个QS调控基因（即 lasI、lasR、rhlI、rhlR、lasA、lasB、phzA1和rhlA）的表达。柚皮素还显著降低了由lasI和rhlI基因产物驱动的酰基高丝氨酸内酯3-oxo-C12-HSL和C4-HSL的产生（Vandeputte et al., 2011）。槲皮素、柚皮素、芹菜素和山奈酚对哈维弧菌BB120和大肠杆菌O157: H7具有抗生物膜形成活性（Truchado et al., 2015; Vikram et al., 2010）。黄酮-3-醇儿茶素能减少QS介导的毒力因子的产生（如吡氰蓝和弹性蛋白酶），以及铜绿假单胞菌PAO1的生物膜形成（Rasamiravaka et al., 2013; Vandeputte et al., 2010）。葡萄柚汁和罗勒根部产生的迷迭香酸可破坏大肠杆菌生物膜的形成（Vattem et al., 2007）。可食用植物和水果的提取物，如车前草、人心果、菠萝和圣罗勒，对紫色色素杆菌产生的紫色素和铜绿假单胞菌PAO1产生的吡氰蓝色素，葡萄球菌蛋白酶和弹性蛋白酶，以及其生物膜形成能力表现出QQ活性（Musthafa et al., 2010）。十字花科植物提取物和成分可抑制QS相关基因的表达，在体外和体内下调大肠杆菌O157: H7的毒力特征；表明它们具有作为抗感染药物的潜力（Lee et al., 2011）。含有没食子酸基团的多酚化合物（如表没食子酸儿茶素，鞣花酸和丹宁酸）可通过阻断AHL介导细菌之间的通信来干扰AHL介导的信号转导（Sarabhai et al., 2013）。在石榴和浆果中包含鞣花素，如安石榴苷和鞣花酸，被肠道微生物水解为鞣花酸，随后，被代谢成尿石素-A和尿石素-B。它们能降低小肠结肠炎耶尔森菌的QS相关过程高达40%，并降低AHLs产生的水平（Giménez-Bastida et al.，2012）。4′,5′-O-二咖啡酰奎宁酸可作为泵抑制剂，在大量革兰氏阳性人类致病菌中具有靶向外排系统的潜力（Fiamegos et al.，2011）。绿原酸和香草酸以及芦丁、犬尿酸和二甲基槲皮素都可以作为QS抑制的阳性对照（Leach et al.，2007）。黄烷醇和原花青素与水果作物病原真菌的孢子和菌丝形成复合物（Feucht et al.，2000），而酚醛聚合物的沉积与细菌增殖速率的降低有关。植物提取物是控制细菌发病机制和微生物调控很有前景的工具。这些活动的知识提供了更深入的见解，解释了为什么这些植物提取物可以在未来用于对抗细菌感染（Vandeputte et al.，2010）并在食物保存方面表现出优越性（Hwa in Lee et al.，2017；Vazquez-Armenta et al.，2017），从而有效保护人类健康。

3 从海洋生物中提取的具有抗菌活性和QQ活性的多酚

海洋和陆地生物产生的次生代谢物能够保护它们免受细菌感染。令人意想不到的是，这些分子对人类和动物的许多病原体表现出活性。自古以来，海洋藻类提取物在民间医学和生药学中被广泛使用。由于海洋和陆地生物面临着截然不同的环境挑战，它们的结构特征及代谢物的药理活性存在显著差异。一些海洋来源（如藻类）的抗菌化合物在抗菌效果方面优于陆地来源的化合物，这主要是因为海水中细菌细胞的数量显著多于空气中的细菌数量，同时海洋生物需要通过固着防止海洋表面生物污染（Bixler and Bhushan，2012）。近几十年来，从海洋生物中寻找生物活性化合物的研究产生了大量具有制药和工业应用的提取物。在过去的10年里，已经有超过1 000多种活性化合物被研究；其中许多已被证明对抗高血压、高胆固醇和其他疾病有效，同时也显示出潜在的抗菌、抗真菌、抗癌药物效率。最后一个方面尤为重要：细菌对现有抗生素的耐药性急剧增加，必然刺激和引导对来自海洋中新的天然候选生物活性分子的研究，这些分子可能具有防污剂和食品保鲜剂的作用。海洋代表了一种令人难以置信的生物多样性，包括大型和微型藻类、刺胞动物、浮游植物、软体动物、海绵、珊瑚、被囊动物和苔藓动物等生物。研究这些生物的初级和次级代谢物，旨在鉴定具有抗菌活性并能阻断或限制微生物通信机制的化合物。在进化过程中，一些海洋生物，如微型和大型藻类（如硅藻类和海藻），已形成成熟的本土系统，以对抗致病细菌和其他环境微生物。因此，这些海洋生物不仅应被考虑用于食品或生产水胶体，还具有提取健康活性的生物活性分子的潜力。现代筛选方法鉴定出多种藻类中具有抗菌性能的次级代谢物，如红藻纲（红色）、绿藻纲（绿色）、褐藻纲（棕色）、金藻纲（金色）和硅藻纲（硅藻；图8-3；Shannon and Abu-Ghannam，2016）。

具有抗菌活性的官能团包括褐藻多酚类、肽、萜烯、脂肪酸、聚乙炔、多糖、吲哚生物碱、甾醇、芳香有机酸、莽草酸、聚酮、对苯二酚、醇、醛、酮、卤代呋喃酮、烷烃和烯烃（Shannon and Abu-Ghannam，2016）。

藻类的生态功能仍未被完全理解。所有与藻类生长和繁殖相关的现象都可以被视作一个重要因素，这一因素与环境条件（通常是不利的条件）密切相关，具体包括对藻类生存的障碍，如食草动物的捕食、与其他生物竞争空间、菌丝体受伤、生物膜形成、渗透胁迫、高水平的紫外线照射、氧气和盐度。此外，每毫升海水中大约有100万个细菌细胞。藻类为了抵抗这些胁迫而产生的一些化学物质具有抗菌潜力。很多化学物质由于海水的稀释效应和恶劣环境而表现出很高的生物活性。例如，绿藻和红藻的乙醇提取物富含多酚，能抑制多种病原体的生长，包括粪肠球菌、溶藻弧菌、霍乱弧菌、金黄色葡萄球菌、鼠伤寒沙门菌和大肠杆菌（Vijayavel and Martinez，2010）。

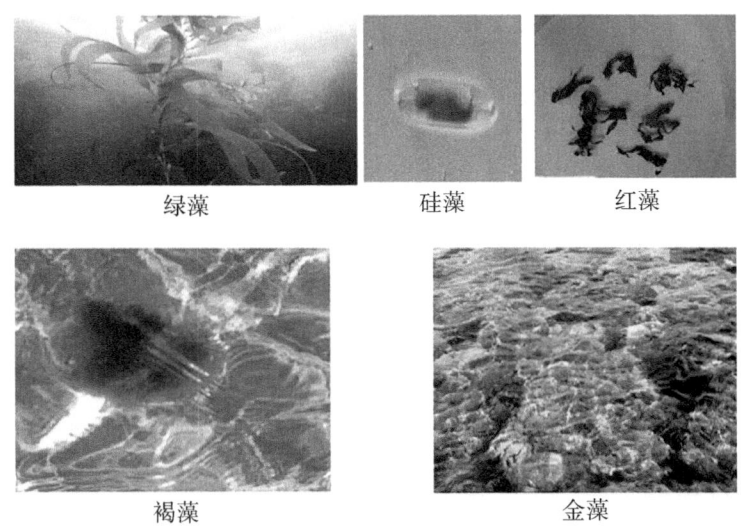

图 8-3 不同类型的海藻

3.1 褐藻多酚的结构、表征和生物活性

海藻中存在的天然酚类化合物包括一个或多个羟基，并直接与芳香烃基团结合。酚类化合物与金属离子螯合需要具备两个邻羟基（o-二酚）结构（Hermund，2018）。它们与自由基反应的能力与酚环和儿茶酚结构（o-二酚）的数量密切相关（Capitani et al.，2009）。在褐藻方面，关于褐藻素含量和生物活性方面的研究最多，然而，只有少数研究探索了红藻和绿藻的化学防御机制。

褐藻，尤其是墨角藻，产生大量的多酚次级代谢物，即褐藻多酚。这些多酚类物质表现出与单宁的相似化学特性；像单宁一样，它们很可能与蛋白质和碳水化合物结合（Hermund，2018）。由于褐藻多酚结构复杂，其具有潜在的多功能天然抗氧化剂特性，即具有初级和次级抗氧化性能。褐藻多酚是水溶性的，由间苯三酚（1,3,5-三羟基苯）形成的寡聚体或聚合分子，在海藻的细胞核周区域的高尔基体中通过醋酸-丙二酸途径生物合成。它们是高度亲水性的成分，储存在被称为藻类的囊泡中，在藻类细胞中执行不同的生物任务，并占叶状体干重的 1%~15%，在某些情况下甚至达到 25%（Hermund，2018）。

通常，褐藻多酚不会分泌到细胞外，仅在细胞受损后才会被释放。褐藻多酚的含量受到多种因素的影响，例如海藻的大小、年龄、组织类型、营养水平，以及盐度和光强度。此外，水温、季节和食草动物的强度也会影响它们的存在。

在褐藻中，这种化合物的含量在不同物种之间可能会有所不同，而墨角藻目的物种通常富含这种类型的化合物（Shibata et al.，2004）。在热带大西洋中，它们的浓度可以达到最高（在那里它们占褐藻干质量的 20%），而在热带太平洋和印度-太平洋地区则最低。天然存在的褐藻多酚类化合物的分子量变化在 0.126~650 kDa，尽管最常见的观察范围在 10~100 kDa（Hermund，2018）。

根据 Martínez 和 Castañeda（2013）的研究，褐藻多酚根据间苯三酚单元（PGUs）的连接方式可分为三类：富科尔、邻苯二酚和富科邻苯二酚。这些分子结构的多样性和复杂性随着 PGUs 数量和连接方式（线性或分支或两者兼有）的增加而增加。褐藻多酚的提取方法会影响其产量和由此产生的生物学特性，包括抗菌特性。所谓的环境友好技术，如超临界水提取（SWE；Plaza et al., 2010），可用于从褐藻中提取酚类化合物，获得高产的褐藻多酚。

从海藻中纯化几种褐藻多酚，如厄氏褐藻、昆布、海藻、良布海藻、羽叶藻、铁钉菜以及鹿角菜均表现出显著的药用活性。

钝马尾藻主要含有呋喃酚、羟基呋喃酚和褐藻多酚。其他物种，如齿缘墨角藻、墨角藻、伸长海条藻、结茎囊链藻，则含有低分子量的褐藻多酚（相当于 4~12 个具有不同位置异构程度的间苯三酚单体），被认为是其主要成分（Panzella and Napolitano, 2017）。

厄氏褐藻也含有多种褐藻多酚，如 6,60-双鹅掌菜酚、8,80-双鹅掌菜酚、8400 o-二鹅掌菜酚、二苯杂二氧系统、fucodiphlorethol G、根富鹅掌菜酚 A 和 triphlorethol-A 等。其他化合物，如鹅掌菜酚、根富鹅掌菜酚、二鹅掌菜酚和 8,80-双鹅掌菜酚已从昆布和羽叶褐藻中分离出来；6,60-双鹅掌菜酚、二苯杂二氧系统和间苯三酚已从棕藻铁钉菜中分离出来（Eom et al., 2012）。褐藻多酚混合物以及聚合物二鹅掌菜酚、根富鹅掌菜酚 A 和 8,8-双鹅掌菜酚能强烈抑制草食性海螺五脏六腑内的消化酶（Shibata et al., 2004）。

褐藻多酚及其衍生物为食品和医药应用提供了一种潜在有用的天然抗菌剂来源。

通常情况下，纯化提取的褐藻多酚对革兰氏阳性和革兰氏阴性菌株均具有抑制活性，其中革兰氏阳性细菌对革兰氏阴性细菌更为敏感，而这种效应随着间苯三酚的聚合而增强（Nagayama et al., 2002）。革兰氏阴性和革兰氏阳性细菌之间的物理差异可能是褐藻多酚提取物的行为基础。革兰氏阴性细菌被高脂多糖含量的外膜包围，这赋予细菌对多种合成和天然抗生素产生更大的抗性。Lopes 等（2012）证明，提取物对革兰氏阳性细菌，特别是葡萄球菌的活性最强，其中表皮葡萄球菌和金黄色葡萄球菌是最敏感的菌株，尤其是从黑角藻和结茎囊链藻中提取的褐藻多酚含量最高，这样的结果具有重要意义。

金黄色葡萄球菌和表皮葡萄球菌与人类感染相关，其中金黄色葡萄球菌是主要的发病和死亡的原因，也参与宿主的免疫功能受损以及生物膜的形成有关，这是一种病原性的信号。这些提取物对革兰氏阴性细菌的作用较小，但也具有活性。此外，结茎囊链藻是唯一一种具有一定能力的海藻，特别是抑制鼠伤寒沙门菌、奇异变形杆菌和大肠杆菌的生长，这三种是胃肠道的主要感染源，其发展会导致膀胱炎、肾结石和膀胱结石，以及腹泻、由结石引起的导管阻塞、急性肾盂肾炎和发热。其中一些菌株会产生一种或多种类型的毒素，严重损害消化系统和肾脏的黏膜，导致痢疾和肾功能衰竭，尤其是在年幼儿童或免疫系统受损的患者中。

物种属于囊藻属和岩藻属的藻类可能含有高分子量的褐藻多酚；同时，它们通常含有更多羟基团的游离形式的褐藻多酚类物质，并对所有研究的细菌显示出更低的抑菌浓度。在囊藻属中，虽然海葡萄藻含有最高量的褐藻多酚类物质，但其提取物对细菌的抑制效果不如结茎囊链藻和长须囊藻，可能是由于低分子量多酚的存在或可发生反应的羟基较少所致。细菌蛋白质与褐藻多酚类物质之间的相互作用在褐藻多酚类物质的杀菌作用中起着重要作用（Eom et al., 2012）。

3.2 褐藻多酚具有抗真菌和 QQ 活性

抗真菌药物的最重要靶点是真菌的细胞膜和细胞壁，其负责与环境进行交流，因此，在代谢过程中起着关键作用。麦角甾醇是真菌细胞膜中主要的甾醇，其负责保持细胞的完整性活力及其正常功能和生长。因此，真菌细胞壁也是抗真菌作用的靶点，其影响真菌细胞壁几丁质和 β-葡聚糖的合成，这两者是维持真菌结构和正常细胞生长的基本成分。抗真菌药物还能影响酵母菌的芽管形成和附着能力，并与线粒体的呼吸链相互作用。抑制酵母菌的芽管形成被认为是几种抗真菌化合物降低微生物毒力的机制，同时也抑制或降低了细菌间的通信机制（Nazzaro et al., 2017）。事实上，通过影响酵母菌的二态转变，抑菌化合物降低了微生物对目标上皮细胞的附着，减少了感染的进程，使其更容易被克服。一些抗真菌化合物也影响线粒体呼吸链的功能，作为潜在的细胞生长抑制剂，并能触发细胞死亡。海藻特别有吸引力，不仅具有丰富的工业价值，而且还发现其具有多样的药理特性的次生代谢产物（Thomas and Kim, 2011）。褐藻多酚类物质特别有趣，因为它们具有重要的生物活性，对真核细胞没有毒性。褐藻多酚提取物对真菌的活性似乎比对细菌的活性要低。纯化的提取物对真菌深红发癣菌（*Trichophyton rubrum*）和白假丝酵母菌显示出抗真菌活性。在测试条件下，曲霉菌通常与孢子吸入相关的过敏症状相关，相反，能抵抗所有提取物的作用。深红发癣菌对褐藻多酚纯化提取物特别敏感，尤其是从黑角藻、结茎囊链藻、长须囊藻、普通马尾藻和海葡萄藻提取的褐藻多酚。结茎囊链藻中褐藻多酚对白假丝酵母菌具有显著的抑制活性，白假丝酵母菌是一种共生酵母，通常定殖在大多数健康人的黏膜上，但不会引起组织损伤。然而，作为机会性病原体，白假丝酵母菌属在各种适宜条件下可引起疾病：其细胞可以从黏膜和肠道传播，并在导管水平上导致侵袭性感染和生物膜形成。生物膜形成能力极大地增加了白假丝酵母菌从共生阶段转变为致病菌的能力（Nazzaro et al., 2017）。褐藻多酚的作用机制可能是改变了酵母细胞膜中麦角甾醇组成。褐藻多酚似乎能增加线粒体脱氢酶的活性，抑制白假丝酵母菌的二态转变，导致假菌丝的形成，其黏附着于上皮细胞的能力减弱，从而降低了白假丝酵母菌的毒力和攻击宿主细胞的能力（Lopes et al., 2013）。关于海藻多酚所表现的 QQ 活性的研究很少。Liu 等（2014）使用紫色色素杆菌 CV026 的细菌模型，评估了海藻多酚的细菌群体感应（QS）抑制剂活性，并鉴定和检测了酚类化合物的存在。对 25 种海藻进行了研究，证明了其中的 12 种具有 QS 抑制剂活性。特别是从海莴苣、石藻和裙带菜提取的多酚显著降低了紫色色素杆菌 CV026 的紫色素产量，在浓度为 1.0 mg/mL 时，分别抑制紫色素的直径为 25.18 mm、21.61 mm 和 20.43 mm。其他提取物，如 *S. thunbergii*、*Ectocarpus lyngbye* 和 *Spirulina platensis*，尽管程度较小，但也降低了紫色素的产量，这种活性是由于在多酚模式中存在儿茶素和表儿茶素。这些结果表明，使用海藻多酚抑制细菌的 QS 系统，可能会开发出新的 QS 抑制剂。

4 地衣次生代谢物：一般性、抗菌和抗真菌活性

地衣是一种共生生物，由真菌（通常是子囊菌）和光合自养的伴侣组成，伴侣可以是绿藻、和/或蓝藻，以及被称为光合体的光合细菌。此外，嵌入地衣皮层的第三个伙伴被鉴定为担子菌酵母（图 8-4；Millot et al., 2017）。

地衣是宝贵的植物资源，具有广泛的用途。自古以来，地衣不仅在民间医学中被认识和

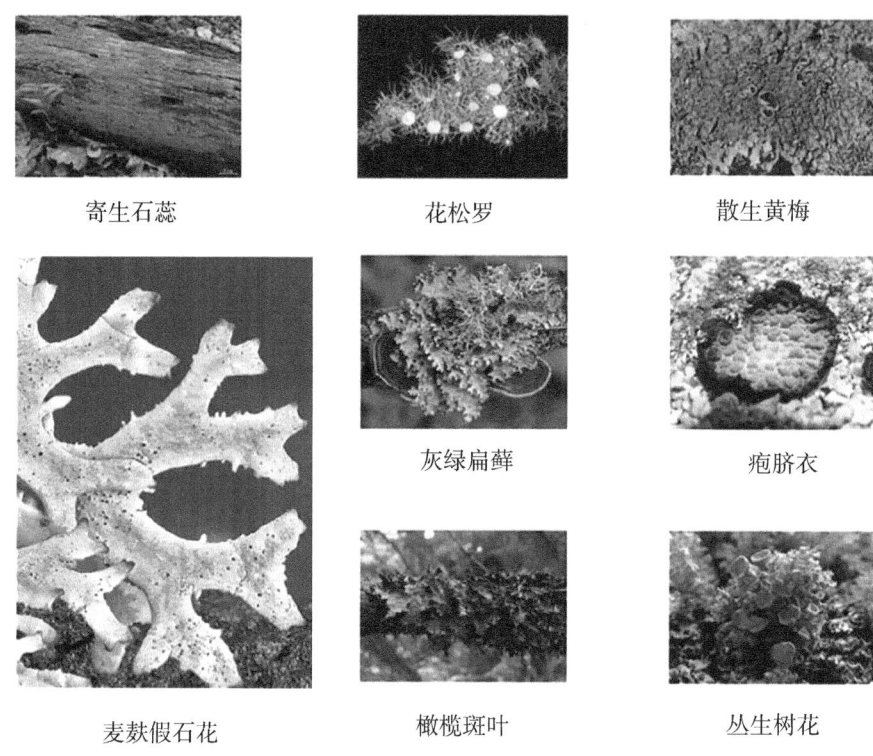

图 8-4 地衣

使用，而且在世界多个国家也被用于食品、化妆品、香料和染料。

地衣的扁平体代表了生活在地衣体内外的其他微生物的支持，包括内生和外生的地衣真菌和细菌。在这个复杂生态系统中，相互作用导致了许多分子的合成，这些分子参与了群落成员之间的相互作用，并影响整体群落的内稳态和生存。地衣中已发现超过 1 000 多种次生代谢物（Shukla et al., 2010）。

这些化合物大多数来自不同的生物合成途径，产生各种化合物。乙酰-聚丙二酰生物合成途径产生次生脂肪酸、酯和相关衍生物，以及聚酮衍生的芳香族化合物，这些化合物由真菌合成，仅与藻类（地衣共生）相关，并对地衣结合的成功起重要作用。通过这一途径，可以合成不同的化合物，这些化合物是由两到三个苔黑素、β-苔黑素型酚类物质通过不同的醚、酯和 C-C 键连接形成。众所周知，这些分子不仅可以在共生体和外部环境之间建立更好的相互作用，还能保护地衣免受氧化应激（Shukla et al., 2010）。其他化合物，尤其是三萜（如藻酮）和少量的二萜，主要是通过甲羟戊酸途径产生的。莽草酸途径则导致三联苯醌和枕酸衍生物的形成，主要发生在海藻科地衣中。

多类聚酮次生代谢产物包括单环酚（苔色酸和 β-苔色酸），由酯键连接的双环或三环酚（缩酚酸，三缩酚酸类和苄酯），以及含有酯键和醚键的化合物（缩酚酸环醚、氨苯砜和二苯醚）。此外，还包括呋喃杂环化合物（二苯并呋喃酸、地衣酸和衍生物），蒽醌类化合物，色酮，萘醌和氧杂蒽酮（Nomura et al., 2013）。其他酚类化合物则包括地几酮、地弗地衣酸、二元酸、培拉托酸、补骨脂酸、原鳔酸和去甲酸（Boustie et al., 2011）。

地衣中的代谢物可以占到菌丝体干重的 20%，通常为 5%~10%。这些代谢物大部分以晶体形式积累在菌丝体（外源物质）的外表面上，或在内部菌丝体髓质中。一般来说，"地衣代谢物"的1~2种皮层会积累较高的数量，如前所述，这些代谢物在有机溶剂中占可提取物质的5%~20%。

地衣在生产植物化学物质时所消耗的能源和碳源成本表明，这些化学物质可能涉及地衣抵御环境压力的防御机制，特别是生物压力，并有助于维持和保护地衣共生体。因此，这种复合分子群落可代表新药的潜在来源。从地衣中分离出大量化合物显示出不同的生物活性。其中，几种地衣植物化学物质展示出抗病毒、抗氧化、抗肿瘤和酶抑制等特性（Gökalsın and Sesal，2016）。

地衣提取物及其代谢物以其抗菌和抗真菌特性而闻名。

Kosanic 和 Rankovic（2011）证明，地衣 *C. furcata*、*H. physodes* 和 *U. polyphylla* 的甲醇提取物对多种革兰氏阳性细菌，如蕈状芽孢杆菌（*B. mycoides*）、枯草芽孢杆菌（*B. subtilis*）和金黄色葡萄球菌以及革兰氏阴性细菌阴沟肠杆菌（*E. cloacae*）、大肠杆菌和肺炎克雷伯菌具有抗菌活性。这些提取物对黄曲霉（*A. flavus*）、烟曲霉、灰霉病菌、白假丝酵母菌、尖孢镰孢菌、大毛霉（*Mucor mucedo*）、宛氏拟青霉（*Paecilomyces variotii*）、变紫拟青霉（*P. purpurescens*）、疣梗拟青霉和哈茨木霉菌（*Trichoderma harzianum*）表现出有效的抑制活性。这些提取物的抗菌活性显示出较高的多酚含量，特别相关的是，考虑到阻止某些物种生长的重要性，如金黄色葡萄球菌、大肠杆菌和白假丝酵母菌，它们能够激活细胞间通信的 QS 系统，从而触发导致其毒力的大部分事件。从地衣中分离出的酚类化合物，如苔黑素衍生物和来自一些缩酚酸的乙基-β-奥尔塞酸酯水解衍生物，分别从白腹地卷、珊瑚枝地衣和石耳属中分离得到，表现出抗真菌和抗菌活性（Boustie et al.，2011）。

4.1 地衣多酚的 QS 和抗生物膜活性

地衣提取物及其代谢物的抗菌特性已被广泛研究，但其 QQ 和抗生物膜潜力仍未得到深入探索。

一些常见的地衣，如灰绿扁藓和麦麸假石花（图8-4），以及它们的丙酮、乙酸乙酯或甲醇提取物，表现出抑制某些病原体（如金黄色葡萄球菌）生物膜的能力，生物膜抑制浓度范围为 0.63~1.25 mg/mL（Mitrovic et al.，2014）。Chang 等（2012）证明从地衣 *Lobaria kurokawae* 中分离出的次生代谢物老龙皮酸 B 具有抗菌生物膜活性，能抑制感染小鼠白假丝酵母菌菌丝转变，并与氮唑类药物协同作用，不仅阻断了真菌的生长，还阻止了该微生物的生物膜形成。姜黄素和邻苯三酚对白假丝酵母菌生物膜具有有效活性；另一方面，天然蒽醌色素紫红素对白假丝酵母菌菌丝形成和生物膜发育也有重要作用。此外，一些地衣的丙酮提取物（属于九科，Parmeliaceae 和 Cladoniaceae 两个科）对游离和附着的白假丝酵母菌细胞有效（Millot et al.，2017）。控制白假丝酵母菌生物膜的发展在临床上具有重要的意义。白假丝酵母菌是人类胃肠道的共生菌，作为一种机会性病原体，尤其通过其形成生物膜的能力，常涉及感染，并增加酵母对抗微生物和免疫防御的抵抗力，已经证明，生物膜内的酵母细胞对传统的抗真菌药物和宿主免疫因子更具抵抗力，因此，大多数与生物膜相关的白假丝酵母菌感染更难治疗。

地衣褐藻多酚的生物作用似乎也取决于其结构和地衣提取物的数量：事实上，一些代谢物似乎并未参与所观察到的抗饱和效应。以萜类为主的地衣（如球藻属）大多数是缺乏活性的。二苯并呋喃类胎座二醇、鞣酸和苔酸（分别分布在微绒枝、膜癞屑衣和厚叶石蕊中）；脱萘素

(分布在麦麸假石花、橄榄斑叶中)，萘乙酸（寄生石蕊、鳞片石蕊和花松萝中，图8-4）；呋喃二羧酸和过络酸的沉淀；三缩的旋光酸和旋光酸；去甾酮（主要在碟形皮叶中）；己糖酸（主要在皱衣和灰绿扁藓中）；罗塞落酸（分布在膜癞屑衣中）；最后，报道表明，三萜醇（存在于微绒枝）在非活性提取物中也有发现。在活性和非活性提取物中发现了一些化合物（乌苏酸、富马原四氢呋喃酸、原四氢呋喃酸、柳氮酸和角鲨烯酸），在这种情况下，提取物的生物活性或非活性可能与单一代谢物的数量有关。例如，地衣提取物具有高比例的地衣酸，这种代谢物的弱活性已被证明。地衣酸占花松萝提取物中40%的干重和皱衣提取物20%的干重；相对而言，这些提取物对白假丝酵母菌生物膜发育和成熟表现出较差的活性。在其他情况下，活性可能与其他因素有关，如异构化速率。例如，地衣酸的活性可能与其异构化有关：事实上，根据结构的不同，地衣酸可以对细菌生物膜（如葡萄球菌）以及真菌生物膜产生作用（例如，通过抑制酵母菌丝转换和降低成熟生物膜厚度，或还原胞外多糖层中的不同糖）。此外，对耐唑和敏感的白假丝酵母菌具有显著的生物膜抑制作用，分别为71.08%和87.84%，（Millot et al., 2017；Nithyanand et al., 2015；Pompilio et al., 2016)。一些化合物，如黏附酸、扁枝衣二酸和槲皮素，主要存在于散生黄梅、丛生树花和石黄衣的活性提取物中；这表明它们在白假丝酵母菌生物膜的抗成熟过程中具有重要意义。特别是，扁枝衣和丛生树花提取物对生物膜和24 h龄生物膜拥有最高的抑制活性。它们的主要成分是扁枝衣二酸（Culberson, 1963；Ristic et al., 2016）。研究评估了扁枝衣二酸对细菌生物膜的影响，缩酚酸已被证明具有抑制铜绿假单胞菌QS系统的作用（Gökalsın and Sesal, 2016）。Mitrovic 等（2014）观察到，从散生黄梅中分离的黏胶酸对白假丝酵母菌生物膜具有活性。在其他情况下，暗腹黄梅中的地衣酸和藻纹苔酸之间可能存在协同机制，进而解释提取物的抗膜活性。另一个缩酚酸、鳞片酸与寸石蕊提取物中的地衣酸相关，其结构与地衣酸相似，可能涉及抗生物膜活性（Millot et al., 2017）。从两种地衣属中分离出酚类和二聚体，属于梅衣科（灰绿扁藓和麦麸假石花），表现出对 *Legionella* 的抗生物膜活性，因为它们可以阻断这种细菌的细胞间通信机制。地衣还含有其他已知的多酚，具有确定的QQ和抗膜活性。槲皮素是从一种可食用地衣（长松萝）中分离出来的，确定了其对诱导白假丝酵母菌对氟康唑的敏感能力，且在对健康无害的剂量下增强其活性。从地衣中提取的槲皮素具有抑制毒力武器产生的能力，包括生物膜的形成、菌丝发育、磷脂酶、蛋白酶、酯酶和溶血活性，这些都与酵母的QS和生物膜的形成有关。其活性也可能与法尼醇的增加相关，法尼醇已知共同调节菌丝发育，生物膜形成和毒力因子的产生（Singh et al., 2015）。

5　结论

一些研究清楚地表明，来自陆地和海洋植物的植物化学物质作为生物活性成分发挥着关键作用，同时对植物和人类的健康也至关重要。设计新的药物来支持对抗病原微生物及其相关的所有毒力现象（如细胞通信过程、黏附和生物膜产生）的可能性，是一个有前景且值得进一步探索的研究方向。

术语表

藻类　一组生物，不属于系统分类单元，具有简单的结构，既可以是自养的单细胞或多细胞，通过光合作用产生化学能和氧气，但在真实组织中没有明显的区别。

抗真菌剂　一种天然的或合成的化学物质，能抑制真菌生物的生长，如酵母和霉菌。

抗菌剂　一种天然的或合成的，能杀死微生物或抑制微生物生长的化学物质。

硅藻类　光合作用的单细胞藻类，大小为 0~500 μm。

革兰氏阴性细菌和革兰氏阳性细菌　革兰氏染色后保持粉红色的细菌定义为革兰氏阴性细菌。它们与革兰氏阳性细菌相反，其革兰氏染色后呈蓝紫色。这是由于革兰氏阴性细菌的细胞壁很薄，由不超过5%的肽聚糖组成（不像革兰氏阳性细菌，其肽聚糖占细胞壁本身的50%~90%），使染料渗透细胞并着色。

Las　铜绿假单胞菌的群体感应调节系统之一，负责人体肺组织中弹性蛋白含量的降解，随后肺出血与铜绿假单胞菌的感染相关。

Lux　基因组，称为 *lux* 操纵子，编码革兰氏阴性生物发光细菌费氏弧菌中的荧光素-荧光素酶细菌系统。

***N*-酰基高丝氨酸内酯（缩写为 AHLs 或 *N*-AHLs）**　一类参与细菌群体感应的信号分子。

褐藻多酚　植物中存在的天然抗氧化剂，可能有助于防止脂蛋白的氧化，对心脏、衰老、癌症和微生物感染有积极作用。

群体淬灭　通过合成或天然化合物阻断或限制 QS 机制的活性，使细菌和真菌能够限制参与细胞-细胞通信和致病性中特定基因的表达。

群体感应　一种依赖于微生物细胞密度的转录调控系统，是同一物种许多细菌细胞用来彼此相互交流的一种机制。

紫色杆菌素　一种天然色素（MW 343.33），为"紫蓝莓"颜色，由细菌如紫色色素杆菌和青菌（*Janthinobacterium lividum*）产生。这些微生物产生紫色杆菌素作为辅助分子，可能会增加它们适应和抵抗不利环境条件的能力。在一些测试中其被用于评价 QS 活性。

缩写词

ABC	ATP-结合盒
AHLs	*N*-酰基高丝氨酸内酯
AI-1	自诱导物 1
AI-2	自诱导物 2
LAB	乳酸菌
ncRNAs	非编码 RNAs
PGUs	间苯三酚单元
QQ	群体淬灭
QS	群体感应
QSI	群体感应抑制剂
QSM	群体感应分子

参考文献

ABACHI S, LEE S, RUPASINGHE H P, 2016. Molecular mechanisms of inhibition of *Streptococcus* species by phytochemicals. Molecules, 21 (2): 215.

ADONIZIO A, KONG K F, MATHEE K, 2008. Inhibition of quorum sensing controlled virulence factor production in *Pseudomonas aeruginosa* by South Florida plant extracts. Antimicrob. Agents Chemother., 52: 198-203.

AL-HUSSAINI R, MAHASNEH A R, 2009. Microbial growth and quorum sensing antagonist activities of herbal plants extracts. Molecules, 14: 3425-3435.

AL-REFI M R, 2016. Antimicrobial, Anti-Biofilm, Anti-Quorum Sensing and Synergistic Effects of Some Medicinal Plants Extracts. Master thesis, The Islamic University, Gaza, pp: 1-150.

AURA A M, 2008. Microbial metabolism of dietary phenolic compounds in the colon. Phytochem. Rev., 7: 407-429.

BACHA K, TARIKU Y, GEBREYESUS F, et al., 2016. Antimicrobial and anti-quorum sensing activities of selected medicinal plants of Ethiopia: implication for development of potent antimicrobial agents. BMC Microbiol., 16 (1): 139.

BASSLER B L, 2002. Small talk: cell-to-cell communication in bacteria. Cell, 109: 421-424.

BÄUMLER A J, SPERANDIO V, 2016. Interactions between the microbiota and pathogenic bacteria in the gut. Nature, 535: 85-93.

BIALONSKA D, KASIMSETTY S G, SCHRADER K K, et al., 2009. The effect of pomegranate (*Punica granatum* L.) by-products and ellagitannins on the growth of human gut bacteria. J. Agric. Food Chem., 57: 8344-8349.

BIXLER G D, BHUSHAN B, 2012. Biofouling: lessons from nature. Philos. Trans. R. Soc. A Math. Phys. Eng. Sci., 370: 2381-2417.

BORRIELLO S P, SETCHELL K D R, AXELSON M, et al., 1985. Production and metabolism of lignans by the human faecal flora. J. Appl. Bacteriol., 58: 37-43.

BOUSTIE J, TOMASI S, GRUBE M, 2011. Bioactive lichen metabolites: alpine habitats as an untapped source. Phytochem. Rev., 10: 287-307.

BRACKMAN G, HILLAERT U, VAN CALENBERGH S, et al., 2009. Use of quorum sensing inhibitors to interfere with biofilm formation and development in *Burkholderia multivorans* and *Burkholderia cenocepacia*. Res. Microbiol., 160: 144-151.

CAPITANI C D, CARVALHO A C L, RIVELLI D P, et al., 2009. Evaluation of natural and synthetic compounds according to their antioxidant activity using a multivariate approach. Eur. J. Lipid Sci. Technol., 111: 1090-1099.

CERDÁ B, ESPIN J C, PARRA S, et al., 2004. The potent in vitro antioxidant ellagitannins from pomegranate juice are metabolized into bioavailable but poor antioxidant hydroxy-6H-dibenzopyran-6-one derivatives by the colonic microflora in healthy humans. Eur. J. Nutr., 43: 205-220.

CHANG W, LI Y, ZHANG L, et al., 2012. Retigeric acid B enhances the efficacy of azoles combating the virulence and biofilm formation of *Candida albicans*. Biol. Pharm. Bull., 35: 1794-1801.

CHANG C Y, KRISHNAN T, WANG H, et al., 2014. Non-antibiotic quorum sensing inhibitors acting against *N*-acyl homoserine lactone synthase as druggable target. Sci. Rep., 4: 7245.

CHEN F, GAO Y, CHEN X, et al., 2013. Quorum quenching enzymes and their application in degrading signal molecules to block quorum sensing-dependent infection. Int. J. Mol. Sci., 14: 17477-17500.

CHERNIN L S, WINSON M K, THOMPSON J M, et al., 1998. Chitinolytic activity in *Chromobacterium violaceum*: substrate analysis and regulation by quorum sensing. J. Bacteriol., 180: 4435-4441.

CULBERSON C F, 1963. The lichen substances of the genus *Evernia*. Phytochemistry, 2: 335-340.

DERYABIN D G, TOLMACHEVA A A, 2015. Antibacterial and anti-quorum sensing molecular composition derived from *Quercus cortex* (oak bark) extract. Molecules, 20: 17093-17108.

DONG Y H, WANG L H, ZHANG L H, 2007. Quorum quenching microbial infections: mechanisms and implications. Philos. Trans. R. Soc. Lond. B, 362 (1483): 1201-1211.

EL-HAMID M IA, 2016. A new promising target for plant extracts: inhibition of bacterial quorum sensing. J. Mol. Biol. Biotechnol., 1: 1.

EOM S H, KIM Y M, KIM S K, 2012. Antimicrobial effect of phlorotannins from marine brown algae. Food Chem. Toxicol., 50: 3251-3255.

ESPIN J C, LARROSA M, GARCIA-CONESA M T, et al., 2013. Biological significance of urolithins, the gut microbial ellagic acid-derived metabolites: the evidence so far. Evid. Based Complement. Alternat. Med., 2013: 27042718.

FEUCHT W, SCHWALB P, ZINKERNAGEL V, 2000. Complexation of fungal structures with monomeric and prooligomeric flavanols. J. Plants Dis. Prot., 107: 106-110.

FIAMEGOS Y C, PANAGIOTIS L, KASTRITIS X, et al., 2011. Antimicrobial and efflux pump inhibitory activity of caffeoylquinic acids from Artemisia absinthium against gram-positive pathogenic bacteria. PLoS One, 6: e18127.

FRATIANNI F, COPPOLA R, SADA A, et al., 2010. A novel functional probiotic product containing phenolics and anthocyanins. Int. J. Probiotics Prebiotics, 5: 85-90.

FRATIANNI F, COPPOLA R, NAZZARO F, 2011. Phenolic composition and antimicrobial and antiquorum sensing activity of an ethanolic extract of peels from the apple cultivar Annurca. J. Med. Food, 14: 957-963.

FRATIANNI F, NAZZARO F, MARANDINO A, et al., 2013. Biochemical composition, antimicrobial activities, and anti-quorum-sensing activities of ethanol and ethyl acetate extracts from *Hypericum connatum* Lam. (*Guttiferae*). J. Med. Food, 16: 454-459.

GAO M, TEPLITSKI M, ROBINSON J B, et al., 2003. Production of substances by *Medicago truncatula* that affect bacterial quorum sensing. Mol. Plant-Microbe Interact., 16: 827-834.

GIMENEZ-BASTIDA J A, TRUCHADO P, LARROSA M, et al., 2012. Urolithins, ellagitannin metabolites produced by colon microbiota, inhibit quorum sensing in *Yersinia enterocolitica*: phenotypic response and associated molecular changes. Food Chem., 132: 1465-1474.

GÖKALSIN B, SESAL N C, 2016. Lichen secondary metabolite evernic acid as potential quorum sensing inhibitor against *Pseudomonas aeruginosa*. World J. Microbiol. Biotechnol., 32: 150.

GONTHIER M P, REMESY C, SCALBERT A, et al., 2006. Microbial metabolism of caffeic acid and its esters chlorogenic and caftaric acids by human faecal microbiota in vitro. Biomed. Pharmacother., 60: 536-540.

GRANATO D, KOOT A, SCHNITZLER E, et al., 2015. Authentication of geographical origin and crop system of grape juices by phenolic compounds and antioxidant activity using chemometrics. J. Food Sci., 80: C584-C593.

GUPTA S, ABU-GHANNAM N, 2011. Recent developments in the application of seaweeds or seaweed extracts as a means for enhancing the safety and quality attributes of foods. Innov. Food Sci. Emerg. Technol., 12: 600-609.

HERMUND D B, 2018. Antioxidant properties of seaweed-derived substances. In: Qin Y (Ed.), Bioactive

Seaweeds for Food Applications. Elsevier, Amsterdam, pp: 201-221.

HOLLMAN P C, KATAN M B, 1999. Dietary flavonoids: intake, health effects and bioavailability. Food Chem. Toxicol., 37: 937-942.

HORNBY J M, JENSEN E C, LISEC A D, et al., 2001. Quorum sensing in the dimorphic fungus *Candida albicans* is mediated by farnesol. Appl. Environ. Microbiol., 67: 2982-2992.

HWA IN LEE S, BARANCELLI G V, MENDES DE CAMARGO T, et al., 2017. Biofilm-producing ability of *Listeria monocytogenes* isolates from Brazilian cheese processing plants. Food Res. Int., 91: 88-91.

KARLSSON P C, HUSS U, JENNER A, et al., 2005. Human fecal water inhibits COX-2 in colonic HT-29 cells: role of phenolic compounds. J. Nutr., 135: 2343-2349.

KEPPLER K, HUMPF H U, 2005. Metabolism of anthocyanins and their phenolic degradation products by the intestinal microflora. Bioorg. Med. Chem., 13: 5195-5205.

KOH C L, SAM C K, YIN W F, et al., 2013. Plant-derived natural products as sources of anti-quorum sensing compounds. Sensors, 13: 6217-6228.

KORUKLUOGLU M, GULGOR G, 2017. Anti quorum sensing activity of plants. In: Mendez-Vilas, A. (Ed.), Antimicrobial research: Novel Bioknowledge and Educational Programs. Formatex Research Center, Badajoz, pp: 529-535.

KOSANIC M, RANKOVIC B, 2011. Antioxidant and antimicrobial properties of some lichens and their constituents. J. Med. Food, 14: 1-7.

KROON P A, WILLIAMS G, 1999. Hydroxycinnamates in plants and food: current and future perspectives. J. Sci. Food Agric., 79: 355-361.

KUHNLE G, SPENCER J P, SCHROETER H, et al., 2000. Epicatechin and catechin are O-methylated and glucuronidated in the small intestine. Biochem. Biophys. Res. Commun., 277: 507-512.

KUIPERS O P, DE RUYTER P G G A, KLEEREBEZEM M, et al., 1998. Quorum sensing-controlled gene expression in lactic acid bacteria. J. Biotechnol., 64: 15-21.

LEACH J E, LLOYD L A, MCGEE J D, et al., 2007. Trafficking of plant defense response compounds. In: Keen N T, Mayama S, Leach J E, Tsuyumu S (Eds.), Delivery and Perception of Pathogen Signals in Plants. The American Phytopathological Society (APS) Press, St. Paul, MN, pp: 1-268.

LEBEER S, DE KEERSMAECKER S C J, VERHOEVEN T L A, et al., 2007. Functional analysis of luxS in the probiotic strain *Lactobacillus rhamnosus* GG reveals a central metabolic role important for growth and biofilm formation. J. Bacteriol., 189: 860-871.

LEE H C, JENNER A M, LOW C D, et al., 2006. Effect of tea phenolics and their aromatic faecal bacterial metabolites on intestinal microbiota. Res. Microbiol., 157: 876-884.

LEE K M, LIM J, NAM S, et al., 2011. Inhibitory effects of broccoli extract on *Escherichia coli* O157: H7 quorum sensing and in vivo virulence. FEMS Microbiol. Lett., 321: 67-74.

LIU Z Y, ZENG H, ZENG M Y, 2014. Primary studies on screening of marine algae polyphenols for quorum sensing inhibitor and their activities. J. Food Saf. Qual., 5: 4097-4101.

LOPES G, SOUSA C, SILVA L R, et al., 2012. Can phlorotannins purified extracts constitute a novel pharmacological alternative for microbial infections with associated inflammatory conditions? PLoS One, 7 (2): e31145.

LOPES G, PINTO E, ANDRADE P B, et al., 2013. Antifungal activity of phlorotannins against dermatophytes and yeasts: approaches to the mechanism of action and influence on *Candida albicans* virulence factor. PLoS One, 8 (8): e72203.

MAHMOUDI E, TARZABAN S, KHODAYGAN P, 2014. Dual behaviour of plants against bacterial quorum sensing: inhibition or excitation. J. Plant Pathol., 96: 295-301.

MANACH C, HUBERT J, LLORACH R, et al., 2009. The complex links between dietary phytochemicals and human health deciphered by metabolomics. Mol. Nutr. Food Res., 53: 1303-1315.

MARTÍNEZ J H I, CASTAÑEDA H G T, 2013. Preparation and chromatographic analysis of phlorotannins. J. Chromatogr. Sci., 51: 825-838.

MESELHY M R, NAKAMURA N, HATTORI M, 1997. Biotransformation of (-)-epicatechin-3-O-gallate by human intestinal bacteria. Chem. Pharm. Bull., 45: 888-893.

MILLOT M, GIRARDOT M, DUTREIX L, et al., 2017. Antifungal and anti-biofilm activities of acetone lichen extracts against *Candida albicans*. Molecules 22: 651-662.

MITROVIC T, STAMENKOVIC S, CVETKOVIC V, et al., 2014. *Platismatia glaucia* and *Pseudevernia furfuracea* lichens as sources of antioxidant, antimicrobial and antibiofilm agents. EXCLI J., 13: 938-953.

MUSTHAFA K S, RAVI A V, ANNAPOORANI A, et al., 2010. Evaluation of anti-quorumsensing activity of edible plants and fruits through inhibition of the N-acyl-homoserine lactone system in *Chromobacterium violaceum* and *Pseudomonas aeruginosa*. Chemotherapy, 56: 333-339.

NAGAYAMA K, IWAMURA Y, SHIBATA T, et al., 2002. Bactericidal activity of phlorotannins from the brown alga *Ecklonia kurome*. Antimicrob. Agents Chemother., 50: 889-893.

NAZZARO F, FRATIANNI F, COPPOLA R, 2013. Quorum sensing and phytochemicals. Int. J. Mol. Sci., 14: 12607-12619.

NAZZARO F, FRATIANNI F, D'ACIERNO A, et al., 2015. Gut microbiota and polyphenols: a strict connection enhancing human health. In: Ravishankar Rai, V. (Ed.), Advances in Food Biotechnology, first ed. John Wiley & Sons, Ltd, New York, pp: 335-349.

NAZZARO F, FRATIANNI F, COPPOLA R, et al., 2017. Essential oils and antifungal activity. Pharmaceuticals, 10: 86.

NITHYANAND P, BEEMA SHAFREEN R M, MUTHAMIL S, et al., 2015. Usnic acid inhibits biofilm formation and virulent morphological traits of *Candida albicans*. Microbiol. Res., 179: 20-28.

NOMURA H, ISSHIKI Y, SAKUDA K, et al., 2013. Effects of oakmoss and its components on biofilm formation of *Legionella pneumophila*. Biol. Pharm. Bull., 36: 833-837.

NOUMI E, SNOUSSI M, MERGHNIA A, et al., 2017. Phytochemical composition, anti-biofilm and anti-quorum sensing potential of fruit, stem and leaves of *Salvadora persica* L. methanolic extracts. Microb. Pathog., 109: 169-176.

PANZELLA L, NAPOLITANO A, 2017. Natural phenol polymers: recent advances in food and health applications. Antioxidants, 6: 30.

PARKAR S G, STEVENSON D E, SKINNER M A, 2008. The potential influence of fruit polyphenols on colonic microflora and human gut health. Int. J. Food Microbiol., 124: 295-298.

PASCUAL-TERESA S, SANCHEZ-BALLESTA M T, 2007. Anthocyanins: from plant to health. Phytochem. Rev., 7: 281-299.

PELEG A Y, HOGAN D A, MYLONAKIS E, 2010. Medically important bacterial-fungal interactions. Nat. Rev. Microbiol., 8: 340-349.

PLAZA M, AMIGO-BENAVENT M, DEL CASTILLO M, et al., 2010. Facts about the formation of new antioxidants in natural samples after subcritical water extraction. Food Res. Int., 43: 2341-2348.

POMPILIO A, RIVIELLO A, CROCETTA V, et al., 2016. Evaluation of antibacterial and antibiofilm mechanisms by usnic acid against methicillin-resistant *Staphylococcus aureus*. Future Microbiol, 11: 1315-1338.

POSSEMIERS S, BOLCA S, EECKHAUT E, et al., 2007. Metabolism of isoflavones, lignans and prenylflavonoids by intestinal bacteria: producer phenotyping and relation with intestinal community. FEMS Microbiol. Ecol., 6: 1372-1383.

QUIRÓS SAUCEDAA E, PACHECO-ORDAZ R, AYALA-ZAVALA J F, et al., 2017. Impact of fruit dietary fibers and polyphenols on modulation of the human gut microbiota. In: Yahia, E. M. (Ed.), Fruit and Vegetable Phytochemicals: Chemistry and Human Health. vol. 1. John Wiley & Sons Ltd, pp: 405-422.

RASAMIRAVAKA T, JEDRZEJOWSKI A, KIENDREBEOGO M, et al., 2013. Endemic Malagasy Dalbergia species inhibit quorum sensing in *Pseudomonas aeruginosa* PAO1. Microbiology, 159 (Pt 5): 924-938.

RASMUSSEN T B, GIVSKOV M, 2006. Quorum sensing inhibitors: a bargain of effects. Microbiology, 152: 895-904.

READING N C, SPERANDIO V, 2006. Quorum sensing: the many languages of bacteria. FEMS Microbiol. Lett., 254: 1-11.

RECHNER A R, SMITH M A, KUHNLE G, et al., 2004. A colonic metabolism of dietary polyphenol: influence of structure on microbial fermentation products. Free Radic. Biol. Med., 36: 212-225.

RISTIC S, RANKOVIC B, KOSANIC M, et al., 2016. Biopharmaceutical potential of two *Ramalina lichens* and their metabolites. Curr. Pharm. Biotechnol., 17: 651-658.

RUDRAPPA T, BAIS H P, 2008. Curcumin, a known phenolic from *Curcuma longa*, attenuates the virulence of *Pseudomonas aeruginosa* PAO1 in whole plant and animal pathogenicity models. J. Agric. Food Chem., 56: 1955-1962.

RUTHERFORD S T, BASSLER B L, 2012. Bacterial quorum sensing: its role in virulence and possibilities for its control. Cold Spring Harb. Perspect. Med., 2: a012427.

SARABHAI S, SHARMA P, CAPALASH N, 2013. Ellagic acid derivatives from *Terminalia chebula* Retz. Down regulate the expression of quorum sensing genes to attenuate *Pseudomonas aeruginosa* PAO1 virulence. PLoS One, 8: e53441.

SAVOIA D, 2012. Plant-derived antimicrobial compounds: alternatives to antibiotics. Future Microbiol, 7: 979-990.

SCALBERT A, WILLIAMSON G, 2000. Dietary intake and bioavailability of polyphenols. J. Nutr., 130: 2073S-2085S.

SCHUSTER M, SEXTON D J, DIGGLE S P, et al., 2013. Acyl-homoserine lactone quorum sensing: from evolution to application. Annu. Rev. Microbiol., 67: 43-63.

SELMA M V, ESPIN J C, TOMAS-BARBERAN F A, 2009. Interaction between phenolics and gut microbiota: role in human health. J. Agric. Food Chem., 57: 6485-6501.

SETCHELL K D R, BROWN N M, ZIMMER-NECHEMIAS L, et al., 2002. Evidence for lack of absorption of soy isoflavone glycosides in humans, supporting the crucial role of intestinal metabolism for bioavailability. Am. J. Clin. Nutr., 76: 447-453.

SHANNON E, ABU-GHANNAM N, 2016. Antibacterial derivatives of marine algae: an overview of pharmacological mechanisms and applications. Mar. Drugs, 14 (81): 1-23.

SHIBATA T, KAWAGUCHI S, HAMA Y, et al., 2004. Local and chemical distribution of phlorotannins in brown algae. J. Appl. Phycol., 16: 291-296.

SHUKLA V, BHATHENA Z, 2016. Broad spectrum anti-quorum sensing activity of tannin-rich crude extracts of Indian medicinal plants. Scientifica, 2016: 1-8.

SHUKLA V, JOSHI G P, RAWAT M, 2010. Lichens as a potential natural source of bioactive compounds: a review. Phytochem. Rev., 9: 303-314.

SINGH B N, UPRETI D K, SINGH B R, et al., 2015. Quercetin sensitizes fluconazole-resistant *Candida albicans* to induce apoptotic cell death by modulating quorum sensing. Antimicrob. Agents Chemother., 59:

2153-2168.

SMITH A H, MACKIE R I, 2004. Effect of condensed tannins on bacterial diversity and metabolic activity in the rat gastrointestinal tract. Appl. Environ. Microbiol., 70: 1104-1115.

STURME M H J, NAKAYAMA J, MOLENAAR D, et al., 2005. An agr-like two-component regulatory system in *Lactobacillus plantarum* is involved in production of a novel cyclic peptide and regulation of adherence. J. Bacteriol., 187: 5224-5235.

SUGA H, SMITH K M, 2003. Molecular mechanisms of bacterial quorum sensing as a new drug target. Curr. Opin. Chem. Biol., 7: 586-591.

THOMAS N V, KIM S K, 2011. Potential pharmacological applications of polyphenolic derivatives from marine brown algae. Environ. Toxicol. Pharmacol., 32: 325-335.

THOMPSON J A, OLIVEIRA R A, DJUKOVIC A, et al., 2015. Manipulation of the quorum-sensing signal AI-2 affects the antibiotic-treated gut microbiota. Cell Rep., 10: 1861-1871.

TOLMACHEVA A A, ROGOZHIN E A, DERYABIN D G, 2014. Antibacterial and quorum sensing regulatory activities of some traditional Eastern-European medicinal plants. Acta Pharma., 64: 173-186.

TRUCHADO P, LARROSA M, CASTRO-IBANEZ I, et al., 2015. Plant food extracts and phytochemicals: their role as quorum sensing inhibitors. Trends Food Sci. Technol., 43: 189-204.

VAN DUYNHOVEN J, VAUGHAN E E, JACOBS D M, et al., 2011. Metabolic fate of polyphenols in the human superorganism. Proc. Natl. Acad. Sci. U. S. A., 108: 4531-4538.

VANDEPUTTE O M, KIENDREBEOGO M, RAJAONSON S, et al., 2010. Identification of catechin as one of the flavonoids from *Combretum albiflorum* bark extract that reduces the production of quorum-sensing-controlled virulence factors in *Pseudomonas aeruginosa* PAO1. Appl. Environ. Microbiol., 76: 243-253.

VANDEPUTTE O M, KIENDREBEOGO M, RASAMIRAVAKA T, et al., 2011. The flavanone naringenin reduces the production of quorum sensing-controlled virulence factors in *Pseudomonas aeruginosa* PAO1. Microbiology, 157: 2120-2132.

VATTEM D A, MIHALIK K, CRIXELL S H, et al., 2007. Dietary phytochemicals as quorum sensing inhibitors. Fitoterapia, 78: 302-310.

VAZQUEZ-ARMENTA F J, BERNAL-MERCADO A T, LIZARDI-MENDOZa J, et al., 2017. Phenolic extracts from grape stems inhibit *Listeria monocytogenes* motility and adhesion to food contact surfaces. J. Adhes. Sci. Technol., 32: 1-19.

VIJAYAVEL K, MARTINEZ J A, 2010. In vitro antioxidant and antimicrobial activities of two Hawaiian marine Limu: *Ulva fasciata* (Chlorophyta) and *Gracilaria salicornia* (Rhodophyta). J. Med. Food, 13: 1494-1499.

VIKRAM A, JAYAPRAKASHA G K, JESUDHASAN P R, et al., 2010. Suppression of bacterial cell-cell signalling, biofilm formation and type III secretion system by citrus flavonoids. J. Appl. Microbiol., 10: 515-527.

WALLE T, HSIEH F, DELEGGE MH, et al., 2004. High absorption but very low bioavailability of oral resveratrol in humans. Drug Metab. Dispos., 32: 1377-1382.

WHITELEY M, DIGGLE S P, GREENBERG E P, 2017. Progress in and promise of bacterial quorum sensing research. Nature, 551: 313-320.

YANG K, DULEY M L, ZHU J, 2018. Metabolomics study reveals enhanced inhibition and metabolic dysregulation in *Escherichia coli* induced by *Lactobacillus acidophilus*-fermented black tea extract. J. Agric. Food Chem., 66: 1386-1393.

第4部分
应用

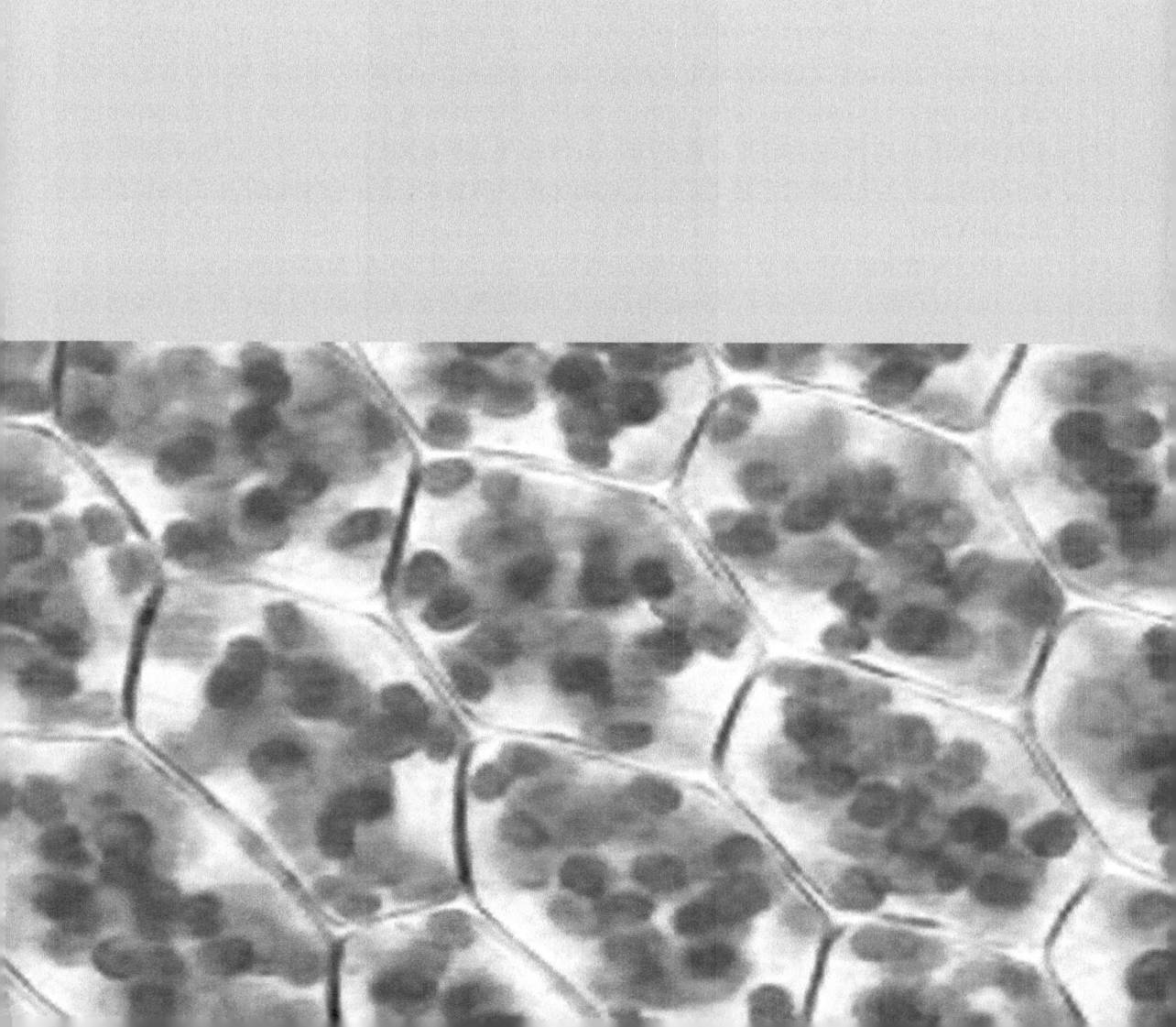

第9章 铜绿假单胞菌群体感应和生物膜抑制

Barış Gökalsın[*,a], Didem Berber[†,a], Nüzhet Cenk Sesal[†,a]

[*] Marmara University, Department of Biology, Institute of Pure and Applied Sciences, Istanbul, Turkey,

[†] Marmara University, Department of Biology, Faculty of Arts and Sciences, Istanbul, Turkey

1 前言

抗生素被发现以来，已被广泛用于治疗感染。然而，这些药物的滥用导致微生物通过获得抗生素耐药性来适应环境。人们主要担心抗生素在短时间内失效、预防和治疗感染将变得不可能，传统感染将重新成为死亡的主要死因之一（Van Hecke et al., 2017）。世界卫生组织报告称，我们缺乏有效治疗抗生素耐药细菌治疗的选择，国际协调必须努力克服这种情况（WHO, 2017）。Laxminarayan 等（2016）估计，平均每年有 214 000 例新生儿败血症死亡是由抗生素耐药病原体引起的。据报道，美国每年大约有 23 000 人死于抗生素耐药感染（CDC, 2017）。住院患者直接遭受抗生素耐药细菌的影响，其中铜绿假单胞菌因多重耐药性而难以治疗。

铜绿假单胞菌是一种革兰氏阴性机会性病原菌，约占所有医院感染的 10%（Diekema et al., 1999）。铜绿假单胞菌可导致肺部、血液、泌尿道和手术部位的感染。虽然它主要感染免疫功能低下的患者，但健康人也可能被感染。多重耐药（MDR）被认为是囊性纤维化（CF）患者高死亡率的主要原因。CF 是一种常见的疾病，全球约有 70 000 人受影响，其通过影响呼吸系统降低肺功能（Rivas Caldas and Boisrame, 2015）。通过集中治疗，CF 患者可延长 35~50 年的寿命。如果不及时治疗，许多 CF 患者可能在年轻时死亡。MDR 生物膜形成使得这种疾病难以用抗生素治疗。

众所周知，当铜绿假单胞菌达到一定的密度时，菌落会产生多糖基质并附着在其表面上，形成宏观上可见且难以去除的生物膜。据报道，这种生物膜对抗生素的耐药性是同一种类游离生物的 1 000~3 000 倍（Olson et al., 2002）。生物膜提供了许多优势，如积累营养物质并保护微生物免受消毒剂、抗生素、紫外线、pH 值、湿度和温度波动，以及以细菌和病毒为食的生物的侵害（Hall-Stoodley et al., 2004）。

对铜绿假单胞菌感染的明显反应是开发新的治疗方法，不仅预防感染还能重组抗生素的使用。抗病毒方法是实现这一目标的主要进展。抗毒性不是直接消灭病原体，而是通过交替抑制其毒力（Fuqua and Greenberg, 2002）。因此，针对抗病毒方法的目标需要特定的生物学信息。许多已知的毒力因子，如细菌毒素、表面蛋白、免疫逃避因子和黏附素。现了解到，除许多行为外，大多数这些因子都受到群体感应（QS）系统的控制，包括生物发光、聚集、偶联、蛋白酶活性以及生物膜形成。

2 群体感应和群体淬灭

已确定浮游细胞不能在环境中自由存在，因为它们必须与其他共存的生物竞争并在极端条件下生存。因此，需要通过 QS 进行通信。较高的细菌种群密度会触发种内、种间或界间的 QS 系统。小而可扩散的化学信号分子被称为自诱导物（AI），被分泌到细菌的局部环境中。已鉴定出三种主要的 AIs 类型：AI-1，也称为 AHLs（N-酰基 L-高丝氨酸内酯），被革兰氏阴性细菌利用；革兰氏阳性细菌使用自诱导肽（AIP）；革兰氏阳性和阴性细菌都使用自诱导物-2（AI-2）进行种间相互作用（Rutherford and Bassler，2012）。QS 系统由五个元素组成，负责 QS 调控：AI 分子、信号合成酶、受体、调节因子和基因。大多数致病细菌通过 QS 系统协调其毒力因子、生物膜形成和抗生素抗性（Li and Tian，2012）。

群体感应一词于 1970 年首次在费氏弧菌和哈维弧菌中提出，这些细菌已知具有发光特征的海洋细菌。通过对这些细菌生物发光特性的研究，揭示了 QS 系统的细节。QS 被定义为细菌的一种感知系统，用于检测其周围环境中种群密度的变化。在细菌生长的滞后阶段，可以很容易地监测到上述细菌的发光特性。在早期对数生长期的培养物中，发光强度达到最大，可将固定相中的无细胞上清液添加到培养基中。AI 分子是可释放到细菌周围环境中的小且可扩散的化合物。在较低的细胞密度下，这些信号分子的浓度保持较低水平；而在较高的细胞密度下，它们会累积到一定的浓度阈值。

在费氏弧菌中，第一个被鉴定的 AI（VAI）是 3-O-C6-HSL（Eberhard et al.，1981）。VAI 的生物合成由 *luxI* 基因以正反馈机制编码。细菌种群密度的增加导致 VAI 分子在周围环境中浓度积累升高。这些信号分子与 *luxR* 基因的相互作用导致 *luxICDABE* 操纵子的转录。随后，由 *luxCDABE* 操纵子编码的荧光素酶使 QS 激活通过发光可见。另外，由于细胞密度降低，低浓度的 VAI 不足以激活 *luxR* 和 *luxI* 基因，因此，荧光素酶不能被编码，这些信息为 QS 机制的研究提供了便利。

据了解，当细菌细胞密度增加，AHL 达到一定阈值时，AHL 分子才能作为 QS 信号发挥作用。为了调控 QS 系统，需要通过 AHL 合酶合成 AHLs，并由于细菌群体密度的增加，这些信号才能在较高浓度下积累。因此，可以通过 AHL 监测来评估 QS 系统。AHL 的积累取决于细菌群落中的物理和化学因素。例如，AHL 分子大部分可通过细胞膜扩散。

尽管使用高剂量的抗生素，一些致病细菌如铜绿假单胞菌仍会导致高死亡率和高发病率，特别是在免疫功能低下的患者中（Borges et al.，2016）。这些致病细菌通过水平基因转移和自发突变等多种方式来避免抗生素的杀菌和抑菌作用（Kalia，2013）。此外，抗生素治疗失败可能是由于群体转移、氧化还原机制或酶水解作用（Hentzer and Givskov，2003）。因此，最近的研究集中在开发替代策略，通过干扰细菌的 QS 来防止细菌对抗生素的耐药性。这些抗 QS 方法被称为群体淬灭（QQ；Hentzer and Givskov，2003）。从植物、动物、真菌、细菌和藻类中发现几种天然或合成的 QS 抑制剂（QSI）化合物，并检测了 QSI 的抑制潜力。此外，计算科学和硅方法的技术进步，允许快速筛选这些化合物。如本章所述，许多研究已经报道了它们的生物学和治疗效果。

2.1 铜绿假单胞菌群体感应系统和生物膜

铜绿假单胞菌的 QS 系统调控关键功能，如毒力、运动性、生物膜形成和次级代谢物的产生。像其他革兰氏阴性细菌一样，铜绿假单胞菌也利用 AHLs 作为其主要的 QS 系统。众所周知，铜绿假单胞菌有四个层次连接的 QS 系统用于种间通信：*las*、*rhl*、*pqs* 和 *iqs*（Daniels et al., 2004；Lee and Zhang, 2015）。

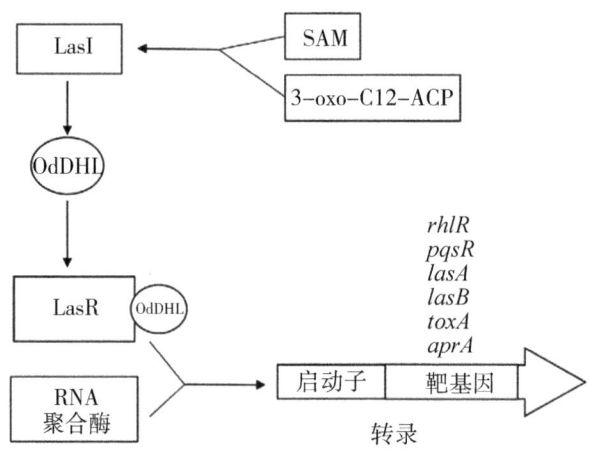

图 9-1 铜绿假单胞菌中 *las* 信号网络的示意

注：*las* 系统通过转录调节因子 LasR 利用信号合成酶 LasI 检测 3-oxo-C12-HSL 的产生，并通过双组分信号转导系统调节毒力因子。

如图 9-1 所示，*las* 系统由转录调节因子 LasR、信号合成酶 LasI 和自诱导物 3-oxo-C12-HSL（OdDHL）组成。*rhl* 系统同样具有 RhlR、RhlI 和自诱导物 C4-HSL（BHL）。*pqs* 系统具有调节因子 PqsR，*pqsABCDE-phnAB* 操纵子和用于信号合成的 PqsH，信号分子 2-烷基-4-喹啉酮（AQs），包括 2-庚基-4-羟基喹啉酮（HHQ）和 2-庚基-3-羟基-4（1H）-喹啉酮，被称为假单胞菌喹啉酮信号（PQS）。最后，最近发现的 *iqs* 系统利用 2-（2-羟基苯基）-噻唑-4-甲醛作为信号分子，并与环境压力有关。

主要的 QS 系统 *las*、*rhl* 和 *pqs* 调控许多毒力因子的产生，如弹性蛋白酶、外毒素 A、鼠李糖脂类、铜绿假单胞菌素、脂肪酶、铁载体、凝集素等。研究表明，MDR 细菌的外排泵系统与 *pqs* 系统有关。在这三个系统中，*las* 负责分层管理其他系统，而 *pqs* 似乎介导 *las* 和 *rhl* 系统，并调节一些毒力因子（Lee and Zhang, 2015）。然而，铜绿假单胞菌的 QS 系统层次结构可能会根据环境胁迫发生改变和适应。例如，在严重的磷酸盐耗竭压力的情况下，*iqs* 系统可以接管 *las* 的功能，或者 *pqs* 系统可以在没有 *las* 系统的情况下激活。因此，毒力途径可根据环境条件和压力而改变。其他调节因子如 QscR 和 RsaL 可以抑制信号产生，从而维持这种复杂的整体信号机制的平衡。

QS 被认为参与了铜绿假单胞菌生物膜的形成。与野生型菌株相比，尽管考虑了培养条件的影响，QS 突变菌株形成扁平且薄弱的生物膜。人们普遍认为 QS 在生物膜形成中起到重要作用，但目前所提出的一些机制仍存在争议。

生物膜形成的第一步是细菌附着在表面上。鞭毛运动（第Ⅳ型菌毛）和黏附素是这个阶段的重要因素。在不可逆的附着之后，微团聚体形成，增加了 QS 信号通信。生物膜的成熟随之开始。成熟生物膜的 3D 结构会根据环境条件和细菌产生的因子的数量而变化。这些因子包括胞外多糖（EPS）如藻酸盐、结构性 DNA、铁螯合剂吡啶，以及鼠李糖脂等表面活性剂，其中大部分直接受到 QS 代谢的调控。适量的鼠李糖脂分泌对于成熟的生物膜和后续的扩散至关重要。鼠李糖脂的过量表达会导致生物膜分散，从而使细菌能够在其他表面上定殖（Boyle et al., 2013）。

在这些阶段，铜绿假单胞菌的 QS 系统在生物膜形成中的作用似乎是显而易见的。然而，一直存在矛盾和多样化的结果和观点。QS-生物膜的关系通常是采用流式细胞系统来研究的。这些系统具有小通道，介质在其中不断循环。细菌菌株在这些通道中形成生物膜结构，并通过激光共聚焦显微镜进行监测。与其野生型对照组相比，这些菌株是 QS 缺陷突变体。关于 QS 与生物膜形成的不同研究结果，导致了一种普遍观点，即培养和环境条件对生物膜结构有重要影响。然而，众所周知，QS 系统对生物膜形成具有重要影响，正如之前解释的，从治疗角度来看，抑制 QS 系统似乎是一个合理的方法（Joo and Otto, 2012）。

2.2 群体感应抑制剂的筛选

QSI 分子必须是高效、稳定、实用的化合物且具有低分子量和高特异性的信号调节因子。同时，对细菌和宿主引起不良效果是很重要的。此外，这些化合物不能受到宿主水解酶的影响。另外，一些化合物结合到受体并激活它们，作为激动剂引起 QS 相关基因的上调。QSIs 优先在其抑制靶点表现出拮抗效应。

群体淬灭（QQ）是一个通用术语，用于描述所有针对抑制 QS 系统的过程。QQ 方法的目的是在不杀死细菌或阻止细菌生长的情况下，破坏细菌之间的通信。QS 抑制有几个靶点。以下是铜绿假单胞菌的 QQ 方法总结，如图 9-2 所示。

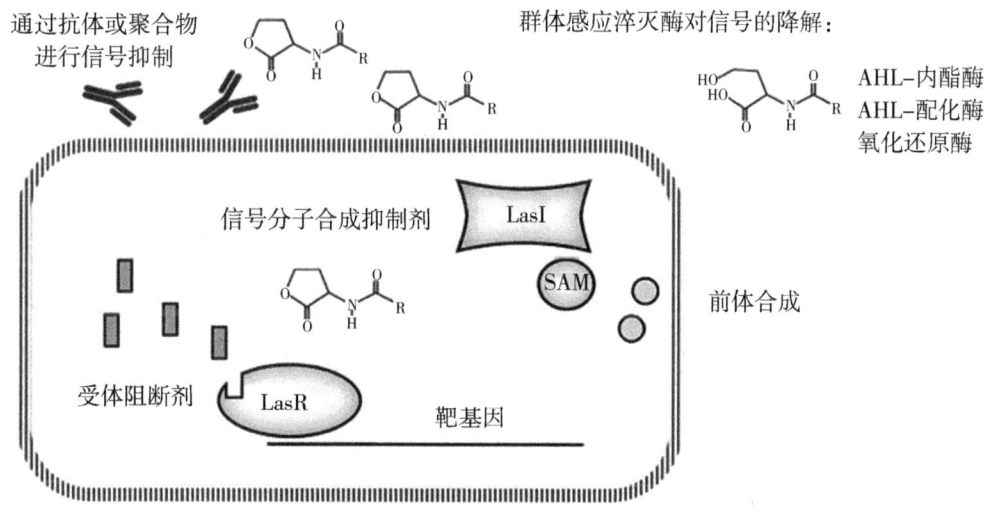

图 9-2　铜绿假单胞菌 las 系统的 QQ 方法和 QSIs

注：这些方法主要集中在抑制自诱导信号的合成，包括其前体的合成、抑制或降解信号分子和阻断信号分子的检测。

2.2.1 群体感应信号生物合成的抑制

抑制 QS 信号生物合成是对治疗铜绿假单胞菌的 QQ 方法之一。在革兰氏阴性细菌中，烯酰基载体蛋白（ACP）还原酶（ENR）和S-腺苷甲硫氨酸（SAM）可能成为N-酰基高丝氨酸内酯（AHL）合成的靶点（Dong et al.，2007）。在AI-2 QS抑制中，利用LuxS酶裂解S-核糖基-L-同型半胱氨酸（SHR）形成4,5-二羟基2,3-戊二酮（DPD）的合成，也可以成为QQ的靶点（Galloway et al.，2011）。

由于细菌抗生素的抗药性是一个重要的全球医疗问题，已对铜绿假单胞菌中的QS系统进行了详细研究。大多数病原菌如铜绿假单胞菌通过QS系统协调其致病性。负责AHL合成和积累的基因已被作为一种替代方法进行靶向研究，揭示了AHL合酶作为抗菌靶点的潜力。为此，已利用抑制基因来减少luxI同源基因的转录。Branny等（2001）从铜绿假单胞菌中分离出rhlI的抑制基因dksA。rhlI基因负责C4-HSL合酶的转录。另外，抑制基因qscR靶向lasI基因，并调控铜绿假单胞菌中QS信号的合成以及毒力因子。据报道，qscR基因的突变导致信号和后续转录的过早发生。其他已知的QS信号合成抑制基因，如铜绿假单胞菌中的rsaL。

NADH依赖的ENR（FabI）负责酰基-ACP合成并形成AHLs的酰基链。据报道，尽管抗菌剂三氯生可抑制FabI酶的产生和C4-HSL的合成，由于外排泵系统的存在，铜绿假单胞菌的耐药性不能被三氯生阻止（LaSarre and Federle，2013）。外排泵系统与QS之间的关系是一个引人注目的课题，进一步的研究可能会揭示有趣的前景。

2.2.2 QS信号降解和失活

QS信号的降解可通过化学、代谢或酶的方式实现。在化学降解中，碱性pH值会导致内酯环打开，而酸性pH值会导致再环化（Rasmussen and Givskov，2006）。然而，大部分QS信号的降解主要由酶来处理，并可使用抗体进行失活。

2.2.3 信号检测抑制

信号分子检测的抑制可以通过竞争拮抗分子在信号分子与受体结合之前来实现。信号受体的失活会抑制毒力因子的表达。大多数天然的QSIs通过抑制LasR、RhlR和PqsR对铜绿假单胞菌具有活性。

2.3 天然群体感应抑制剂

天然QSIs大部分是从天然来源中获得的化合物、提取物、酶和抗体。

2.3.1 群体淬灭酶

已报道的QQ酶有AHL-内酯酶，酰化酶，氧化还原酶，对氧磷酶和2,4-二氧化酶（Hod）。

AHL内酯酶

AHL内酯酶参与金属蛋白质组，通过水解HSL环的酯键形成酰基高丝氨酸内酯。这组酶由于具有高度保守的HSL环，对AHL分子表现出显著的特异性（LaSarre and Federle，2013）。自诱导失活基因（AiiA）首次被描述为内酯酶，在芽孢杆菌属炭疽杆菌（*B. anthracis*）、蜡样芽孢杆菌、蕈状芽孢杆菌、枯草芽孢杆菌、苏云金杆菌（*B. thuringiensis*）、节杆菌属，酸杆菌（*Acidobacteria*），农杆菌属和克雷伯菌属中发现，并被证明能降解AHL（Kalia and Purohit，2011）。

据报道，铜绿假单胞菌和泰国芽孢杆菌中的AiiA等位基因抑制了AHLs的聚集。在植物病原体根瘤农杆菌中发现，AttM与AiiA具有低水平的相似性，尽管它们具有相同保守的

HXDH 基序。此外，AiiB、AhlD、AhlK、AidC、QlcA、BipB01、BipB04、BipB05、BipB07、QsdA、AiiM、AidH 和 QsdH 被报道作为内酯酶。它们在 DNA 序列和对金属离子的依赖性上有所不同（LaSarre and Federle，2013）。

哺乳动物酶对氧磷酶 1、对氧磷酶 2 和对氧磷酶 3（PON1、PON2 和 PON3）表现出内酯酶活性。Hraiech 等（2014）发现了一种新的 SsoPox 变异分子（超耐热内酯酶）。研究证明，在急性肺炎大鼠模型中 SsoPox-I（磷酸三酯酶样内酯酶）具有 QQ 潜力，并在气管内应用 SsoPox-I 可提高其存活率，同时也降低了 lasB 活性、铜绿假单胞菌素合成、蛋白酶活性、生物膜形成和肺组织损伤。

最近，Tang 等（2015）报道了从鼠尾菌中新发现的 AHL 内酯酶 MomL，其抑制铜绿假单胞菌中蛋白酶和铜绿假单胞菌素的产生。研究人员在秀丽隐杆线虫（*Caenorhabditis elegans*）模型中评估了 MomL 对铜绿假单胞菌毒力的影响，并观察到毒力抑制。

酰基转移酶

这组酶能将 HSL 和酰基侧链之间的酰胺键裂解。在这种裂解之后，形成了一个脂肪酸链和一个 HSL 基团。据报道，这些酶的特异性与酰基链的长度和链的第三位的取代有关（LaSarre and Federle，2013）。

假单胞菌对其自身的 AHL 具有酰化酶活性（Grandclement et al.，2016）。在铜绿假单胞菌中发现了编码酰胺酶的基因 *pvdQ*（PA2385），*quiP*（PA1032）和 *hacB*（PA0305）（Huang et al.，2006；Wahjudi et al.，2011；Sio et al.，2006）。

Grover 等（2016）将来自蜂蜜曲霉酰化酶固定在聚氨酯涂层中。这种固定化酶抑制了铜绿假单胞菌 ATCC 10145 和 PAO1 菌株的生物膜形成和铜绿假单胞菌素的产生。Sunder 等（2017）研究了青霉素酰化酶（PVAs）对黑腐果胶杆菌（*P. atrosepticum*）和根瘤农杆菌的影响。研究人员将这些酶转移到铜绿假单胞菌中，并研究其对毒力因子和生物膜形成的抑制，以及提高在昆虫急性感染模型中的存活率。

副磷酸酶类

副磷酸酶是哺乳动物酶，作为 QQ 酶起作用。有 3 种类型的副磷酸酶（PON1，PON2 和 PON3）。PONs 被描述为关于破坏 QS 系统类似的内酯酶（Grandclement et al.，2016）。Stoltz 等（2007）报道在小鼠气管上皮细胞中过表达的 PON2 引起 3OC12-HSL 的降解。

最近，来自糖尿病足溃疡患者的铜绿假单胞菌菌株中检测 SsoPox-W263I。研究报道了对毒力因子（蛋白酶和铜绿假单胞菌素产生）的抑制。研究表明，与 5-氟尿嘧啶和 C-30 相比，SsoPox-W263I 更高效（Guendouze et al.，2017）。此外，人血清对氧磷酶 1（hPON1）表现出对铜绿假单胞菌素、鼠李糖脂、弹性蛋白酶、葡萄球菌 LasA 蛋白酶和碱性蛋白酶活性的降低（Aybey and Demirkan，2016）。

氧化还原酶

另一组酶，氧化还原酶，能氧化或还原 AHL 的酰基侧链而不降解。这些酶通过修饰 AHL 分子脂肪酸侧链的 C3 酮基来发挥 QSIs 的作用。来自宏基因组衍生克隆的 NADP-依赖性氧化还原酶 BpiB09，其使 AHL 介导铜绿假单胞菌生物膜形成失活（Weiland-Brauer et al.，2016）。

2,4-双加氧酶（Hod）

Hod，另一种 QQ 酶，使 PQS 形成邻苯二甲酸和一氧化碳。Pustelny 等（2009）研究将来自节杆菌属 Rue61a 的杂环裂解酶 Hod 外源添加到铜绿假单胞菌培养物中，在植物叶片感染模型中可抑制关键毒力因子和组织损伤。

2.3.2 抗体

QQ 抗体主要针对铜绿假单胞菌中 HSL 信号失活。另外，其他在信号合成中起作用的因素也可以被靶向。

采用单克隆或多克隆抗体的免疫药物治疗方法已被用于减弱 QS、控制毒力调控和生物膜形成，如表 9-1 所示。

表 9-1 群体淬灭抗体

抗体	研究	靶标	参考文献
3-oxo-C12-HSL-牛血清白蛋白复合物	3-oxo-C12-HSL-载体蛋白偶联物抗肺肿瘤坏死因子（TNF）-α 和抗巨噬细胞凋亡	铜绿假单胞菌 C12HSL	Miyairi et al., 2006
HSL-2 和 HSL-4 单克隆抗体（mAbs）	MAbs	铜绿假单胞菌 HSLs	Palliyil et al., 2014
MAbs	靶向 DNABI 蛋白 DNA 结合位点的 MAbs	铜绿假单胞菌生物被膜破坏，MAbs 抗生素治疗	Novotny et al., 2016
RS2-1G9	单克隆抗体	铜绿假单胞菌 C12HSL	Kaufmann et al., 2006
XYD-11G2	单克隆抗体	铜绿假单胞菌 C12HSL	De Lamo Marin et al., 2007

Miyairi 等（2006）研究了对具有 OdDHL 载体蛋白共轭物对患有铜绿假单胞菌肺部感染的小鼠的主动免疫效果。结果表明，血清中具有 OdDHL 特异性抗体的小鼠对铜绿假单胞菌感染的生存率较高。免疫血清能增加 OdDHL 诱导的巨噬细胞凋亡的细胞活力，并提高其存活率。Palliyil 等（2014）利用羊免疫和重组抗体技术制备单克隆抗体（MAbs），并检测铜绿假单胞菌中 HSL 分子。实现了在尿液中对 HSL 的"nmol"级灵敏度检测。通过使用线虫和小鼠模型，对使用 HSL MAbs 单抗治疗的感染组与对照组的存活率进行了比较，使用 HSL MAbs 单抗治疗小鼠的存活率呈现出显著增加。这些研究表明，抗 QS 信号分子的抗体可以作为一种可行的补充治疗方法。

2.3.3 天然群体感应抑制剂化合物和提取物

多种生物产生天然 QSIs，如细菌、藻类、动物、植物或真菌，且其 QS 抑制效果已被许多研究证明。这些抑制剂在生化结构上表现出高度的多样性。不幸的是，目前对 QSIs 的分子结构或化学基团在 QS 介导途径中的功能尚缺乏详细的信息（LaSarre and Federle, 2013）。

细菌

已报道了一些潜在的 QSIs 来自多种细菌门的成员，如放线菌门、厚壁菌门、蓝藻门、拟杆菌门和变形菌门。从其他生物体中分离出细菌并研究其 QSI 活性是常见的做法。考虑到竞争环境，这种方法似乎是合理的。如前所述，许多细菌通过酶显示其 QSI 活性。已知具有 QSI 活性的 AHL 内酯酶有 MomL、*hqiA*、AiiAAI96、*Sso*Pox、SsoPox-W263I、AIiiM、AttM、AiiB、Ahl、AhlD、MLR6805、DlhR、Qsd、AidC、AhlS、Aii20J、QsdA、GKL、MCP、AidH、AiiO、QsdH、QlcA、BpiB01、BpiB04、BpiB05 和 BpiB07。MacQ、青霉素 V 酰化酶、青霉素 G 酰化酶 KcPGA、AiiD、PvdQ QuiP HacA、HacB、AibP、AhlM、AiiC 和 Aac 都是酰

化酶。LsrK 和 LsrG 是 AI-2 介导的 QS 的抑制剂。Hod 和 CarAB 分别靶向 PQS 和 DSF 通路。CYP102A1 是一种氧化还原酶（LaSarre and Federle，2013；Grandclement et al.，2016）。

Devaraj 等（2017）报道了 147 株土壤放线菌的 QSI 潜力，其对铜绿假单胞菌 PAO1 菌群运动和铜绿假单胞菌素产生均表现出阳性结果。他们观察到，三种放线菌（微单孢菌、红球菌和链霉菌）抑制了紫色色素杆菌 CV026 紫色杆菌素的产生，并抑制了铜绿假单胞菌 PAO1 的群体运动和铜绿假单胞菌素的产生。Yaniv 等（2017）检测了红海浮游微生物群中 2 500 条细菌人工染色体（BAC）文库克隆对指示生物紫色色素杆菌的 QSI 潜力。他们发现一种活性化合物 14-A5 可抑制 QS 途径，并降低铜绿假单胞菌生物膜的形成。

Chang 等（2017）对北大西洋地表水的海洋菌株进行筛选，并评估其抗 QS 的潜力。研究发现，根瘤菌属 NAO1 提取物中含有基于 AHL 的 QS 类似物分子。他们观察到，根瘤菌属 NAO1 不仅抑制 QS 系统，还抑制紫色色素杆菌和铜绿假单胞菌 PAO1 生物膜的形成。他们还研究了铜绿假单胞菌 PAO1 的铁载体和弹性蛋白酶活性等毒力因子，并发现其受到一定程度的抑制。研究发现，由于抑制铜绿假单胞菌生物膜的形成，当应用于紫色色素杆菌 NAO1 的次生代谢产物时，对氨基糖苷类抗生素的敏感性增加。

水莱茵海默氏菌（*Rheinheimera aquimaris*）QSI02 乙酸乙酯（EtOAc）提取物被检测出对紫色色素杆菌 CV026 具有 QQ 活性。通过使用生物测定指导的分离方案，从水莱茵海默氏菌中检测到一种活性二酮哌啶因子，环（Trp-Ser）。这种二酮哌啶因子可抑制铜绿假单胞菌 PAO1 中铜绿假单胞菌素、弹性蛋白酶活性和生物膜形成（Sun et al.，2016）。

Muller 等（2014）从土壤中分离并鉴定出红斑红球菌作为一种 PQS 降解菌。他们报道了红斑红球菌 BG43 能将铜绿假单胞菌的 QS 分子 HHQ 和 PQS 降解为氨基苯甲酸，也可以将为 2-庚基-4-羟基喹啉-N-氧化物转化为 PQS 的潜力。在菌株 BG43 的质粒 pRLCBG43 上鉴定出了 aqdA1B1C1 和 aqdA2B2C2，即为 PQS 诱导基因，分别负责编码 HHQ 羟化为 PQS 途径的酶以及 PQS 降解为氨基苯甲酸盐途径中的酶。推测这些基因在铜绿假单胞菌中 PQS 裂解的双加氧酶表达中起着重要作用。AqdC 蛋白首次被鉴定为裂解 PQS 的酶。在铜绿假单胞菌 PAO1 中表达 PQS 双氧酶基因 *aqdC1* 或 *aqdC2*，证明其对铜绿假单胞菌素、鼠李糖脂和铁载体产生的潜在抑制作用（Muller et al.，2015）。

藻类

从大型藻类和微型藻类中可提出多种活性化合物，如具有 QS 抑制作用的褐藻多酚。许多海洋生物中已经报道了海洋来源的 QSIs。此外，一些大型藻类能通过这些化合物来保护自己抵御表面相关的细菌侵害（Saurav et al.，2017）。

Rajamani 等（2008）评估了绿藻 *Chlamydomonas reinhardtii* CC-2137 中的天然化合物光色素对铜绿假单胞菌的影响。光色素是维生素核黄素的衍生物。发现核黄素和光色素能增加 *luxCDABE* 基因的表达，表明这些化合物对 LasR 受体具有特异性，并且光色素能激活 LasR 结合到 LasI 启动子上。

此外，核黄素和发光素可结合到 LasR 的同一结合口袋上，这一点通过分子对接方法得到确认。

澳大利亚大型红藻 *Delisea pulchra*，因其次生代谢物具有抗菌作用而受到关注。它们产生溴化和氯化呋喃酮在结构类上与 AHL 分子相似，可能很容易与信号受体结合（Shannon and Abu-Ghannam，2016）。

动物

从动物中鉴定出几种 QSIs，其中大部分用于抗菌污染。从海绵（*Luffariella variabilis*）中获得的二倍半萜代谢物甘露内酯、甘露内酯单乙酸酯和二甘露内酯，以及海藻虫属点状透明虫（*Hyalinella punctate*）的乙酸乙酯提取物，对铜绿假单胞菌的 QSI 作用而广为人知（Skindersoe et al., 2008；Pejin et al., 2016）。

Costantino 等（2017）从一种名为左旋呋喃内酯中分离出 γ-内酯，该化合物与印度尼西亚海洋海绵（*Plakortis cf. Lita*）提取得到的 LasI/R 系统有关。该化合物在大肠杆菌 pSB401 和紫色色素杆菌 CV026 进行了短酰基链信号的 QQ 潜力检测，但未检测到抑制效果。但他们观察到该化合物能抑制 AHL 诱导的 C6-HSL 检测 pSB401 菌株和 C12-HSL 检测 pSB1075 菌株的生物发光，以及降低铜绿假单胞菌 PAO1 蛋白酶活性。

Skindersoe 等（2008）对大堡礁的 284 个海洋生物样本进行了筛选，以通过两个 QSI 选择系统（QSIS1 和 QSIS2）评估它们的 QSI 活性。从海绵（*Luffariella variabilis*）中分离出的三种 C25 倍半萜代谢物（manoalide, manoalide monoacetate 和 secomanoalide）被证明对 *lasB: gfp* [ASV] 融合具有 QSI 效应。

海藻虫属点状透明虫（*Hyalinella punctate*）和淡水苔藓虫，对其抗生物膜和抗 QS 活性进行了检测（Pejin et al., 2016）。点状透明虫的乙酸乙酯提取物对铜绿假单胞菌 PAO1 具有显著的抗生物膜活性。这些提取物对抽搐运动也表现出良好的效果，并有效地抑制铜绿假单胞菌 PAO1 铁载体的产生。

Quintana 等（2015）评估 26 种海绵、7 种软珊瑚、5 种藻类和 1 种珊瑚虫的 39 种提取物的 QSI 活性。研究发现，*Eunicea laciniata*、*Svenzea tubulosa*、*Ircinia felix* 和 *Neopetrosia carbonaria* 的粗提取物对 QS 抑制作用相当有效。研究人员从海绵 *I. felix* 的粗提取物中分离出呋喃酯类三萜化合物，具有中等抗 QS 潜力。

Mai 等（2015）从 *Leucetta chagosensis* 的粗提物中分离出 isonaamine A、isonaamidine A、isonaamine D、leucettamine D 和 di-isonaamidine A 化合物。isonaamine D 和 isonaamidine A 抑制哈维弧菌的三条 QS 途径。其中，Isonaamidine A 对 AI-2 生物传感器的抑制作用最大。

利用紫色色素杆菌 CV017 对来自海洋生物（海绵、藻类、真菌、海鞘、蓝藻、陆地植物）的 78 种天然产物 QQ 潜力进行评估。研究发现，脱甲氧基恩西卡林、中间酰胺、苦脲酸、菌核消毒素、微胞素 A 和 B 以及曲酸是有效且丰富的 QS 抑制化合物。中间酰胺和苦脲酸对大肠杆菌 pSB401 和大肠杆菌 pSB1075 具有毒性。脱甲氧基恩西卡林和菌核消毒素降低了大肠杆菌 pSB1075（C12-HSL 监测）的 QS 依赖性发光，而菌核消毒素、脱甲氧基恩西卡林、微胞素 A 和 B 以及曲酸降低了大肠杆菌 pSB401 的 QS 依赖性发光（Dobretsov et al., 2011）。

Díaz 等（2015）从软珊瑚中分离出 5 种脂质化合物，从褐珊瑚中分离出 3 种萜类化合物和甾醇。这些化合物从 *Ochrobactrum pseudogringnonense*、*Alteromonas macleodii*、哈维弧菌、铜绿假单胞菌和金黄色葡萄球菌被检测出来。它们的有效性取决于细菌，效果取决于细菌的种类。发现 batyl alcohol 1 和 fuscoside E peracetate 6 具有强抗生物膜效果，并且毒性较小。

植物

许多天然 QSIs 可以从各种植物中获得，其中一些来源于蔬菜和可食用水果。这些植物来源的 QSI 分子可以作为激动剂或拮抗剂。植物可以控制 QS 系统的观点源于植物没有任何类似于人类的免疫机制。这一想法认为，它们可通过抗 QS 化合物与其他 QS 病原体作斗争，其中大部分是次生代谢产物（Kalia，2013）。这些分子可以从植物组织（根、叶等）中提取，由于发育阶段不同，其产量在植物之间有所差异。这些分子的化学性质有时与 QS 信号分子相似。为了确定这些分子的安全性，还必须考虑毒性参数。

从植物中获得的次生代谢物由于其抗菌、抗真菌和抗肿瘤特性（以及其他特性）而具有重要意义。抗菌活性植物化合物有酚类、酚酸类、醌类、皂苷、FLs、单宁、香豆素、萜类化合物和生物碱。此外，植物化合物的抗生物膜活性是由柚皮素、尿皮苷、水杨酸、熊果酸、肉桂醛、甲基丁香酚、大蒜提取物和可食用水果提供（Asfour，2018）。

槲皮素是一种常见的黄酮类化合物，具有抗氧化和补充效果。Gopu 等（2015）评估了槲皮素对铜绿假单胞菌的抗 QS 潜力，并报道了其对几种毒力因子的抑制作用，如海藻酸盐、铜绿假单胞菌素、蛋白酶、弹性蛋白酶和胞外多糖的产生，以及生物膜形成和运动能力。

乌苏酸是一种已知的，无毒且具有多种药理作用的化合物。Ren 等（2005）创建了一个包含 13 000 种化合物的文库，并筛选了它们抑制哈维弧菌和铜绿假单胞菌 PAO1 生物膜形成的潜力。他们观察到，从黑乌木中获得的乌苏酸（10 μg/mL）能抑制生物膜的形成。

对 *Acer palmatum*、*Acer pseudosieboldianum* 和 *Cercis chinensis* 的甲醇叶提取物对 PAO1 生物膜形成、群体运动、铜绿假单胞菌素产生和秀丽隐杆线虫杀伤活性的潜在抑制作用。研究发现，这些提取物能成功地抑制生物膜形成、群体运动和 AI 产生（Niu et al.，2017）。

Jakobsen 等（2012a）从大蒜提取分离出的大蒜素（4,5,9-三硫代十二烯-1,6,11-三烯-9-氧化物），是一种含硫化合物。研究人员在体内外开发合成了大蒜素，以评估 QS 的潜力。他们指出，这种合成的大蒜素成功地抑制与铜绿假单胞菌相关的毒力基因。此外，有报道称，大蒜素与妥布霉素在铜绿假单胞菌生物膜中具有协同作用。

Packiavathy 等（2014）研究了来自姜黄的姜黄素对大肠杆菌、铜绿假单胞菌 PAO1、奇异变形杆菌和黏质沙雷菌的抗 QS 活性。研究发现，姜黄素抑制生物膜形成、胞外多糖（EPS）和海藻酸盐产生和群体运动，此外，生物膜对常规抗生素的敏感性增强。

Yang 等（2010）检测 ENR 抑制剂、三氯生和绿茶表没食子儿茶素没食子酸酯（EGCG）对 QS 抑制潜力。据报道，EGCG 对铜绿假单胞菌的 ENR 具有较高的结合亲和力，表明其具有 QQ 抑制效应。此外，该化合物还抑制了铜绿假单胞菌的群体运动和生物膜形成。

Vandeputte 等（2011）评估了商业可获得的黄酮类化合物（芹菜素、芹菜素类黄酮醇、山萘酮、槲皮素、杨梅素、柚皮素、柚皮苷、柚皮酮、槲皮酮、反式苯甲酰乙酮）对铜绿假单胞菌 PAO1 和紫色色素杆菌 CV026 的抑制作用。研究发现，柚皮素抑制了铜绿假单胞菌 PAO1 的弹性蛋白酶活性，生物膜形成，*lasB*、*lasI*、*lasR*、*rhlA*、*rhlI*、*rhlR* 基因表达，以及 3-oxo-C12-HSL 和 C4-HSL 的产生。此外，本研究还发现花旗松素能抑制铜绿假单胞菌 PAO1 铜绿假单胞菌素产生和弹性蛋白酶活性。

苯乙酸是一种以其抗真菌、抗氧化和抗炎特性而闻名的化合物。Musthafa 等（2012b）评估了苯基乙酸抗 QS 潜力，观察到其对 PAO1 的蛋白酶和弹性蛋白酶活性、EPS 产生以及群体运动具有抑制作用。

Kumar 等（2015）研究表明，来自生姜根的姜酮对铜绿假单胞菌群体运动和抽搐运动、生物膜形成、鼠李糖脂产生、弹性蛋白酶、蛋白酶和铜绿假单胞菌素具有抑制作用。此外，通过分子对接评估了姜酮的抗 QS 潜力，该化合物还可通过与信号受体结合（TraR、LasR、

RhlR 和 PqsR）来阻断 QS 信号。

真菌

真菌产生了许多具有 QSI 潜力的次生代谢产物，这种代谢产物的产生会根据环境条件的不同而变化。真菌的 QSI 活性对于药物和食品工业非常重要。Rasmussen 等（2005）报道了来自根赤青霉菌（*P. radicicola*）和粪生青霉菌（*P. coprobium*）的赤霉素和青霉酸次生代谢产物具有 QSI 潜力。这些真菌毒素被发现能够调控铜绿假单胞菌中与 QSI 相关的基因，表明其具有 QSI 活性（赤霉素和青霉酸的抑制作用分别为 45%和 60%）。此外，他们还发现棒曲霉素能增强妥布霉素对生物膜形成的潜力。与对照组相比，这种化合物在小鼠肺部感染模型中清除铜绿假单胞菌效果也相当好。

同时，巴西蘑菇水提取物对铜绿假单胞菌的 PAO1 QS 调控的毒力因子和生物膜形成进行了评估。在亚最小抑菌（Sub-MIC）浓度（不抑制细菌生长）下，提取物显示出降低铜绿假单胞菌的毒力因子，如铜绿假单胞菌素的产生、抽搐和游动性。此外，在 Sub-MIC 值下，生物膜形成也以浓度依赖的方式受到抑制（Soković et al., 2014）。

从白假丝酵母菌中分离出的倍半萜醇法尼醇和对苯乙醇苷类酪醇被报道为 QQ 分子和生物膜抑制剂。几种细菌，如假单胞菌属分泌吩嗪化合物，这些化合物在细菌致病性中起重要作用。例如，铜绿假单胞菌中的铜绿假单胞菌素在患有囊性纤维化的患者肺部定殖。Cugini 等（2010）报道法尼醇增加了 *lasR* 突变株中 PQS 和 C4HSL 的水平。因此，在白假丝酵母菌和铜绿假单胞菌共培养中，高水平的 PQS 导致铜绿假单胞菌中吩嗪（铜绿假单胞菌素）产生增加。PQS 的增加依赖于 *rhlR*、*rhlI* 和 *pqsH*。在 *lasR* 突变株中，法尼醇通过激活 RhlR 诱导 *pqsH* 的转录。此外，他们观察到在低细胞密度时，法尼醇抑制 PqsR 的活性。

sRNAs

可能通过靶向小 RNAs（sRNA）实现 QS 抑制，即转录后调节。Sonnleitner 等（2011）鉴定出 PhrS 作为 PqsR 合成的激活因子。Jakobsen 等（2017）证明先前已知的 QSI 大蒜素也可能作为铜绿假单胞菌中 sRNAs 的抑制剂 RsmY 和 RsmZ。sRNAs 被认为是抗病毒方法的新靶点。

从天然来源发现了大量抗铜绿假单胞菌的 QSIs 化合物，其中许多在表 9-2 中列出。

表 9-2 抗铜绿假单胞菌的天然 QSI 化合物

QSI	抑制作用	参考文献
来自生姜（*Zingber officinale*）中 [6]-姜酚、[6] 姜烯酚和姜酮	生物膜形成，铜绿假单胞菌素产生	Kumar et al., 2015 和 Kim et al., 2015
来自大蒜的大蒜新素和大蒜素	毒力、LasI、LasR、sRNAs RsmY、RsmZ	Jakobsen et al., 2012a
黄芩素（类黄酮）黄芩	生物膜形成	Zeng et al., 2008
黄芩苷水合物、桂皮醛和金缕梅鞣素	增加生物膜对抗生素的敏感性	Brackman et al., 2011
二萜化合物巴豆	生物膜形成	Carneiro et al., 2010

续表

QSI	抑制作用	参考文献
Cassipourol、β-谷甾醇和α-香树脂醇，萜类化合物来自桔梗	生物膜形成，EPS的产生，运动	Rasamiravaka et al.，2017
儿茶酚	铜绿假单胞菌素产量、弹性蛋白酶活性、生物膜产生	Vandeputte et al.，2010
大黄酚、紫花前胡内酯、紫草素和大黄素来源于传统中草药	生物膜形成	Ding et al.，2011
香豆酸酯（酚类化合物）来自三叶草	生物膜形成、铜绿假单胞菌素、LasB弹性蛋白酶产生、蛋白水解活性、运动性	Rasamiravaka et al.，2013
姜黄素来自姜黄	生物膜的形成，EPS和藻酸盐的产生，泳动和群集运动，生物膜对抗生素的敏感性增强	Packiavathy et al.，2014
Delfia tsurwhatensis	生物膜形成	Singh et al.，2017
二羟基香柠檬素和香柠檬素	生物膜形成	Girennavar et al.，2008
鞣花酸衍生物	生物膜形成	Sarabhai et al.，2013
绿茶中的表没食子儿茶素没食子酸盐，三氯生（5-氯-2-（2,4-二氯苯氧基）苯酚）	生物膜的形成和群集运动，显示出与环丙沙星的协同活性	Yang et al.，2010
圣草酚	铜绿假单胞菌素产量、弹性蛋白酶活性	Vandeputte et al.，2011
丁香酚来源于丁香科	生物膜形成	Zhou et al.，2013
煤地衣二酸	LasB、RhIA、生物膜形成	Gokalsin and Sesal，2016
法尼醇	产生花青素，PQS	Cugini et al.，2010
lberin	RhIA、LasB	Jakobsen et al.，2012b
来自肉豆蔻的 Malabaricone C	铜绿假单胞菌素产生、弹性蛋白酶活性、生物膜形成	Chong et al.，2011
马内酯，马内酯单乙酸酯来自海绵	生物膜形成	Skindersoe et al.，2008
从茴香中提取甲基丁香酚	生物膜形成，运动性	Packiavathy et al.，2012
柚皮素（黄酮）商品	生物膜形成、弹性蛋白酶活性 *lasB*、*lasl*、*lasR*、*rhlA*、*rhll*、*rhlR* 基因表达、AHL产生	Vandeputte et al.，2011
对香豆素基-羟基-豆酸（三萜的香豆素酯）*Diospyros dendo* Welw.	生物膜形成	Hu et al.，2006
棒曲霉素（呋喃吡喃酮）和青霉酸（呋喃酮）来自青霉菌	LasR，RhlR	Rasmussen et al.，2005

续表

QSI	抑制作用	参考文献
苯乙酸	铜绿假单胞菌素产量、弹性蛋白酶活性、生物膜产量、泳动性	Musthafa et al., 2012b
核黄素及其衍生物发光色素	LasR	Rajamani et al., 2008
迷迭香酸	生物膜形成	Walker et al., 2004
迷迭香酸、柚皮苷、绿原酸、桑色素和芒果苷	LasR、RhlR、生物膜形成、蛋白酶、弹性蛋白酶、溶血素产生	Annapoorani et al., 2012
水杨酸	AHL 的产生、抽搐和游动性、蛋白酶活性、LasR、Rhl、POS 活性、鼠李糖脂的产生、铜绿假单胞菌素的产生、生物膜的形成	Yang et al., 2009 和 Bandara et al., 2006
水杨酸、单宁酸和反式肉桂醛	游动性和铜绿假单胞菌素产生	Chang et al., 2014
人参皂苷、人参皂苷和人参多糖	AHL 合成，LasA，LasB	Song et al., 2010
倍半萜绿花醇和三萜，乌索里酸和白蛋白酸，来自地钱	生物膜形成和促弹性溶解活性	Gilabert et al., 2015
Solenopsina A（生物碱）*Solenopsis invicta*（昆虫；蚂蚁）	生物膜形成	Park et al., 2008
紫杉叶素	铜绿假单胞菌素的产生，弹性蛋白酶活性	Vandeputte et al., 2011
乌索酸（三萜）*Diospyros dendo* Welw	生物膜形成	Ren et al., 2005
玉米黄素（类胡萝卜素）	LasB、RhlA、生物膜形成	Gokalsin et al., 2017

2.3.4 噬菌体

细菌通过噬菌体感染被裂解，因此发展了针对这些病毒的各种防御机制，如聚集的有规则间隔短回文重复序列（CRISPRs）。尽管关于 CRISPR-Cas 调控的信息有限，但一些研究表明，QS 系统发挥了积极的作用（Patterson et al., 2016；Hoyland-Kroghsbo et al., 2017）。

Hoyland-Kroghsbo 等（2017）报道，QS 调节剂可以激活或抑制铜绿假单胞菌中的 CRISPR-Cas 系统。因此，他们提到了 QSI-噬菌体联合多疗法的可能性。Qin 等（2017）也报道 QS 参与铜绿假单胞菌对噬菌体 K5 感染的防御机制。相反，在大蜡螟的竞争种群中，温和噬菌体 D3112 和 JBD30 被确定为更倾向于 QS 能力强的铜绿假单胞菌，而不是 QS 缺陷菌株。然而，Mumford 和 Friman（2017）揭示了溶菌性 PT7 噬菌体会导致 LasR 缺陷的铜绿假单胞菌种群减少，而在竞争对手金黄色葡萄球菌和嗜麦芽狭窄营养单胞菌（*Stenotrophomonas maltophilia*）存在的情况下，PAO1 菌株则增加。

2.4 合成群体感应调节器

虽然大多数 QSIs 是从天然产物中获得的,但另一种方法是开发合成 QSIs。这类化合物有时来源于天然配体。此外,还有 HSLs 的结构模拟和结构不相关的化合物(表 9-3)。

AHL 分子在其结构上有一个头部和一个尾部:一个带有 N-酰基残基尾部的 HSL 基团。通常,合成的 AHL 模拟物有一个部分完整,另一个部分为衍生结构,旨在创建一个具有拮抗效应的更强大的分子。Biswas 等(2017)合成并检测了几种 AHL 模拟物,如乙酰氧基葡萄糖胺、羟基葡萄糖酰胺和 3-氧基葡萄糖胺对铜绿假单胞菌 MH602 菌株的作用。研究表明,最强的 QSI 化合物 9b(一种羟基葡萄糖酰胺)对铜绿假单胞菌的 las 系统抑制达 79.1%。对接研究还揭示了这些化合物的结合方式。Morkunas 等(2012)合成了一系列非生物 OdDHL 模拟物,并表明其中一些能够抑制 QS 和铜绿假单胞菌素的产生。Hodgkinson 等(2012)合成并评估 OdDHL 模拟物,发现了一些可以调节铜绿假单胞菌的 las 系统的新化合物。

AHL 类似物的研究主要集中在与受体活性位点结合,但阻止 QS 信号的检测。这个想法是合成作为拮抗剂的化合物。然而,一些修饰可能导致激动效应。例如,间位溴硫内酯(meta-bromo-thiolactone, mBTL)在没有 AHLs 的情况下作为激动剂,但在天然 AIs 存在时,对铜绿假单胞菌中的 LasR 和 RhlR 起到拮抗作用(Oloughlin et al., 2013)。

研究表明,PQS 信号通路负责铜绿假单胞菌中毒力因子的产生和生物膜成熟。前体 HHQ 是由邻氨基苯甲酸盐和 β-酮脂肪酸产生,随后转化为 PQS(LaSarre and Federle, 2013)。基于配体或基于片段的方法,研究人员使用激动剂/拮抗作为其 QSI 特性。以邻氨基苯甲酸盐为靶点可作为抑制 PQS 的有效方法。甲基氨基苯甲酸酯和卤代邻氨基苯甲酸酯类似物被证明能抑制 PQS 的生物合成(Calfee et al., 2001)。另一种选择方法是靶向 PqsD,其是 HHQ 和 PQS 生物合成中的一个关键酶。最近,S-苯基-L-半胱氨酸亚砜可抑制犬尿氨酸酶,这种酶催化犬尿氨酸裂解成邻氨基苯甲酸和 3-羟基邻氨基苯甲酸(Kasper et al., 2016)。

表 9-3 人工合成的 QSI 类化合物,包括纳米颗粒

化合物	抑制作用	参考文献
(Z)-5-亚辛基噻唑烷-24-二酮(TZD-C8)	LasI	Lidor et al., 2015
2,5-哌嗪二酮	LasR	Musthafa et al., 2012a
2-氨基酸噁二唑类化合物	PqsR	Zender et al., 2013
2-硝基苯类衍生物	PqsD	Storz et al., 2013
3-硝基苯乙酰基-高丝氨酸内酯	LasR	Geske et al., 2008a
4-硝基-苯乙酰-L-高丝氨酸内酯	LasR	Geske et al., 2007
4-硝基吡啶-N-氧化物	毒力	Rasmussen et al., 2005
5-芳基脲基噻吩-2-羧酸	PqsD	Sahner et al., 2013

续表

化合物	抑制作用	参考文献
苯甲酰胺基苯甲酸	PqsD	Hinsberger et al., 2014 和 Weidel et al., 2013
氯代吡啶类药效团	LasR	Marsden et al., 2010
化合物 1	PqsR	Lu et al., 2014
呋喃 C-30	毒力因子	Hentzer et al., 2003
呋喃酮 F2、F3 和 F4	OscR	Liu et al., 2010
镰刀酸类似物	*las*、*rhl* 系统	Tung et al., 2017
羟基氨基葡萄糖胺	*las*、*rhl* 系统	Biswas et al., 2017
长链 4-氨基喹啉类化合物	PQS，生物膜形成	Aleksic et al., 2017
间溴硫代内酯	LasR 和 RhlR	O'loughlin et al., 2013
菌胶法制备纳米银	生物膜形成、Las A 蛋白酶、Las B 弹性蛋白酶、铜绿假单胞菌素、铁载体、铜绿假单胞菌螯铁蛋白、鼠李糖脂、褐藻胶	Singh et al., 2015
N-癸酰环戊酰胺	LasR 和 RhlR	Ishida et al., 2007
OdDHL 模拟物	LasR、铜绿假单胞菌素产生、生物膜形成	Morkunas et al., 2012
OdDHL 模拟物包含（杂）芳香族	LasR	Hodgkinson et al., 2012
苯丙酰基-高丝氨酸内酯	LasR	Geske et al., 2008b
喹唑啉酮（OZN）	PqsR	Ilangovan et al., 2013
含蜜源化学物质的硒纳米颗粒	LasR、生物膜形成、弹性蛋白酶	Prateeksha et al., 2017
纳米银	生物膜形成	Barapatre et al., 2016
硅纳米颗粒浸渍敷料	生物膜形成	Velazquez-Velazquez et al., 2015
纳米氧化锌	生物膜形成、弹性蛋白酶、铜绿假单胞菌素	Garcia-Lara et al., 2015

也有与 AHLs 无关的合成调节器。例如，Lidor 等（2015）合成并研究了噻唑烷二酮型分子，观察到被称为（Z）-5-辛烯噻唑烷-2,4-二酮（TZD-C8）的化合物具有较强的运动性和生物膜抑制特性。此外，他们还探索了 QSI 特性和硅结构的亲和力，发现 TZD-C8 对 las 系统具有显著的抑制潜力。

Tung 等（2017）通过微波辅助合成法合成了 40 种新的镰孢菌酸类似物，研究了其对铜绿假单胞菌中 *las* 和 *rhl* Qs 系统的 QSI 潜力。他们发现其中一种类似物能够抑制 *las* 系统和相关的毒力因子。

此外，几项研究调查了具有潜在的纳米颗粒的 QSI 能力，并取得了令人期待的结果。

Singh 等（2015）利用少根根霉（*Rhizopus arrhizus*）BRS-07 代谢物，评价了真菌银纳米颗粒（Ag-NPs）对铜绿假单胞菌的 QS 抑制特性。结果显示，这些 Ag-NPs 可以抑制 QS 调控的毒力因子，包括 LasA 蛋白酶、LasB 弹性蛋白酶、铜绿假单胞菌素、铁载体、铜绿假单胞菌螯铁蛋白、鼠李糖脂和藻酸盐。Barapatre 等（2016）通过木质素降解真菌黄曲霉和构巢裸壳孢菌（*Emericella nidulans*）生物合成 Ag-NPs，观察到了抗菌效果以及强抗生物膜效果。另一方面，Velazquez-Velazquez 等（2015）将 Ag-NPs 浸渍在敷料中，并检测其对抗铜绿假单胞菌生物膜的效果，认为 Ag-NP 浸渍敷料可以减少或预防伤口环境中的细菌生长。

2017 年，Prateeksha 等研究含有蜂蜜植物化学物质的硒纳米颗粒（Se-NPs）对铜绿假单胞菌生物膜和 QS 的作用。利用 Se-NPs 作为药物传递系统的载体。结果表明，与其他对照组相比，纳米支架在体外和体内均表现出更强的 QS 和生物膜抑制作用。分子对接研究表明，纳米支架有助于蜂蜜植物化学物质与 LasR 受体结合。

预料之中，候选化合物的数量可能过多而难以处理，并且它们可以通过各种方式进行修饰。如何进行这些修饰有许多选择，可以通过高通量和虚拟筛选小分子来确定许多新合成的基于结构支架的 QSIs。另一个调控 QS 的思路是通过衍生其中间产物的合成类似物来抑制 AHL 合成酶，如前所述。考虑到 SAM 既是 AI1 又是 AI2 的前体，其是一个有趣的中间体（Kalia，2013）。

3　结论与意见

由于细菌抗生素耐药性的增加越来越受到关注，抗生素对常见感染的效果可能不久的将来会失效。因此，随着 QQ 的领先应用，抗毒力策略将成为应对的主要方向。迄今为止，已经发现大量对铜绿假单胞菌的 QSIs，包括丰富的天然化合物，在本章中对它们进行了详细的介绍。毫无疑问，还有许多天然化合物及其合成衍生物等待研究其 QSI 潜力。现代技术使我们能通过使用精心开发的监测菌株来快速进行筛选。在硅方法中，通过计算机辅助模拟构建用于虚拟筛选的化合物以提高筛选速度。这些方法节省了实验室的时间、成本和劳动力，因此，分析和处理它们的局限性是必要的。

与浮游细菌相比，生物膜形成具有复杂结构。因此，应当假设 QS 信号的产生和浓度分布是不均匀的。信号分子的扩散也会发生变化，导致一些细菌检测到更多的 AHL 分子。此外，生物膜的扁平结构或蘑菇结构必然会产生不同的信号梯度。为了更好地理解系统的流动性并相应地选择抑制剂，应对所有这些因素进行深入研究。

目前，许多研究都集中于使用 QQ 方法的多重疗法上。多重疗法可以与两种或更多种 QSIs、酶、抗体或抗生素结合使用。已知 QSIs 能够增强细菌对抗生素和噬菌体感染的敏感性。此外，针对多物种和菌种的感染模型和应用使这种方法更加实用。当然，QQ 疗法也存在一些缺点和未知参数，细菌甚至可能获得对 QSIs 的耐药性。进一步的研究将增强我们有效利用 QSIs 的信心，并解决这些缺点。

术语表

酰基高丝氨酸内酯　一种参与群体感应机制的小而可扩散的信号分子。
抗生素耐药性　微生物抵抗抗生素作用的能力，这些抗生素以前可以成功地治疗它们。

抗病毒治疗 抑制细菌的毒力因子而不需要消灭任何病原体。

生物膜 附着在表面并分泌胞外多糖的一层原核微生物。

囊性纤维化 一种经常遇到的疾病并影响全球约70 000人，通过影响呼吸系统来降低肺功能。

多药耐药 微生物对多种抗菌药物表现出的耐药性。

群体淬灭 在不杀死细菌或阻止细菌生长的情况下，破坏细菌通信的一种替代策略。

群体感应抑制剂 抑制群体感应机制的天然或合成化合物。

群体感应系统 细菌通信系统可通过小且可扩散的化学信号分子称为自体诱导物来调控多种行为，如生物发光、聚集、偶联、蛋白酶活性和生物膜形成等。

虚拟筛选 一种计算机辅助技术，在药物发现中用于搜索化学库，包括具有特殊性质的化合物。

缩写词

3-O-C6-HSL	N-3-(氧己烷)-高丝氨酸内酯
3-oxo-C12-HSL，OdDHL	N-（3-氧-十二酰）-高丝氨酸内酯
Ag-NP	银纳米颗粒
AHL	N-酰基-L-高丝氨酸内酯
AI	自诱导物
AI-2	自诱导物-2
AiiA	自诱导失活基因
AIP	自诱导肽
AQ	2-烷基-4-喹诺酮类
BAC	细菌人工染色体
C4-HSL，BHL	N-丁酰-L-同丝氨酸内酯
CDC	美国疾病控制和预防中心
CF	囊性纤维化
CRISPR	聚集有规律的间隔短回文重复
DPD	4,5-二羟基-2,3-戊二酮
EGCG	表儿茶素
ENR	烯酰基载体蛋白（ACP）还原酶
EPS	胞外多糖
FabI	NADH 依赖的 ENR
HHQ	2-庚基-4-羟基喹啉酮
Hod	2,4-双加氧酶
hPON1	人血清对氧磷酶1
mBTL	间溴硫内酯
MDR	多重耐药
PON1	对氧磷酶1
PON2	对氧磷酶2

PON3	对氧磷酶 3
PQS	喹诺酮假单胞菌信号
PVA	青霉素 V 酰化酶
QQ	群体淬灭
QS	群体感应
QSI	群体感应抑制剂
QSIS	群体感应抑制剂选择器系统
SAM	S-腺苷酰蛋氨酸
Se-NP	硒纳米颗粒
sRNA	小 RNA
SsoPox	超耐热内酯酶
TZD-C8	(Z)-5-辛基噻唑烷-2,4-二酮
VAI	费氏弧菌人工智能

致谢

作者感谢土耳其科学与技术研究理事会的支持（TÜBİTAK-315S092）。

参考文献

ALEKSIĆ I, ŠEGAN S, ANDRIĆ F, et al., 2017. Long-chain 4-aminoquinolines as quorum sensing inhibitors in *Serratia marcescens* and *Pseudomonas aeruginosa*. ACS Chem. Biol., 12 (5): 1425-1434.

ANNAPOORANI A, UMAMAGESWARAN V, PARAMESWARI R, et al., 2012. Computational discovery of putative quorum sensing inhibitors against LasR and RhlR receptor proteins of *Pseudomonas aeruginosa*. J. Comput. Aided Mol. Des., 26 (9): 1067-1077.

ASFOUR H Z, 2018. Antiquorum sensing natural compounds. J. Microsc. Ultrastruct., 6 (1): 1-10.

AYBEY A, DEMIRKAN E, 2016. Inhibition of quorum sensing-controlled virulence factors in *Pseudomonas aeruginosa* by human serum paraoxonase. J. Med. Microbiol., 65 (2): 105-113.

BANDARA M B, ZHU H, SANKARIDURG P R, et al., 2006. Salicylic acid reduces the production of several potential virulence factors of *Pseudomonas aeruginosa* associated with microbial keratitis. Invest. Ophthalmol. Vis. Sci., 47 (10): 4453-4460.

BARAPATRE A, AADIL K R, JHA H, 2016. Synergistic antibacterial and antibiofilm activity of silver nanoparticles biosynthesized by lignin-degrading fungus. Bioresour. Bioprocess., 3 (1): 8.

BISWAS N N, YU T T, KIMYON O, et al., 2017. Synthesis of antimicrobial glucosamides as bacterial quorum sensing mechanism inhibitors. Bioorg. Med. Chem., 25 (3): 1183-1194.

BORGES A, ABREU A C, DIAS C, et al., 2016. New perspectives on the use of phytochemicals as an emergent strategy to control bacterial infections including biofilms. Molecules, 21 (7): 877.

BOYLE K E, HEILMANN S, VAN DITMARSCH D, et al., 2013. Exploiting social evolution in biofilms. Curr. Opin. Microbiol., 16 (2): 207-212.

BRACKMAN G, COS P, MAES L, et al., 2011. Quorum sensing inhibitors increase the susceptibility of bacterial biofilms to antibiotics in vitro and in vivo. Antimicrob. Agents Chemother., 55 (6): 2655-2661.

BRANNY P, PEARSON J P, PESCI E C, et al., 2001. Inhibition of quorum sensingby a *Pseudomonas aerug-*

inosa dksA homologue. J. Bacteriol., 183 (5): 1531-1539.

CALFEE M W, COLEMAN J P, PESCI E C, 2001. Interference with *Pseudomonas quinolone* signal synthesis inhibits virulence factor expression by *Pseudomonas aeruginosa*. Proc. Natl. Acad. Sci. U. S. A., 98 (20): 11633-11637.

CARNEIRO V A, SANTOS H S, ARRUDA FV, et al., 2010. Casbane diterpene as a promising natural antimicrobial agent against biofilm-associated infections. Molecules, 16 (1): 190-201.

CENTERS FOR DISEASE CONTROL AND PREVENTION, 2017. Available from: https://www.cdc.gov/drugresistance/about.html (Accessed 2017).

CHANG C Y, KRISHNAN T, WANG H, et al., 2014. Non-antibiotic quorum sensing inhibitors acting against *N*-acyl homoserine lactone synthase as druggable target. Sci. Rep., 4: 7245.

CHANG H, ZHOU J, ZHU X, et al., 2017. Strain identification and quorum sensing inhibition characterization of marine-derived *Rhizobium* sp. NAO1. R. Soc. Open Sci., 4 (3): 170025.

CHONG Y M, YIN W F, HO C Y, et al., 2011. Malabaricone C from Myristica cinnamomea exhibits anti-quorum sensing activity. J. Nat. Prod., 74 (10): 2261-2264.

COSTANTINO V, DELLA SALA G, SAURAV K, et al., 2017. Plakofuranolactone as a quorum quenching agent from the Indonesian sponge *Plakortis* cf. lita. Mar. Drugs., 15 (3).

CUGINI C, MORALES D K, HOGAN D A, 2010. *Candida albicans* - produced farnesol stimulates *Pseudomonas quinolone* signal production in LasR-defective *Pseudomonas aeruginosa* strains. Microbiology, 156 (Pt 10): 3096-3107.

DANIELS R, VANDERLEYDEN J, MICHIELS J, 2004. Quorum sensing and swarming migration in bacteria. FEMS Microbiol. Rev., 28 (3): 261-289.

DE LAMO MARIN S, XU Y, MEIJLER M M, et al., 2007. Antibody catalyzed hydrolysis of a quorum sensing signal found in Gram-negative bacteria. Bioorg. Med. Chem. Lett., 17 (6): 1549-1552.

DEVARAJ K, TAN G Y A, CHAN K G, 2017. Quorum quenching properties of Actinobacteria isolated from Malaysian tropical soils. Arch. Microbiol., 199 (6): 897-906.

DÍAZ Y M, LAVERDE G V, GAMBA L R, et al., 2015. Biofilm inhibition activity of compounds isolated from two *Eunicea* species collected at the Caribbean Sea. Rev. Bras, 25: 605-611.

DIEKEMA D J, PFALLER M A, JONES R N, et al., 1999. Survey of bloodstream infections due to gram-negative bacilli: frequency of occurrence and antimicrobial susceptibility of isolates collected in the United States, Canada, and Latin America for the SENTRY Antimicrobial Surveillance Program, 1997. Clin. Infect. Dis., 29 (3): 595-607.

DING X, YIN B, QIAN L, et al., 2011. Screening for novel quorum-sensing inhibitors to interfere with the formation of *Pseudomonas aeruginosa* biofilm. J. Med. Microbiol., 60 (Pt 12): 1827-1834.

DOBRETSOV S, TEPLITSKI M, BAYER M, et al., 2011. Inhibition of marine biofouling by bacterial quorum sensing inhibitors. Biofouling, 27 (8): 893-905.

DONG Y H, WANG L Y, ZHANG L H, 2007. Quorum-quenching microbial infections: mechanisms and implications. Philos. Trans. R. Soc. B Biol. Sci., 362 (1483): 1201-1211.

EBERHARD A, BURLINGAME A L, EBERHARD C, et al., 1981. Structural identification of autoinducer of *Photobacterium fischeri* luciferase. Biochemistry, 20 (9): 2444-2449.

FUQUA C, GREENBERG E P, 2002. Listening in on bacteria: acyl-homoserine lactone signalling. Nat. Rev. Mol. Cell Biol., 3 (9): 685-695.

GALLOWAY W R, HODGKINSON J T, BOWDEN S D, et al., 2011. Quorum sensing in Gramnegative bacteria: small-molecule modulation of AHL and AI-2 quorum sensing pathways. Chem. Rev., 111 (1): 28-67.

GARCIA-LARA B, SAUCEDO-MORA M A, ROLDAN-SANCHEZ J A, et al., 2015. Inhibition of quorum-sensingdependent virulence factors and biofilm formation of clinical and environmental *Pseudomonas aeruginosa* strains by ZnO nanoparticles. Lett. Appl. Microbiol., 61 (3): 299-305.

GESKE G D, O'NEILL J C, MILLER D M, et al., 2007. Modulation of bacterial quorum sensing with synthetic ligands: systematic evaluation of N-acylated homoserine lactones in multiple species and new insights into their mechanisms of action. J. Am. Chem. Soc., 129 (44): 13613-13625.

GESKE G D, MATTMAN M E, BLACKWELL H E, 2008a. Evaluation of a focused library of N-aryl L-homoserine lactones reveals a new set of potent quorum sensing modulators. Bioorg. Med. Chem. Lett., 18 (22): 5978-5981.

GESKE G D, O'NEILL J C, MILLER D M, et al., 2008b. Comparative analyses of N-acylated homoserine lactones reveal unique structural features that dictate their ability to activate or inhibit quorum sensing. ChemBioChem, 9 (3): 389-400.

GILABERT M, MARCINKEVICIUS K, ANDUJAR S, et al., 2015. Sesqui- and triterpenoids from the liverwort *Lepidozia chordulifera* inhibitors of bacterial biofilm and elastase activity of human pathogenic bacteria. Phytomedicine, 22 (1): 77-85.

GIRENNAVAR B, CEPEDA M, SONI K A, et al., 2008. Grapefruit juice and its furocoumarin inhibits autoinducer signaling and biofilm formation in bacteria. Int. J. Food Microbiol., 125 (2): 204-208.

GOKALSIN B, SESAL N C, 2016. Lichen secondary metabolite evernic acid as potential quorum sensing inhibitor against *Pseudomonas aeruginosa*. World J. Microbiol. Biotechnol., 32 (9): 150.

GOKALSIN B, AKSOYDAN B, ERMAN B, et al., 2017. Reducing virulence and biofilm of *Pseudomonas aeruginosa* by potential quorum sensing inhibitor carotenoid: Zeaxanthin. Microb. Ecol., 74 (2): 466-473.

GOPU V, MEENA C K, SHETTY P H, 2015. Quercetin influences quorum sensing in food borne bacteria: in-vitro and in-silico evidence. PLoS One, 10 (8): e0134684.

GRANDCLEMENT C, TANNIERES M, MORERA S, et al., 2016. Quorum quenching: role in nature and applied developments. FEMS Microbiol. Rev., 40 (1): 86-116.

GROVER N, PLAKS J G, SUMMERS S R, et al., 2016. Acylase-containing polyurethane coatings with anti-biofilm activity. Biotechnol. Bioeng., 113 (12): 2535-2543.

GUENDOUZE A, PLENER L, BZDRENGA J, et al., 2017. Effect of quorum quenching lactonase in clinical isolates of *Pseudomonas aeruginosa* and comparison with quorum sensing inhibitors. Front. Microbiol., 8: 227.

HALL-STOODLEY L, COSTERTON J W, STOODLEY P, 2004. Bacterial biofilms: from the natural environment to infectious diseases. Nat. Rev. Microbiol., 2 (2): 95-108.

HENTZER M, GIVSKOV M, 2003. Pharmacological inhibition of quorum sensing for the treatment of chronic bacterial infections. J. Clin. Investig., 112 (9): 1300-1307.

HENTZER M, WU H, ANDERSEN J B, et al., 2003. Attenuation of *Pseudomonas aeruginosa* virulence by quorum sensing inhibitors. EMBO J., 22 (15): 3803-3815.

HINSBERGER S, DE JONG J C, GROH M, et al., 2014. Benzamidobenzoic acids as potent PqsD inhibitors for the treatment of *Pseudomonas aeruginosa* infections. Eur. J. Med. Chem., 76: 343-351.

HODGKINSON J T, GALLOWAY W R, WRIGHT M, et al., 2012. Design, synthesis and biological evaluation of non-natural modulators of quorum sensing in *Pseudomonas aeruginosa*. Org. Biomol. Chem., 10 (30): 6032-6044.

HOYLAND-KROGHSBO N M, PACZKOWSKI J, MUKHERJEE S, et al., 2017. Quorum sensing controls the *Pseudomonas aeruginosa* CRISPR-Cas adaptive immune system. Proc. Natl. Acad. Sci. U. S. A.,

114 (1): 131-135.

HRAIECH S, HIBLOT J, LAFLEUR J, et al., 2014. Inhaled lactonase reduces *Pseudomonas aeruginosa* quorum sensing and mortality in rat pneumonia. PLoS One, 9 (10): e107125.

HU J F, GARO E, GOERING M G, et al., 2006. Bacterial biofilm inhibitors from Diospyros dendo. J. Nat. Prod., 69 (1): 118-120.

HUANG J J, PETERSEN A, WHITELEY M, et al., 2006. Identification of QuiP, the product of gene PA1032, as the second acyl-homoserine lactone acylase of *Pseudomonas aeruginosa* PAO1. Appl. Environ. Microbiol., 72 (2): 1190-1197.

ILANGOVAN A, FLETCHER M, RAMPIONI G, et al., 2013. Structural basis for native agonist and syntheticinhibitor recognition by the *Pseudomonas aeruginosa* quorum sensing regulator PqsR (MvfR). PLoS Pathog., 9 (7): e1003508.

ISHIDA T, IKEDA T, TAKIGUCHI N, et al., 2007. Inhibition of quorum sensing in *Pseudomonas aeruginosa* by N-acyl cyclopentylamides. Appl. Environ. Microbiol., 73 (10): 3183-3188.

JAKOBSEN T H, VAN GENNIP M, PHIPPS R K, et al., 2012a. Ajoene, a sulfur-rich molecule from garlic, inhibits genes controlled by quorum sensing. Antimicrob. Agents Chemother., 56 (5): 2314-2325.

JAKOBSEN T H, BRAGASON S K, PHIPPS R K, et al., 2012b. Food as a source for quorum sensing inhibitors: iberin from horseradish revealed as a quorum sensing inhibitor of *Pseudomonas aeruginosa*. Appl. Environ. Microbiol., 78 (7): 2410-2421.

JAKOBSEN T H, WARMING A N, VEJBORG R M, et al., 2017. A broad range quorum sensing inhibitor working through sRNA inhibition. Sci. Rep., 7 (1): 9857.

JOO H S, OTTO M, 2012. Molecular basis of in vivo biofilm formation by bacterial pathogens. Chem. Biol., 19 (12): 1503-1513.

KALIA V C, 2013. Quorum sensing inhibitors: an overview. Biotechnol. Adv., 31 (2): 224-245.

KALIA V C, PUROHIT H J, 2011. Quenching the quorum sensing system: potential antibacterial drug targets. Crit. Rev. Microbiol., 37 (2): 121-140.

KASPER S H, BONOCORA R P, WADE J T, et al., 2016. Chemical inhibition of kynureninase reduces *Pseudomonas aeruginosa* quorum sensing and virulence factor expression. ACS Chem. Biol., 11 (4): 1106-1117.

KAUFMANN G F, SARTORIO R, LEE S-H, et al., 2006. Antibody interference with N-acyl homoserine lactone-mediated bacterial quorum sensing. J. Am. Chem. Soc., 128 (9): 2802-2803.

KIM H S, LEE S H, BYUN Y, et al., 2015. 6-Gingerol reduces *Pseudomonas aeruginosa* biofilm formation and virulence via quorum sensing inhibition. Sci. Rep., 5: 8656.

KUMAR L, CHHIBBER S, KUMAR R, et al., 2015. Zingerone silences quorum sensing and attenuates virulence of *Pseudomonas aeruginosa*. Fitoterapia, 102: 84-95.

LASARRE B, FEDERLE M J, 2013. Exploiting quorum sensing to confuse bacterial pathogens. Microbiol. Mol. Biol. Rev., 77 (1): 73-111.

LAXMINARAYAN R, MATSOSO P, PANT S, et al., 2016. Access to effective antimicrobials: a worldwide challenge. Lancet, 387 (10014): 168-175.

LEE J, ZHANG L, 2015. The hierarchy quorum sensing network in *Pseudomonas aeruginosa*. Protein Cell, 6 (1): 26-41.

LI Y H, TIAN X, 2012. Quorum sensing and bacterial social interactions in biofilms. Sensors (Basel), 12 (3): 2519-2538.

LIDOR O, AL-QUNTAR A, PESCI E C, et al., 2015. Mechanistic analysis of a synthetic inhibitor of the *Pseudomonas aeruginosa* LasI quorum-sensing signal synthase. Sci. Rep., 5: 16569.

LIU H B, LEE J H, KIM J S, et al., 2010. Inhibitors of the *Pseudomonas aeruginosa* quorum-sensing regulator, QscR. Biotechnol. Bioeng., 106 (1): 119-126.

LU C, MAURER C K, KIRSCH B, et al., 2014. Overcoming the unexpected functional inversion of a PqsR antagonist in *Pseudomonas aeruginosa*: an in vivo potent antivirulence agent targeting pqs quorum sensing. Angew. Chem. Int. Ed., 53 (4): 1109-1112.

MAI T, TINTILLIER F, LUCASSON A, et al., 2015. Quorum sensing inhibitors from Leucetta chagosensis Dendy, 1863. Lett. Appl. Microbiol., 61 (4): 311-317.

MARSDEN D M, NICHOLSON R L, SKINDERSOE, et al., 2010. Discovery of a quorum sensing modulator pharmacophore by 3D small-molecule microarray screening. Org. Biomol. Chem., 8 (23): 5313-5323.

MIYAIRI S, TATEDA K, FUSE E T, et al., 2006. Immunization with 3-oxododecanoyl-L-homoserine lactone-protein conjugate protects mice from lethal *Pseudomonas aeruginosa* lung infection. J. Med. Microbiol., 55 (Pt 10): 1381-1387.

MORKUNAS B, GALLOWAY W R, WRIGHT M, et al., 2012. Inhibition of the production of the *Pseudomonas aeruginosa* virulence factor pyocyanin in wild-type cells by quorum sensing autoinducer-mimics. Org. Biomol. Chem., 10 (42): 8452-8464.

MULLER C, BIRMES F S, NIEWERTH H, et al., 2014. Conversion of the *Pseudomonas aeruginosa* quinolone signal and related alkylhydroxyquinolines by Rhodococcus sp. strain BG43. Appl. Environ. Microbiol., 80 (23): 7266-7274.

MULLER C, BIRMES F S, RUCKERT C, et al., 2015. *Rhodococcus erythropolis* BG43 genes mediating *Pseudomonas aeruginosa* quinolone signal degradation and virulence factor attenuation. Appl. Environ. Microbiol., 81 (22): 7720-7729.

MUMFORD R, FRIMAN V P, 2017. Bacterial competition and quorum-sensing signalling shape the eco-evolutionary outcomes of model in vitro phage therapy. Evol. Appl., 10 (2): 161-169.

MUSTHAFA K S, BALAMURUGAN K, PANDIAN S K, et al., 2012a. 2, 5-Piperazinedione inhibits quorum sensingdependent factor production in *Pseudomonas aeruginosa* PAO1. J. Basic Microbiol., 52 (6): 679-686.

MUSTHAFA K S, SIVAMARUTHI B S, PANDIAN S K, et al., 2012b. Quorum sensing inhibition in *Pseudomonas aeruginosa* PAO1 by antagonistic compound phenylacetic acid. Curr. Microbiol., 65 (5): 475-480.

NIU K, KUK M, JUNG H, et al., 2017. Leaf extracts of selected gardening trees can attenuate quorum sensing and pathogenicity of *Pseudomonas aeruginosa* PAO1. Indian J. Microbiol., 1-10.

NOVOTNY L A, JURCISEK J A, GOODMAN S D, et al., 2016. Monoclonal antibodies against DNA-binding tips of DNABII proteins disrupt biofilms in vitro and induce bacterial clearance in vivo. EBioMedicine, 10: 33-44.

O'LOUGHLIN C T, MILLER L C, SIRYAPORN A, et al., 2013. A quorumsensing inhibitor blocks *Pseudomonas aeruginosa* virulence and biofilm formation. Proc. Natl. Acad. Sci. U. S. A., 110 (44): 17981-17986.

OLSON M E, CERI H, MORCK D W, et al., 2002. Biofilm bacteria: formation and comparative susceptibility to antibiotics. Can. J. Vet. Res., 66 (2): 86-92.

PACKIAVATHY I A, AGILANDESWARI P, MUSTHAFA K S, et al., 2012. Antibiofilm and quorum sensing inhibitory potential of Cuminum cyminum and its secondary metabolite methyl eugenol against Gram negative bacterial pathogens. Food Res. Int., 45 (1): 85-92.

PACKIAVATHY I A, PRIYA S, PANDIAN S K, et al., 2014. Inhibition of biofilm development of uropathogens by curcumin—an anti-quorum sensing agent from Curcuma longa. Food Chem., 148: 453-460.

PALLIYIL S, DOWNHAM C, BROADBENT I, et al., 2014. High-sensitivity monoclonal antibodies specific for homoserine lactones protect mice from lethal *Pseudomonas aeruginosa* infections. Appl. Environ. Microbiol., 80 (2): 462-469.

PARK J, KAUFMANN G F, BOWEN J P, et al., 2008. Solenopsin A, a venom alkaloid from the fire ant *Solenopsis invicta*, inhibits quorum-sensing signaling in *Pseudomonas aeruginosa*. J. Infect. Dis., 198 (8): 1198-1201.

PATTERSON A G, JACKSON S A, TAYLOR C, et al., 2016. Quorum sensing controls adaptive immunity through the regulation of multiple CRISPRCas systems. Mol. Cell, 64 (6): 1102-1108.

PEJIN B, CIRIC A, KARAMAN I, et al., 2016. *In vitro* antibiofilm activity of the freshwater bryozoan *Hyalinella punctata*: a case study of *Pseudomonas aeruginosa* PAO1. Nat. Prod. Res., 30 (16): 1847-1850.

PRATEEKSHA SINGH B R, SHOEB M, SHARMA S, et al., 2017. Scaffold of selenium nanovectors and honey phytochemicals for inhibition of *Pseudomonas aeruginosa* quorum sensing and biofilm formation. Front. Cell. Infect. Microbiol., 7: 93.

PUSTELNY C, ALBERS A, BULDT-KARENTZOPOULOS K, et al., 2009. Dioxygenase-mediated quenching of quinolone-dependent quorum sensing in *Pseudomonas aeruginosa*. Chem. Biol., 16 (12): 1259-1267.

QIN X, SUN Q, YANG B, et al., 2017. Quorum sensing influences phage infection efficiency via affecting cell population and physiological state. J. Basic Microbiol., 57 (2): 162-170.

QUINTANA J, BRANGO-VANEGAS J, COSTA G M, et al., 2015. Marine organisms as source of extracts to disrupt bacterial communication: bioguided isolation and identification of quorum sensing inhibitors from Ircinia felix. Rev. Bras, 25: 199-207.

RAJAMANI S, BAUER W D, ROBINSON J B, et al., 2008. The vitamin riboflavin and its derivative lumichrome activate the LasR bacterial quorumsensing receptor. Mol. Plant-Microbe Interact., 21 (9): 1184-1192.

RASAMIRAVAKA T, JEDRZEJOWSKI A, KIENDREBEOGO M, et al., 2013. Endemic malagasy Dalbergia species inhibit quorum sensing in *Pseudomonas aeruginosa* PAO1. Microbiology, 159 (Pt 5): 924-938.

RASAMIRAVAKA T, NGEZAHAYO J, POTTIER L, et al., 2017. Terpenoids from *Platostoma rotundifolium* (Briq.) A. J. Paton alter the expression of quorum sensingrelated virulence factors and the formation of biofilm in *Pseudomonas aeruginosa* PAO1. Int. J. Mol. Sci., 18 (6).

RASMUSSEN T B, GIVSKOV M, 2006. Quorum sensing inhibitors: a bargain of effects. Microbiology, 152 (Pt 4): 895-904.

RASMUSSEN T B, SKINDERSOE M E, BJARNSHOLT T, et al., 2005. Identity and effects of quorumsensing inhibitors produced by *Penicillium* species. Microbiology, 151 (Pt 5): 1325-1340.

REN D, ZUO R, GONZALEZ BARRIOS A F, et al., 2005. Differential gene expression for investigation of *Escherichia coli* biofilm inhibition by plant extract ursolic acid. Appl. Environ. Microbiol., 71 (7): 4022-4034.

RIVAS CALDAS R, BOISRAME S, 2015. Upper aero-digestive contamination by *Pseudomonas aeruginosa* and implications in cystic fibrosis. J. Cyst. Fibros., 14 (1): 6-15.

RUTHERFORD S T, BASSLER B L, 2012. Bacterial quorum sensing: its role in virulence and possibilities for its control. Cold Spring Harb. Perspect. Med., 2 (11).

SAHNER J H, BRENGEL C, STORZ M P, et al., 2013. Combining in silico and biophysical methods for the development of *Pseudomonas aeruginosa* quorum sensing inhibitors: an alternative approach for structure-based drug design. J. Med. Chem., 56 (21): 8656-8664.

SARABHAI S, SHARMA P, CAPALASH N, 2013. Ellagic acid derivatives from Terminalia chebula

Retz. downregulate the expression of quorum sensing genes to attenuate *Pseudomonas aeruginosa* PAO1 virulence. PLoS One, 8 (1): e53441.

SAURAV K, COSTANTINO V, VENTURI V, et al., 2017. Quorum sensing inhibitors from the sea discovered using bacterial *N*-acyl-homoserine lactone-based biosensors. Mar. Drugs, 15 (3): 53.

SHANNON E, ABU-GHANNAM N, 2016. Antibacterial derivatives of marine algae: an overview of pharmacological mechanisms and applications. Mar. Drugs, 14 (4).

SINGH B R, SINGH B N, SINGH A, et al., 2015. Mycofabricated biosilver nanoparticles interrupt *Pseudomonas aeruginosa* quorum sensing systems. Sci. Rep., 5: 13719.

SINGH V K, MISHRA A, JHA B, 2017. Anti–quorum sensing and anti–biofilm activity of *Delftia tsuruhatensis* extract by attenuating the quorum sensing-controlled virulence factor production in *Pseudomonas aeruginosa*. Front. Cell. Infect. Microbiol., 7: 337.

SIO C F, OTTEN L G, COOL R H, et al., 2006. Quorum quenching by an *N*–acyl–homoserine lactone acylase from *Pseudomonas aeruginosa* PAO1. Infect. Immun., 74 (3): 1673-1682.

SKINDERSOE M E, ETTINGER-EPSTEIN P, RASMUSSEN T B, et al., 2008. Quorum sensing antagonism from marine organisms. Mar. Biotechnol. (NY), 10 (1): 56-63.

SOKOVIĆ M, ĆIRIĆ A, GLAMOČLIJA J, et al., 2014. Agaricus blazei hot water extract shows anti quorum sensing activity in the nosocomial human pathogen *Pseudomonas aeruginosa*. Molecules, 19 (4): 4189-4199.

SONG Z, KONG K F, WU H, et al., 2010. *Panax ginseng* has anti-infective activity against opportunistic pathogen *Pseudomonas aeruginosa* by inhibiting quorum sensing, a bacterial communication process critical for establishing infection. Phytomedicine, 17 (13): 1040-1046.

SONNLEITNER E, GONZALEZ N, SORGER-DOMENIGG T, et al., 2011. The small RNA PhrS stimulates synthesis of the *Pseudomonas aeruginosa* quinolone signal. Mol. Microbiol., 80 (4): 868-885.

STOLTZ D A, OZER E A, NG C J, et al., 2007. Paraoxonase–2 deficiency enhances *Pseudomonas aeruginosa* quorum sensing in murine tracheal epithelia. Am. J. Phys. Lung Cell. Mol. Phys., 292 (4): L852-L860.

STORZ M P, BRENGEL C, WEIDEL E, et al., 2013. Biochemical and biophysical analysis of a chiral PqsD inhibitor revealing tight–binding behavior and enantiomers with contrary thermodynamic signatures. ACS Chem. Biol., 8 (12): 2794-2801.

SUN S, DAI X, SUN J, et al., 2016. A diketopiperazine factor from *Rheinheimera aquimaris* QSI02 exhibits anti-quorum sensing activity. Sci. Rep., 6: 39637.

SUNDER A V, UTARI P D, RAMASAMY S, et al., 2017. Penicillin V acylases from gram-negative bacteria degrade *N*-acylhomoserine lactones and attenuate virulence in *Pseudomonas aeruginosa*. Appl. Microbiol. Biotechnol., 101 (6): 2383-2395.

TANG K, SU Y, BRACKMAN G, et al., 2015. MomL, a novel marine-derived *N*-acyl homoserine lactonase from Muricauda olearia. Appl. Environ. Microbiol., 81 (2): 774-782.

TUNG T T, JAKOBSEN T H, DAO T T, et al., 2017. Fusaric acid and analogues as Gram-negative bacterial quorum sensing inhibitors. Eur. J. Med. Chem., 126: 1011-1020.

VAN HECKE O, WANG K, LEE J J, et al., 2017. Implications of antibiotic-resistance for patients' recovery from common infections in the community: a systematic review and meta–analysis. Clin. Infect. Dis., 65 (3): 371-382.

VANDEPUTTE O M, KIENDREBEOGO M, RAJAONSON S, et al., 2010. Identification of catechin as one of the flavonoids from Combretum albiflorum bark extract that reduces the production of quorum-sensing-controlled virulence factors in *Pseudomonas aeruginosa* PAO1. Appl. Environ. Microbiol., 76 (1): 243-253.

VANDEPUTTE O M, KIENDREBEOGO M, RASAMIRAVAKA T, et al., 2011. The flavanone naringenin reduces the production of quorum sensingcontrolled virulence factors in *Pseudomonas aeruginosa* PAO1. Microbiology, 157 (Pt 7): 2120-2132.

VELAZQUEZ-VELAZQUEZ J L, SANTOS-FLORES A, ARAUJO-MELENDEZ J, 2015. Anti-biofilm and cytotoxicity activity of impregnated dressings with silver nanoparticles. Mater. Sci. Eng. C Mater. Biol. Appl., 49: 604-611.

WAHJUDI M, PAPAIOANNOU E, HENDRAWATI O, et al., 2011. PA0305 of *Pseudomonas aeruginosa* is a quorum quenching acylhomoserine lactone acylase belonging to the Ntn hydrolase superfamily. Microbiology, 157 (Pt 7): 2042-2055.

WALKER T S, BAIS H P, DEZIEL E, et al., 2004. *Pseudomonas aeruginosa*-plant root interactions. Pathogenicity, biofilm formation, and root exudation. Plant Physiol., 134 (1): 320-331.

WEIDEL E, DE JONG J C, BRENGEL C, et al., 2013. Structure optimization of 2-benzamidobenzoic acids as PqsD inhibitors for *Pseudomonas aeruginosa* infectiongns and elucidation of bindi mode by SPR, STD NMR, and molecular docking. J. Med. Chem., 56 (15): 6146-6155.

WEILAND-BRAUER N, KISCH M J, PINNOW N, et al., 2016. Highly effective inhibition of biofilm formation by the first metagenome-derived AI-2 quenching enzyme. Front. Microbiol., 7: 1098.

WORLD HEALTH ORGANIZATION, 2017. Antibacterial agents in clinical development. World Health Organization, Geneva.

YANG L, RYBTKE M T, JAKOBSEN T H, et al., 2009. Computer-aided identification of recognized drugs as *Pseudomonas aeruginosa* quorum-sensing inhibitors. Antimicrob. Agents Chemother., 53 (6): 2432-2443.

YANG L, LIU Y, STERNBERG C, et al., 2010. Evaluation of enoyl-acyl carrier protein reductase inhibitors as *Pseudomonas aeruginosa* quorum-quenching reagents. Molecules, 15 (2): 780-792.

YANIV K, GOLBERG K, KRAMARSKY-WINTER E, et al., 2017. Functional marine metagenomic screening for anti-quorum sensing and anti-biofilm activity. Biofouling, 33 (1): 1-13.

ZENDER M, KLEIN T, HENN C, et al., 2013. Discovery and biophysical characterization of 2-amino-oxadiazoles as novel antagonists of PqsR, an important regulator of *Pseudomonas aeruginosa* virulence. J. Med. Chem., 56 (17): 6761-6774.

ZENG Z, QIAN L, CAO L, et al., 2008. Virtual screening for novel quorum sensing inhibitors to eradicate biofilm formation of *Pseudomonas aeruginosa*. Appl. Microbiol. Biotechnol., 79 (1): 119-126.

ZHOU L, ZHENG H, TANG Y, et al., 2013. Eugenol inhibits quorum sensing at sub-inhibitory concentrations. Biotechnol. Lett., 35 (4): 631-637.

第10章 神经性疾病中的环肽：cyclo(His-pro)的情况

Ilaria Bellezza，Matthew J. Peirce，Alba Minelli

Department of Experimental Medicine，University of Perugia，Perugia，Italy

1 前言

细菌群体感应（QS）是一种基于信号的产生、分泌和检测的细胞间通信系统，信号分子被称为自诱导物。这些信号的产生是为了响应种群水平的变化，如细菌种群的密度和组成物种的变化，同时调节基因表达和协调集体行为。通过这种方式，不同的细菌群落获得了作为一个群体的能力，开启毒力因子的产生、生物膜形成和耐药性的发展。事实上，如果仅通过单个细菌细胞单独作用，许多微生物过程将是无效的（Ng and Bassler, 2009; Bassler and Vogel, 2013; Cornforth et al., 2014; Schluter et al., 2016）。研究表明，QS过程在革兰氏阴性细菌和革兰氏阳性细菌中均存在，且许多有趣的"跨界"信号的例子（即在原核细胞和真核细胞之间的跨越分类学边界的信号）已被发现。从一开始，原核生物和真核生物通过感知和响应彼此信号的产生和释放来生存和共存（Rosier et al., 2016）。这些信号一旦被合成，其浓度与种群密度成正比，并会被分泌出来。当信号水平超过阈值时，信号会被QS受体检测，该受体参与下游转导级联，启动高细胞密度的基因表达程序。

在革兰氏阴性细菌中，AIs分子属于几种化学类别，如酰基高丝氨酸内酯（AHLs）、烷基喹诺酮类、α-羟基酮和可扩散信号因子（脂肪酸类化合物），它们由普通代谢物通过单一合成酶或一系列酶促反应合成（Hawver et al., 2016; Ryan et al., 2015）。AIs分子可以自由穿过细胞膜，当它们的胞外浓度足够高时，通过结合胞质受体以自分泌的方式扩散回细胞内。随后，作为转录因子，它们与AIs结合后调节QS基因的表达。

通常，革兰氏阴性细菌使用QS系统与发光（Lux）控制操纵子（LuxI/LuxR系统）同源系统，该系统最初在发光海洋细菌费氏弧菌中发现。LuxI同源物作为AI的合成酶，而LuxR同源物则是胞质转录因子AI受体，通过与AI结合，LuxR获得反转录激活能力从而介导靶基因的转录（Hawver et al., 2016; Ryan et al., 2015）。

革兰氏阳性细菌的同源系统稍微复杂一些，由短寡肽介导。自诱导肽（AIPs）由AIP合成酶产生，在蛋白水解后通过转运体蛋白主动分泌，通常是ATP结合盒转运体。当它们的细胞外浓度达到一个阈值水平时，就会被运回到细胞质中，然后，被QS转录因子检测到。在高细胞密度时，AIPs一旦通过转运体释放到细胞外空间，就会经过翻译后修饰并与跨膜受体结合，从而触发磷酸化级联反应，控制下游的QS反应（Cook and Federle, 2014; Hawver et al., 2016; Monnet et al., 2016）。革兰氏阳性QS受体由Rap、NprR、PlcR和PrgX四个成员组成一个家族，统称为RNPP（Rocha-Estrada et al., 2010）。其中，Rap是天冬氨酸磷酸酶和转录激活蛋白，而NprR、PlcR和PrgX则是DNA结合转录因子。此外，

NprR 是一个中性蛋白酶调节因子，PlcR 是一种磷脂酶 C 调节因子，PrgX 调节质粒接合转移（Rocha-Estrada et al., 2010; Zouhir et al., 2013; Cook and Federle, 2014）。第三类分子是呋喃糖基硼酸二酯，也称为自诱导物-2（AI-2），其代表物种间交流的通用信号（Hense and Schuster, 2015）。

因此，尽管具体的细节各不相同，但一般来说，QS 系统将可扩散的信使分子释放到特定的靶向调控和上下相关的基因表达程序中。QS 使合作成为可能，这与传统的旁分泌信号不同，后者发送和接收细胞往往是不同的，而在 QS 中，发送和接收通常是同一个细胞。此外，这些 QS 行为具有独特的生态功能和进化特性（Diggle et al., 2007）。QS 系统整合和处理来自物理环境的信息，以在群落水平上优化其活动，从而使不同细胞群体的合作和协调行为成为可能。因此，QS 系统广泛分布于多种分类群中，如真菌、植物（藻类）、动物，甚至是病毒，其中宿主细胞的裂解可能取决于病毒浓度（Hallmann, 2011; Hogan, 2006; Moussa et al., 2012; Sprague and Winans, 2006; Weitz et al., 2008）。

2 环肽

在细菌中发现许多具有生物活性的多肽，这些多肽由典型的 20 个氨基酸或非蛋白氨基酸组成（Clardy and Walsh, 2004）。这些线性或环肽可以在核糖体中合成，随后通过肽合成酶独立于核糖体进行加工或生产（Clardy et al., 2006）。肽和蛋白质一样，都是生物活性分子，为合理的药物设计提供了有吸引力的初始线索（Tapeinou et al., 2015）。然而，线性肽易被蛋白降解，在历史上并未成为制药行业的首选分子。不过，通过引入修饰可使这些分子更具吸引力。事实上，环化可导致稳定性增加，N-甲基化可增加膜通透性和稳定性，非天然氨基酸的加入增加了特异性和稳定性，聚乙二醇化（聚乙二醇（PEG）聚合物链的附着或合并）能够降低清除能力。此外，各种结构限制（如二硫键）以及"装订"肽的最新进展，也有助于提高肽的效力和特异性。目前，这些修饰被视为未来治疗很有前途的新模式（Tapeinou et al., 2015; Verdine and Hilinski, 2012）。特别地，环化后的肽表现出优越的结合亲和性，并降低与受体结合相关的熵成本。事实上，将肽限制在环状结构中会降低其母体线性结构的构象自由度，从而增强其代谢稳定性、生物利用度和特异性。自然产生的环肽具有广泛且不寻常的强大生物活性；其中大多数化合物中控制着种内和种间细菌的毒力以及细菌-宿主相互作用，这些化合物通过被动扩散穿透细胞，有些化合物，如临床上重要的药物环孢素 A，表现出良好的口服生物利用度（Bockus et al., 2013）。环二肽类（CDPs），如恩镰孢菌素，可由植物病原菌产生，通过干扰宿主细胞的生理行为而发挥真菌毒素的作用。此外，它们还显示出多种生物活性，如杀虫、抗真菌和抗菌作用。事实上，这些化合物对人类和动物细胞系产生强大的细胞毒性作用，被认为是潜在的抗癌药物（Prosperini et al., 2017）。因此，在广泛的临床环境中，环肽被认为是很有前途的药物开发先导化合物，并具有吸引力的生物合成靶点。

文献中描述的 QS 肽被存储在策划的 Quorumpeps 数据库中（http://quorumpeps.ugent.be），该数据库详细介绍了这些多肽及其 QS 过程（受体、活性、种类）（Wynendaele et al., 2015）。对 Quorumpeps 中列出的 231 条 QS 肽的分析表明，环肽在 QS 肽空间中占据了一个明显的部分，这一发现是根据其大小分布、致密性、亲脂性/疏水性、芳香族氨基酸的存在以及硫原子的存在。另外，物种分布的结果表明，大多数革兰氏阳性细菌合成了化学性质相似的多肽。值得注意的是，环肽被称为θ-防御素，在恒河猴和狒狒的白细胞中也发现了环肽（Lehrer et al., 2012），其参与一系列病毒的防御包括HIV-1、HSV、A型流感病毒和冠状病毒（严重急性呼吸系统综合征）以及细菌病原体炭疽芽孢杆菌，因此，环肽代表了一个有趣的潜在治疗药物。

2.1 环肽类

CDPs，也被称为2,5-二酮哌嗪，是一类小的生物活性分子，主要作为QS效应子，其含有家族定义的CDP核心/支架结构（图10-1），并由变形杆菌物种以及人类产生（Bellezza et al., 2014a, b；Borthwick, 2012；Cornacchia et al., 2012；Minelli et al., 2008；Mishra et al., 2017；Prasad, 1995）。

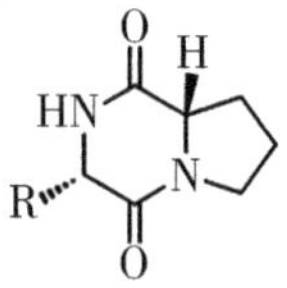

图10-1 含组氨酸的环状二肽家族界定核心/支架结构

CDPs是一类由两个氮原子的哌嗪六元环形成酰胺键的环状有机化合物。CDPs的命名是由两个氨基酸的三个字母代码表示，并加上一个前缀来指定绝对构型（例如，cyclo（L-Xaa-L-Yaa）。CDPs可同时存在顺式和反式两种异构体，但以顺式构型为主（Eguchi and Kakuta，1974）。各种氨基酸修饰赋予其不同的化学和生物学功能。CDPs由于具有更高的稳定性、蛋白酶抗性和构象刚性，这些因素增加了它们与生物靶标特异性相互作用的能力，因此，CDPs表现出比线性对应物更好的生物活性（Liskamp et al., 2011；Menegatti et al., 2013）。CDPs构成了细菌、真菌、植物和动物产生的一大类次级代谢产物（Borthwick, 2012；Giessen and Marahiel, 2014；Huang et al., 2010；Mishra et al., 2017；Prasad, 1995）。事实上，大约90%的CDP生产者是细菌（Giessen and Marahiel, 2014）。CDP支架可以通过纯化学方法使用不同的固相合成，也可以在溶液回流条件下合成（Borthwick, 2012；Gonzalez et al., 2012），或更自然地，通过生物合成酶，例如非核糖体肽合成酶（NRPSs）和CDP合成酶（CDPSs）（CDPs；Belin et al., 2012；Giessen and Marahiel, 2014）。CDPs常见的化学合成包括将单个氨基酸在高温下进行缩合。在一端被胺取代的二肽和在另一端被酯取代的二肽也可以自发地环化形成CDP。然而，为了促使环化反应并限制外消旋作用，条件必须进行优化。这是化学合成CDP最常用的方法。氨基二肽酯的环化反应也可以在热条件下进行，通常在甲苯或二甲苯等高沸点溶剂中回流24 h（Borthwick, 2012）。此外，在加工食品和饮料中，CDPs往往是不必要的副反应产物或寡聚和多肽的降解产物（Borthwick and Da Costa, 2017；Prasad, 1995）。它们经常在焙烤咖啡、炖牛肉和啤酒等产品的化学降解过程中形成（Chen et al., 2009；Gautschi et al., 1997；Ginz and Engelhardt, 2000）。

非酶促过程也可以导致各种生物体中功能性CDPs的形成，例如cyclo（L-His-L-Pro）的描述（Bellezza et al., 2014a, b；Minelli et al., 2008）。在哺乳动物中，环（组氨酸-脯氨酸）（CHP）是由焦谷氨酸氨基肽酶对促甲状腺激素释放激素（TRH，pGlu-His-Pro）的作用获得的。随后，所得到的二肽被非酶环化形成CHP。脯氨酸诱导的限制促进了组氨酸和脯氨酸之间肽键的顺式构象，从而促进环化反应，进而生成 CDP 支架。

在 CDP 形成的酶学途径中，有两个不相关的生物合成酶家族催化 CDPs 的形成：NRPSs 和 CDPSs。研究表明，CDP 支架可以通过一个或多个特定的 NRPSs 合成，或通过特定的生物合成途径，或通过链延伸过程中提前释放的二肽基中间体。某些多肽的 NRPS 基因在原核生物中通常被组织在一个操纵子中，而在真核生物中则被组织在一个基因簇中（Schwarzer et al., 2003）。NRPSs 是一种大模块化酶，它同时作为模板和生物合成机制。每个模块负责将一个氨基酸合并入最终肽段中，并在肽段合成过程中催化特定合成步骤的结构域（Felnagle et al., 2008）。在每个模块中，NRPSs 由三个必要的结构域组成：一个腺苷化（A）结构域；一个硫化（T）结构域，翻译后被 4′-磷酸泛酰巯基（4′-Ppant）臂修饰，也称为肽基载体蛋白（PCP）结构域；以及一个缩合（C）结构域。这些结构域由大约 15 个氨基酸的短间隔区分隔开。A 结构域选择、激活并将单体加载到 PCP 结构域上。在这里，T 结构域的 4′-Ppant 臂上的硫醇基团介导了腺苷化氨基酸的亲核攻击。随后，由 C 结构域催化的两个相邻的 T-结合氨基酰基中间体之间形成肽键（Belin et al., 2012）。另一个重要的 NRPS 催化单元是位于 C 端的硫酯酶（TE）结构域，其通过水解或大环化催化肽的释放。此外，修饰域整合到不同位置的 NRPS 模块中，对氨基酸进行修饰合并。异构化和 N-甲基转移酶结构域分别催化 D-和甲基化氨基酸的生成（Koglin and Walsh, 2009; Strieker et al., 2010）。NRPSs 不仅依赖于 20 个典型的氨基酸，而且还使用了几种不同的构建模块，包括非蛋白质原性氨基酸，这有助于非核糖体肽的结构多样性及其不同的生物活性（Koglin and Walsh, 2009）。CDPs 一旦由 NRPSs 合成，就可以通过剪裁酶进一步修饰，通常由与 NRPS 基因聚类的基因编码。大多数已知的 NRPS 来源的 CDP 是由真菌产生的，而很少有细菌被认为是 NRPS 来源的 CDP 的生产者。

许多 CDPs 可以通过特定的 NRPS 途径形成，如 brevianamide F、erythrochelin、麦角胺、roquefortine C、acetylaszonalenin、thaxtomin A、胶黏毒素、西洛菌素 PL 等（Balibar and Walsh, 2006; Correia et al., 2003; García-Estrada et al., 2011; Gardiner et al., 2004; Healy et al., 2002; Maiya et al., 2006; Lazos et al., 2010; Yin et al., 2009）。在少数情况下，NRPSs 可在合成较长肽段的过程中形成 CDPs，作为截短的副产物，如在生物合成 cyclo (D-Phe-L-Pro) 和 cyclomarazine A 的过程中（Gruenewald et al., 2004; Schultz et al., 2008）。CDPs 的生物合成也可以由 CDPS 介导；CDPS 是一类依赖于 tRNA 肽键形成酶家族，并不需要氨基酸充电。CDPSs 具有一个共同的结构，类似于 Ic 类氨基酸 tRNA 合成酶（aaRSs）的催化结构域，如 TyrRS 和 TrpRS（Sauguet et al., 2011）。CDPSs 和 class-Ic aaRSs 都具有高度保守的罗斯曼折叠结构域，包含与核苷酸结合相关的结构特征，如黄素腺嘌呤二核苷酸、烟酰胺腺嘌呤二核苷酸（NAD^+）和烟酰胺腺嘌呤二核苷酸磷酸（$NADP^+$），以及一个螺旋连接多肽 1（CP1）亚结构域。然而，class-IcaaRSs 具有参与 ATP 的结合签名基序（HIGH 和 KMSKS 序列），而这些在 CDPSs 中并不存在。此外，CDPSs 不具有明显的 tRNA 结合域，而是含有位于螺旋 α4 中的大量带正电荷的残基，这些残基对于氨酰-tRNA 底物的结合非常重要。所有这些观察到的 CDPSs 与它们祖先的 aaRSs 之间的差异导致了 CDP 生物合成的独特酶特性。CDPSs 以氨基酸 tRNAs 为底物，催化 CDP 肽键的形成（Belin et al., 2012; Giessen and Marahiel, 2012; Giessen et al., 2013; Gondry et al., 2009），将两个氨酰-tRNA 从其在核糖体蛋白质合成中的重要作用中分离出来用作底物，并催化 CDP 形成所需的两个肽键的形成（Lahoud and Hou, 2010）。合成过程是由第一个氨酰基底物结合启动，可能涉及带负电荷的核糖-磷酸 tRNA 骨架与螺旋 α4 中的正电荷之间的离子相互作用（Bonnefond et al., 2011; Sauguet et al., 2011）。因此，通过使用氨酰-tRNA 作为底物，CDPSs 代表了初级和次级代谢之间的直接联系。CDPSs 的催化机理可以用乒乓模型来描述。所有的 CDPSs 都具有两个表面可及的口袋，其中含有对底物选择和催化至关重要的活性位点残基。两种 aa-tRNA 底物的不同氨酰基结合位点被称为口袋 1（P1）和口袋 2（P2）。在第一个底物特异性识别后，第一个氨酰基转移到 P1 保守的丝氨酸残基上。在这里，tRNA 部分与 α4 螺旋中的碱性残基相互作用，生成一个氨酰基酶中间体（Moutiez et al., 2014）。同时，第二个 aa-tRNA 的氨酰基部分通过 α6~α7 环与 P2 相互作用。最后，氨酰基-酶中间体与第二个 aa-tRNA 反应，生成二肽基-酶中间体，其通过保守酪氨酸的参与发生分子内环化，导致 CDP 骨架形成最终产物。这些 CDPs 可以通过紧密相连的剪接酶进行修饰。到目前为止，约有 163 个假定的 CDPS 基因被鉴定，其中 150 个在细菌中被报道，分布在 6 个门（放线菌门、拟杆菌门、衣原体门、蓝细菌门、厚壁菌门和变形菌门）。大多数已知的 CDPSs 存在于放线菌门中，迄今已报道的 CDPSs 有 77 个。12 个 CDPS 分布在 4 个真核生物门（子囊菌门、环节菌门、纤毛菌门、刺胞菌门），1 个伊斯巴尼亚盐陆生菌中也 CDPS 报道（Belin et al., 2012; Giessen and Marahiel, 2014; Tommonaro et al., 2012）。一些细菌 CDPSs 已经被充分鉴定，如营养链霉菌中的白诺氏菌素、枯草芽孢杆菌中的普切明，以及结核分枝杆菌（*Mycobacterium tuberculosis*）中的麦考环素（mycocyclosin）（Belin et al., 2012; Giessen et al., 2013）。

生物合成酶通常是物理上的，正如之前提到的，在转录过程中涉及的剪裁酶负责特异性地修饰含有 CDP 的天然产物。在几乎所有的 NRPS 和 CDPS 基因簇中，都可以找到修饰初始组装的 CDP 骨架的假定剪裁酶，并负责引入对 CDP 生物活性至关重要的官能团。在 CDPS 依赖的途径中，发现了大量不同的修饰酶与各自的 CDPS 基因密切相关（Belin et al., 2012; Giessen and Marahiel, 2014），包括不同类型的氧化还原酶、水解酶、转移酶和连接酶。CDPS 簇中最常见的假定剪裁酶是环二肽氧化酶（CDOs）。CDOs 由两个不同的小亚基组成，形成一个明显的兆道尔顿蛋白复合物。根据底物的不同，CDO 可以依次进行一个或两个脱氢反应。尽管已提出了三种不同的方案：直接脱氢、α-羟基化随后失水和亚胺形成与随后烯胺重排，但对此的精确反应机理尚未阐明（Gondry et al., 2001）。已知的 CDO 包括至少 7 种不同的 P450 酶、5 种不同类型的 α-酮戊二酸/Fe II 依赖性加氧酶，以及 3 种不同的含黄素单加氧酶。除氧化还原酶外，在 CDPS 基因簇中还发现了大量不同的 *C*-、*N*-和 *O*-氧甲基转移酶，α/β-水解酶，肽连接酶和酰基辅酶 A 转移酶，其中观察到属于 LuxR 和 MarR 家族的不同转录因子。它们通常参与调节响应环境刺激的各种过程，如对有毒化学物质和抗生素的响应，这可能暗示了 CDPS 依赖的修饰 CDPs 的生物学功能（Ellison and Miller, 2006）。在关于 NRPS 依赖途径的研究中，已报道了类似的修饰酶，同样地，调节 CDP 支架和侧链氧化的酶也是最多的（Belin et al., 2012）。真菌 NRPS 基因簇的一个显著特点是普遍存在不同的异戊烯基转移酶，其在含有色氨酸的 CDP 支架的不同位置进行异戊烯基化和逆异戊烯基化（Yu et al., 2012）。根据 NRPS 和 CDPS 基因簇中发现各种假定修饰酶，推测高度修饰的 CDPs 代表了一个具有不同功能的微生物天然产物家族。

此外，值得注意的是，CDP 核心不仅使这些分子抵抗蛋白水解，还能通过肠道屏障和血脑屏障（BBB; Beck et al., 2012; Teixidó et al., 2009）。因此，灵活性和稳定性相结合为 CDP 分子提供了生物学特性和广泛治疗的可能性（Bellezza et al., 2014a, b）。首先发现

CDPs 的生物学作用具有抑制纤溶酶原激活物抑制剂-1（PAI-1）的能力，可干预心血管疾病和凝血功能（Einholm et al.，2003），之后又发现 CDPs 具有抗菌作用（Rhee，2004），抗肿瘤（Nicholson et al.，2006），抗真菌和抗病毒活性（Kwak et al.，2013；Kwak et al.，2014；Mishra et al.，2017）。韩国发酵蔬菜泡菜是一种富含 Pro-based CDPs 的食品，具有抗多重耐药菌活性（Liu et al.，2017），用植物乳杆菌（Lactobacillus plantarum）LBP-K10 发酵蔬菜产生的化合物 cyclo（L-Val-L-Pro）和 cyclo（L-Phe-L-Pro）可抑制白假丝酵母菌的生长（Kwak et al.，2014）。cyclo（D-Tyr-D-Phe）从发酵改良的芽孢杆菌营养肉汤中提取，与横纹肌类昆虫病原线虫相关的 N 菌株中表现出对 A549 细胞具有显著的抗肿瘤活性，而对正常成纤维细胞无细胞毒性（Kumar et al.，2013）。CDPs 的多效性作用反映在它们结合一系列靶点的能力上：通过与催产素受体高亲和力结合，作为拮抗剂，CDPs 可以抑制射精（Borthwick et al.，2012；Clément et al.，2013）。冷水海洋海绵释放的 CDPs 能协同发挥化学防御作用（Sjögren et al.，2011）。此外，CHP 是 TRH（促甲状腺释放激素）的分解代谢产物，可控制血糖水平（Choi et al.，2012；Jung et al.，2011，2016；Koo et al.，2011；Lee et al.，2015，2013；Park et al.，2012）。CHP 与锌有关，作为一种抗糖尿病药物，对人类没有副作用，已在美国获得专利（Uyemura et al.，2010）。

2.2 QS 系统中的 CDPs

CDPs 最早是从海洋海绵中分离得到的（Schmitz et al.，1983），同时也发现于革兰氏阴性海洋细菌（Jayatilake et al.，1996）和革兰氏阳性海洋细菌（Stierle et al.，1988）。因此，海洋细菌被认为是这类具有生物活性化合物的潜在来源。众所周知，海洋海绵是新微生物的丰富来源，可产生具有潜在药理活性的化合物（Hentschel et al.，2001）。海洋海绵及其表面和内部空间提供了一个高度专业化的环境生态位，其中含有大量细菌。其数量比海水中所含的细菌高出 2~3 个数量级（Engel et al.，2002；Friedrich et al.，2001）。海绵-细菌伴生菌分布广泛且进化古老，海绵的大小和密度与细菌伴生菌的含量有直接关系。一些数据表明，微生物和海绵的共存是有益的（Proksch et al.，2002）。蓝细菌的自养能力能为宿主海绵提供额外的碳源和固定氮，与异养细菌的特异性关联促进多种有机化合物的代谢，而与相关细菌和蓝细菌群落产生的次级代谢产物则增强了宿主的化学防御（Abbamondi et al.，2014；De Rosa et al.，2003）。研究表明，与海绵相关的微生物区系具有物种特异性（Friedrich et al.，2001；Schmidt et al.，2000；Webster and Hill，2001；Webster et al.，2001），并代表一个稳定的种群（Friedrich et al.，2001；Webster and Hill，2001；Webster et al.，2001），能够与海绵本身进行交流。海洋海绵相关细菌分泌 QS 信号，如 N-酰基高丝氨酸内酯（AHLs；Taylor et al.，2004）和 CDPs（Tommonaro et al.，2012）。Holden 等（1999）报道了一些革兰氏阴性细菌产生和分泌 CDPs，这些 CDPs 可以激活和/或拮抗其他基于 LuxR 的 QS 系统。恶臭假单胞菌 WCS358 能产生和分泌 4 种 CDPs，其中一些能够与 QS 的 LuxI 和 LuxR 同源物相互作用（Degrassi et al.，2002）。从一系列革兰氏阴性细菌中分离出来的一组 CDPs，据报道可调节先前认为对 AHLs 特异性的敏感 AHL 生物传感器菌株中的 LuxR、TraR 或 LasR 活性（Degrassi et al.，2002；Holden et al.，1999；Park et al.，2006）。一个生物体中产生的 QS 信号能调节第二个生物体的行为。因此，从铜绿假单胞菌、荧光假单胞菌、恶臭假单胞菌、产碱假单胞菌（P. alcaligenes）、奇异变形杆菌、成团肠杆菌（E. agglomerans）、创伤弧菌和弗氏枸橼酸杆菌（C. freundii）的培养上清中单独或混合分离的 CDPs 代表了种间和跨界信号

（Campbell et al.，2009）。cyclo（*D-Ala-L-Val*）和 cyclo（*L-Pro-L-Tyr*）能抑制参与 AHL 依赖 QS 调控的 LuxR 型调控蛋白的活性（Campbell et al.，2009；Galloway et al.，2011）。此外，lasI 依赖的 QS 系统抑制 CDP 的生物合成，这与拟南芥植物相互作用的决定因素有关，因为 cyclo（*L-Pro-L-Tyr*）、cyclo（*L-Pro-L-Val*）和 cyclo（*L-Pro-L-Phe*）似乎模仿了生长素的生物学作用，后者是一种天然的植物激素（Ortiz-Castro et al.，2011）。

在特定的海绵-细菌关联菌组合中，相关细菌可能会随着海绵材料一起繁殖或至少存活。对居蟹皮海绵无菌细胞培养的核糖体 RNA 研究显示，细菌具有特异性的 16S rRNA 基因条带（Thakur et al.，2003）。海绵肽中 D-氨基酸和异常氨基酸的存在支持了海绵肽的微生物来源（Fusetani and Matsunaga，1993）。CDPs 归属于苔海绵（Schmitz et al.，1983），其均由伴生菌微球菌属（*Microboccus* sp.）产生（Stierle et al.，1988）。从海绵 *Theonella swinhoei* 中分离得到的环状糖肽，包含在不可培养的共生菌病原菌"假丝酵母菌-帕劳蒂壳内菌（*Candidatus Entotheonella palauensis*）"中（Schmidt et al.，2000）。此外，从与海洋海绵居蟹皮海绵相关的鲁杰氏菌属的变形菌，与变异鸢尾相关的葡萄球菌属和芽孢杆菌属菌株，以及海洋海绵贪婪掘海绵相关的海洋细菌弧菌属中，分离得到了一些调控细菌-海绵相互作用的 CDPs（De Rosa et al.，2003；Mitova et al.，2004）。

众所周知，细菌与真核宿主建立致病或共生关系，如铜绿假单胞菌作为一种众所周知的人类和植物病原体，在根际这个受根系分泌物影响的狭窄土壤区域中增殖。为了克服宿主防御，铜绿假单胞菌产生毒素、黏附素、铜绿假单胞菌素和其他毒力因子（Battle et al.，2008；de Abreu et al.，2014）。通过一种由 CDPs 发挥关键作用的 QS 机制（Gonzaález et al.，2017）。铜绿假单胞菌的 QS 系统较为复杂，las 和 rhl 系统分别依赖于 3-oxo-C12-*L*-HSL 和 C4-*L*-HSL。这些化合物由 *lasI* 和 *rhlI* 基因编码的酰基-*L*-高丝氨酸内酯合成酶合成（de Kievit and Iglewski，2000；Fuqua and Greenberg，2002；Lee and Zhang，2015）。第三类 QS 系统包括 *pqs* 基因簇编码的 2-庚基-3-羟基-4-（1H）-喹诺酮和 2-庚基-4-羟基喹诺酮（Gallagher et al.，2002；Lee and Zhang，2015）。所有这些系统都将信号转导连接到 LysR 型的转录因子，即 LasR、RhlR 和 PqsR，它们特异性地响应同源信号分子并驱动数百个基因的表达（Lee and Zhang，2015）。信号层次结构，位于 pqs 和 rhl 系统的上游，由信号分子 2-（2-羟基苯基）-噻唑-4-甲醛（IQS，也称为铜绿醛）进一步定义，该信号分子由 *ambBCDE* 基因簇合成，通过产生铜绿假单胞菌螯铁蛋白铁载体在致病过程中发挥重要作用（Dandekar and Greenberg，2013；Lee et al.，2013；Lee and Zhang，2015）。*ambBCDE* 基因簇编码 *L*-2-氨基-4-甲氧基-反式-3-丁烯酸（AMB）生物合成酶，通过非核糖体肽合酶（NRPS）途径发生，对原核生物和真核生物具有毒性作用（Rojas Murcia et al.，2015）。利用生物信息学和功能学方法，González 及其合作者（2017）最近从铜绿假单胞菌 PAO1 野生型（WT）菌株中鉴定了 NRPS，并研究了 CDPs 在细菌生理中的作用及其与植物的相互作用。研究发现，在假定的 MM-NRPS 缺陷的突变体中，CDPs 的产生发生了变化，这些变化虽然对毒力无效，但在很高的浓度下，通过与同源 AHL 的结合位点相互作用，干扰了 QS 系统。通过使用细菌-植物互作系统（即铜绿假单胞菌-拟南芥共培养），观察到所选择的 WT 和 NRPS 突变体抑制根生长或促进根分支与 AHL 依赖的 QS 状态有关，并在体内受到 CDP 水平的调控。CDPs 也会造成食品腐败，因为 QS 系统控制着食品腐败生态系统中的细菌行为（Gu et al.，2013；Skandamis and Nychas，2012）。大黄鱼是中国重要的海水养殖鱼类之一，由于消化酶和微生物活性在死后短时间内，即使在冷藏条件下，也极易腐败变质。微生物生长及其代谢副产物导致产生三甲胺、有机酸、醇类、硫化物、生物胺、醛类和酮类等具有令人不愉快和不可接受的异味（Gram and Dalgaard, 2002）。冷冻鱼的微生物腐败主要与革兰氏阴性蛋白水解嗜营养细菌的存在有关，主要包括希瓦氏菌属、假单胞菌属和肠杆菌科（Gram and Dalgaard, 2002; Skandamis and Nychas, 2012），每个物种通过产生、分泌和响应小的扩散分子来激活或抑制特定的靶基因表达，进而调节细胞间通信。在变质的牛奶、肉类、蔬菜和水产品中已经报道了包括 AHLs 和 AI-2 在内的各种信号化合物（Blana and Nychas, 2014; Liu et al., 2006; Rash et al., 2005）。一些研究者（Gu et al., 2013; Zhu et al., 2016）通过从腐败鱼中分离出的大黄鱼（*P. crocea*）特异性腐败微生物（SSO），研究了 SSO 的 QS 系统的作用。他们发现，希瓦氏菌属主要是波罗的海希瓦氏菌（*S. baltica*）和腐败希瓦氏菌（*S. outrefaciens*），是大黄鱼（*P. crocea*）货架期结束时的优势菌属。在无细胞的波罗的海希瓦氏菌培养物中，检测到 AI-2 和两个 CDPs，即 cyclo-(*L*-Pro-*L*-Leu) 和 cyclo-(*L*-Pro-*L*-Phe)。在 cyclo-(*L*-Pro-*L*-Leu) 存在下，扰流板中生物膜的产生、三甲胺和腐胺的产量显著增加，而在 cyclo-(*L*-Pro-*L*-Phe) 和 4,5-二羟基-2,3-戊二酮（AI-2 前体）存在的情况下则没有这种现象。外源 cyclo-(*L*-Pro-*L*-Leu) 暴露可上调 luxR、torA 和 ODC 的转录水平。在鱼类匀浆中，冷藏条件下，外源性 cyclo-(*L*-Pro-*L*-Leu) 提高了优势菌和 H_2S 产生菌的生长速率，而外源性 AI-2 前体则抑制了肠杆菌科等竞争细菌的生长。cyclo-(*L*-Pro-*L*-Leu) 刺激匀浆腐败过程中代谢物的积累，从而证实了波罗的海希瓦氏菌在大黄鱼中的腐败潜力是受 CDPs 介导的 QS 调控的。最后，由于细菌微生物组/细菌组和真菌微生物组/真菌组的重要性日益增长，CDPs 作为口腔细菌和真菌属之间的中介物的作用也已被研究（Brown et al., 2015）。口腔假丝酵母菌病是 HIV 感染的主要并发症（Shiboski et al., 2001; Shiboski, 2002; Thompson et al., 2010）。Brown 等（2015）的研究重点关注口腔菌群中假丝酵母菌与其他分类群的相互作用。口腔代谢产物是宿主、口腔细菌微生物组（细菌组）和口腔真菌微生物组（真菌组）的产物。细菌组和真菌组的功能变化导致了健康口腔环境和口腔念珠菌病之间的差异，而疾病和对照样本之间的相关性显著变化，表明生态系统中存在潜在的代谢变化。通过对整个口腔代谢组进行分析，并使用相关差异概率网络分析，证明了环单肽和二肽作为口腔细菌和真菌属之间的 QS 介质的重要作用，并推测了 CDPs 可能与口腔念珠菌病因学有关。Marchesan 等（2015）通过分析牙周炎患者的微生物群落组成，发现互养菌门（Synergistetes）中存在几种牙周致病菌。作者证明互养菌门与两个新的代谢物——cyclo(Leu-Pro) 和 cyclo(Phe-Pro)——密切相关，它们作为 QS 分子可引起牙周菌群失调和牙周病。

内源性或益生菌株已被证明可以减轻细菌病原体产生的毒力因子。事实上，Li 等（2011）证明了阴道常驻罗伊氏乳杆菌（*L. reuteri*）RC-14 能产生 CDPs，包括 cyclo(*L*-Phe-*L*-Pro) 和 cyclo(*L*-Tyr-*L*-Pro)，其干扰葡萄球菌的 QS 系统 agr，这是毒力基因的关键调节因子。这导致负责月经相关中毒性休克综合征的典型月经金黄色葡萄球菌菌株抑制中毒性休克综合征毒素-1 的表达（Li et al., 2011）。

3 CHP 在神经系统疾病中的作用

3.1 背景

促甲状腺素释放激素（TRH）是由 pGlu-His-Pro-NH_2 形成的三肽，通过焦谷氨酰基肽酶的作用在下丘脑后生成，随后转化为线性二肽（His-Pro-NH_2），最后在 37℃ 的非酶过程中环化形成 CHP，也被称为组氨酰脯氨酸二酮哌嗪（Minelli et al.，2008；Prasad and Peterkofsky，1976）。20 世纪 70 年代，广泛的研究表明 CHP 在中枢神经系统中普遍存在，这一发现引发了对 CDP 生物学作用的大量研究。动物服用外源性 CHP 后，伴随着多种生物活性，如减弱氯胺酮麻醉、延长戊巴比妥诱导的睡眠以及减轻酒精的药理作用等（Prasad，2001）。此外，CHP 在调节食物摄入量、身体核心温度和疼痛感知方面发挥重要作用，同时作为内分泌效应物，抑制催乳素的分泌（Morley et al.，1981；Prasad，1995，2001）。所有这些作用似乎都与多巴胺机制有关。Faden 等（1981）的研究表明，TRH 和 TRH 类似化合物可改善脊柱创伤后的神经功能恢复，增强认知功能。尽管表现出强大的内分泌、镇痛和自主神经作用，这些特性限制了 TRH 的治疗应用。然而，令人惊讶的是，研究还发现，TRH 的代谢产物 CHP 保留了所有的药理活性，而没有已知的副作用（Faden et al.，2004，2005）。Taubert 等（2007）研究表明，CDP 是有机阳离子转运体 2（OCT2）的特异性底物，OCT2 是一种钠离子依赖的转运体多巴胺能脑结构中高表达，典型的靶向帕金森病尤其是黑质致密部，这进一步支持了 CHP 参与脑功能和对神经系统疾病的潜在影响。(Taubert et al.，2007）。CHP 不仅在这些区域存在共定位现象，进一步研究表明，它可以保护神经元免受盐碱糖醇诱导的细胞毒性影响，这是一种与帕金森病有关的 L-DOPA 代谢产物。

3.2 当前的理解

过去 10 多年来，Minelli 及其同事一直致力于探究 CHP 在大脑中的作用。事实上，2006 年底发现的第一个线索是，这种分子可能用于治疗神经系统疾病，当发现只有在试验条件的情况下，CHP 能有效保护多巴胺使 PC12 细胞免受凋亡。此外，研究表明，CDPs 处理能激活两种热休克蛋白（Hsp），即 hsp27 和 α-B-晶状蛋白（Minelli et al.，2006），这两种蛋白在蛋白的正确折叠过程中发挥重要作用。然而，Hsps 通过减轻由蛋白质错误折叠引起的细胞凋亡，在神经退行性疾病中起着重要作用。尽管当时这种效果几乎未被注意到，但现在可以将化合物的细胞保护抗凋亡作用与增强管理代谢应激的能力（如蛋白质错误折叠反应）联系在一起，这可能具有相当重要的意义。CHP 通过触发 NF-E2 相关因子-2（Nrf2）核积累，降低活性氧（ROS）的产生，防止谷氨酸、鱼藤酮、百草枯和 β-淀粉样蛋白等应激源引起的谷胱甘肽（GSH）耗竭。Nrf2 是一种转录因子，可上调抗氧化/亲电反应元件（ARE-EpRE）相关基因（详见下文）。基于这些发现，CHP 作为大脑可调节的 Nrf2 通路的选择性激活剂，可能是一种有前途的神经保护剂，通过诱导 II 期基因发挥作用（Minelli et al.，2009a，b）。氧化应激是指 ROS 的产生超过细胞缓冲能力的一种状态。ROS 活性极强，可对蛋白质、核酸、脂质等大分子造成不可修复的损伤，从而导致细胞死亡或基因突变。神经元是终末分化细胞，因此，对氧化应激极为敏感。事实上，神经元在很大程度上依赖于周围神经胶质细胞对 GSH 的可用性（Hsu et al.，2005；Reynolds et al.，2007）。GSH 是一种非常重要的三肽，可通过氧化还原反应增加 ROS 水平和修复被氧化的细胞大分子。有几种酶参与 GSH 的作用，其中大多数受到 Nrf2 的转录调控（详见后文；Brigelius-Flohé and Flohé，2011；Minelli et al.，2009a，b）。此外，神经胶质细胞作为中枢神经系统的免疫系统，通过增加活性氮（Reactive nitrogen species，RNS）的产生，对神经炎症刺激作出响应，例如，一氧化氮（NO），其是一种非常容易扩散的分子，可与 ROS 反应，尤其是超氧阴离子，产生高反应性和有毒的过氧亚硝酸盐。这种情况被认为是亚硝化应激。因此，大脑对氧化还原状态的变化非常敏感，维持氧化还原平衡对于防止氧化损伤非常重要。当神经胶质细胞遭受极高水平的ROS过度激活时，脑细胞会经历氧化应激和亚硝化应激，这两者相互作用会破坏正常的神经元功能。实际上，氧化应激和亚硝化应激的标志物是所有神经退行性疾病的一个明确的特征，并且有力地证实了ROS/RNS与神经退行性疾病之间的因果关系（Gupta et al., 2014；Leszek et al., 2016；Tsang and Chung, 2009；Valko et al., 2007）。在氧化应激条件下，线粒体和氧化磷酸化产生能量的过程功能失调，导致更高水平的ROS产生，并减少ATP合成。值得指出的是，线粒体功能障碍与神经退行性疾病密切相关。事实上，在线粒体衰竭的情况下，NADPH氧化酶会产生超氧阴离子，并与主要由诱导型一氧化氮合酶产生的NO结合，生成高RNS的过氧亚硝酸盐（Contestabile et al., 2003；Dasuri et al., 2013；Grottelli et al., 2016；Valko et al., 2007）。不适当的氧化应激信号会导致细胞凋亡，而凋亡细胞本身会释放ROS，因此，可以想象ROS诱导的细胞凋亡形成一个自我持续循环，进一步释放ROS，从而导致额外的细胞凋亡。

在哺乳动物中，氧化应激损伤主要由NF-E2相关因子2（Nrf2）控制。Kelch样ECH相关蛋白1（Keap1）系统作为一种抗应激反应是从祖先遗传而来的，旨在维持细胞内稳态（Bellezza et al., 2010；Brigelius-Flohe and Flohe, 2011；Itoh et al., 1999）。在基础条件下，Nrf2被细胞质Keap1隔离，并靶向蛋白酶体降解（Bellezza et al., 2010；Brigelius-Flohé and Flohé, 2011；Itoh et al., 1999）。在氧化应激条件下，Nrf2-Keap1的相互作用以剂量依赖的方式解离，使Nrf2转移到细胞核中，并与其中一个小的Maf蛋白异源二聚化。异源二聚体识别存在于Nrf2靶基因调控区的增强子序列中的抗氧化反应元件（AREs），对招募转录的关键因子至关重要（Suzuki et al., 2013；Suzuki and Yamamoto, 2015）。Nrf2影响近500个基因的表达，这些基因编码的蛋白充当氧化还原平衡因子、解毒酶、应激反应蛋白和代谢酶（Fuse and Kobayashi, 2017；Hahn et al., 2015；Yang et al., 2016），因此，Nrf2可以被视为氧化应激反应的主要调节因子。由此可见，CHP具有激活Nrf2系统的能力，可以被认为是一种抗氧化化合物。这种诱导保护性抗氧化反应的能力可能使CDP成为基于氧化损伤的神经系统疾病的一种有价值的治疗方法。然而，由于神经系统疾病是由内质网（ER）应激、钙负荷、兴奋性毒性和炎症共同作用的多因素疾病，因此，推测二肽的有益作用不能仅仅归因于Nrf2的激活。

事实上，应激条件会导致几种途径的激活，包括未折叠蛋白反应（UPR），UPR是由错误折叠的蛋白质聚集在内质网的腔室中引起的，这种情况被认为是内质网应激。内质网应激通过分子伴侣GRP78/Bip（结合型免疫球蛋白/78 kDa葡萄糖调节蛋白）的解离，导致位于内质网膜的PERK［蛋白激酶R（PKR）样内质网激酶］、ATF6（转录激活因子6）和IRE1（肌醇需求酶1）三种应激传感器蛋白的激活。这导致通过PERK介导的eIF2α磷酸化对蛋白翻译的普遍抑制，以缓解内质网蛋白负荷。此外，通过ATF6和IRE1α分支上调分子伴侣的表达，增加细胞的折叠能力。当应激刺激超过细胞修复能力时，内稳态无法恢复，细胞发生凋亡。事实上，持续的应激条件会诱导转录因子CHOP（C/EBP同源蛋白），从而诱启动细胞凋亡机制。

通过发现CDP可以抵消小胶质细胞中束霉素诱导的内质网应激，从而证实了CHP在UPR调控中的作用（Bellezza et al., 2014a, b）。实际上，CHP通过激活eIF2α和GRP78/Bip来诱导保护性UPR，并通过降低促凋亡蛋白CHOP的表达来保护细胞免受凋亡。这些分子事件显著降低了衣霉素诱导的细胞活力下降（Bellezza et al., 2014a, b）。值得注意的是，UPR的PERK臂激活Nrf2，相反，通过减少氧化应激，可以减少氧化和错误折叠蛋白质的数量。

各种研究表明，在多种实验模型中，Nrf2 在对抗 NF-κB 介导的炎症反应中发挥着重要作用（Bellezza et al., 2010, 2014a, b; Brigelius-Flohé and Flohé, 2011; Sandberg et al., 2014）。术语 NF-κB（活化 B 细胞的核因子 kappa-轻链增强子）指的是一类控制炎症反应的转录因子家族。研究最多的成员是 p50-p65 异源二聚体，它在炎症刺激下诱导促炎介质的表达。NF-κB 通过与其抑制子 IκBα 结合，从而在胞质中维持非活性状态。典型的 NF-κB 激活途径依赖于 IKK（IκB 激酶）蛋白激酶的激活，使 IκBα 磷酸化，进而被蛋白酶体降解。该事件导致 NF-κB 的激活和核转位，随后上调 NF-κB 靶基因（Bellezza et al., 2010）。

在转录水平上，NF-κB 与转录共激活因子 CREB 结合蛋白竞争，从而抑制 Nrf2 信号。此外，NF-κB 通过招募组蛋白去乙酰化酶 3（HDAC3），引起局部去乙酰化，从而降低 Nrf2 信号（Wang et al., 2012）。在这两种转录因子同时出现在细胞核的情况下，NF-κB 可拮抗 Nrf2 诱导的基因转录，通过抑制 NF-κB 信号的化合物来减轻炎症反应并激活 Nrf2 通路（Grottelli et al., 2016; Kim et al., 2013; Li et al., 2008; Minelli et al., 2012）。这一联系最初通过研究表明，Nrf2 缺陷小鼠表现出神经退行性表型（Burton et al., 2006），而 Nrf2 的缺失与细胞因子产生的增加相关（Pan et al., 2012）。在 Nrf2 近端启动子中，存在几个 κB 位点（即 NF-κB 识别并结合的基因组序列）；因此，在促炎性刺激如肿瘤坏死因子 α（TNFα）存在的情况下，一些细胞通过上调 Nrf2 来响应，导致细胞因子基因表达的反馈抑制（Rushworth et al., 2012）。此外，NF-κB 激活可通过小 GTP 酶 RAC1（Ras 相关 C3 肉毒素底物 1）调节 Nrf2 活性，作为一种保护性抗炎机制。一旦被 LPS（脂多糖）激活，RAC1 通过激活 Nrf2 上调 HO-1（血红素加氧酶 1）的表达，使细胞转向更还原的环境，这对于终止 NF-κB 的激活至关重要（Cuadrado et al., 2014）。利用小鼠耳部炎症模型，探究 CHP 在 NF-κB 系统中的分子机制。观察到 CHP 可减轻 12-十四烷基邻苯二酚-13-醋酸酯诱导的水肿。此外，在促炎分子 LPS 刺激的小鼠永生化小胶质 BV2 细胞中，CHP 干扰了抗氧化 Nrf2/HO-1 和促炎 NF-κB 通路之间的串扰。事实上，环加氧酶-2 和基质金属蛋白酶 3 这两个是受 NF-κB 调控的基因产物被 CHP 下调，并在血红素加氧酶 1（HO-1）敲除细胞中上调。基于这些数据，表明 CHP 通过 Nrf2 介导的 HO-1 激活来抑制促炎的 NF-κB 信号，因此，提出 CHP 作为体内抗炎化合物使用（Minelli et al., 2012）。越来越明显的是，神经炎症是所有神经退行性疾病共同特征之一（Bellezza et al., 2014a, b; Dinkova-Kostova et al., 2018; González-Reyes et al., 2017）。通常由外周炎症触发，该术语描述了中枢神经系统细胞产生的各种免疫反应，如小胶质细胞、星形胶质细胞和血脑屏障，这些细胞之间通过动态串扰连接。在长期持续的炎症存在的情况下，神经炎症反应会导致突触损伤，神经元死亡，并最终导致神经退行性变（Boulamery and Desplat-Jego, 2017; Lyman et al., 2014; Rustenhoven et al., 2017）。如前所述，CHP 的作用可以通过两种方式抵消这种致病状态：通过激活 Nrf2 和诱导 HO-1 活性，该化合物可能同时驱动保护性抗氧化反应，减轻氧化应激损伤，同时抑制 NF-κB 信号传导，减少与炎症相关的损伤（Minelli et al., 2012）。出于这些考虑，假设 CDP 抑制胶质细胞炎症。全身给药 CHP 可通过下调全身（肝）和局部（脑）TNFα 的表达，在中枢神经系统发挥抗炎作用，从而抵消 LPS 诱导的胶质增生（Bellezza et al., 2014a, b）。已知这些作用可以降低炎性神经毒素对神经元的有害作用（Catorce and Gevorkian, 2016）。这些数据表明，CHP 在神经炎症环境中的有益作用及其在神经炎性疾病中的潜在治疗效用，研究认为 CDP 可能用于治疗其他神经病理疾病。为了更直接地检验这种可能性，Minelli 和同事检测了 CHP 对 hSOD1G93A 小鼠小胶质细胞的影响。这些转基因小鼠表达 Gly93-Ala 突变的超氧化物歧化酶 1（SOD1）编码人类基因（SOD1G93A），概括了肌萎缩性脊髓侧索硬化症（ALS）的几个方面，并提供了强大的模型系统用于研究疾病的病理生理机制和筛选潜在的治疗化合物（Grottelli et al., 2015, 2016）。在这种情况下，CHP 即使在 SOD1G93A 环境中，也可以作为抗氧化剂。更重要的是，它甚至提供了通过强烈上调神经元生长因子-脑源性神经营养因子的 mRNA 水平，来超越保护神经元的再生，该分子不仅与保存现有的神经元功能有关，而且也与新神经元的生长和分化有关。因此，CHP 可能既抑制 SOD1 突变引起的与氧化应激和小胶质细胞炎症反应相关的神经元损伤，又能直接作用于神经元本身，以保持并可能恢复其功能，这表明其可能作为一种治疗药物用于预防或延缓 ALS 疾病的进展（图 10-2）。

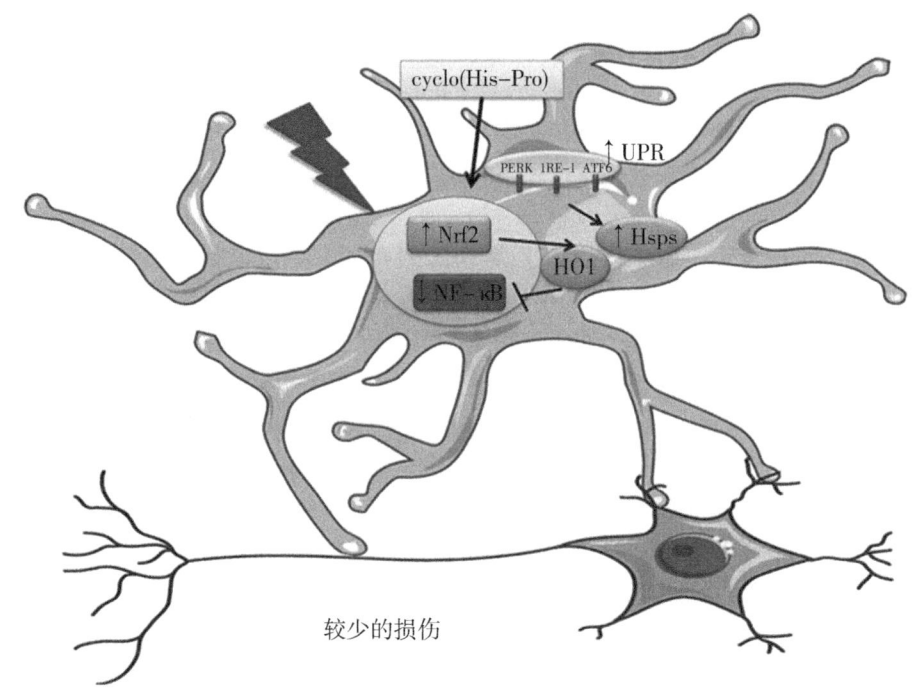

图 10-2　cyclo（His-Pro）作用方案

注：cyclo（His-Pro）通过增加保护性未折叠蛋白反应（UPR），
通过激活 Nrf2 诱导的抗氧化反应，以及通过 NF-κB 抑制小胶质细胞
下调促炎反应，保护神经元细胞免受神经毒性损伤。

3.3　展望

神经退行性疾病是多因素疾病，尽管每种疾病都有不同的病因。然而，常见的致病机制如神经炎症、氧化应激和内质网应激，是神经退行性变的基础。研究表明，CHP 可在几种神经毒素暴露的细胞模型中抵消这些致病途径。迄今为止，只有突变的 SOD1 细胞作为 ALS 的金标准模型被使用。CHP 是否对其他神经退行性疾病特异性细胞和动物模型有效仍需进一步检验。

通过多种途径发挥作用，CHP 保护神经元免受神经毒性损伤。首先，它增加了保护性的未折叠蛋白反应（UPR），帮助细胞应对蛋白质的错误折叠；其次，通过激活 Nrf2，它诱导抗氧化反应，减少氧化应激的损伤。最后，其抑制了 NF-κB 的活性，从而降低了促炎反应，进一步保护神经元细胞。

近期研究指出，肠道微生物群的不平衡可能导致大脑出现病理信号，进而引发促炎反应、氧化应激和细胞退变等多种神经退行性疾病（Noble et al., 2017）。人类胃肠道中细菌细胞数量超过宿主细胞的 10 倍，编码的基因数量超过宿主基因的 100 倍。这些与人类消化道相关的微生物被称为肠道微生物组/微生物群。虽然在不同的研究中，对人类肠道细菌种类数量的评估差异很大，但普遍认为，个体拥有超过 1 000 种微生物，并且它们在物种水平上形成系统发育类型（Lozupone et al., 2012），这些微生物通过 QS 机制进行相互沟通（Bivar Xavier, 2018）。人类肠道微生物组在健康和疾病中的作用一直是广泛研究的课题，而细菌共生菌在各种神经系统疾病中的作用也得到广泛认可（Byrd et al., 2018；Caballero-Villarraso et al., 2017；Cox and Weiner, 2018；Friedland and Chapman, 2017；Ho et al., 2018；Kitai and Tang, 2018；Mariett et al., 2018；Perez-Pardo et al., 2017；Roszyk and Puszczewicz, 2017；Sherwin et al., 2018；Thion et al., 2018；Yang and Duan, 2018）。事实上，胃肠道与中枢神经系统通过肠-脑轴紧密相连，形成了一个神经内分泌信号和免疫因子相互交织的双向网络。研究表明，肠道微生物能将来自所摄入的食物信息传递给中枢神经系统，以获得系统反应（Noble et al., 2017）。在 1995 年，胃肠道中的 CHP 与肠-胰岛轴中肠道肽有关（Prasad, 1995）。目前，由于 CHP 作为 QS 信号的重要作用，能控制细菌群体水平反应的行为和功能，提出了 CHP 作为调控肠道微生物群的新作用。因此，CHP，通过直接作用于中枢神经系统细胞和潜在作用于肠道微生物，被认为是一种治疗神经退行性疾病的潜在新药。然而，关于后者的信息仍不明确，因为，这些信息可能需要通过临床前试验来验证。鉴于这一情况，未来希望能够激发科学界对验证这一假设的兴趣，以开发与肠道微生物功能失调相关的多种疾病的新治疗方式。

4 结论

研究表明，CHP 作为一种内源性 CDP，可减少氧化、内质网应激以及炎症，而炎症被认为是几种神经系统疾病的主要罪魁祸首。因此，即使通过口服给药，这种 CDP 也能穿过血脑屏障，发挥对胶质细胞的有益作用，因为胶质细胞的异常反应是目前导致神经元死亡的原因之一。此外，由于 CHP 可以作为 QS 信号，我们有理由认为这种二肽能调节肠道微生物组，从而在多种与微生物组失衡相关的病理中获得临床益处。

因此，通过直接作用于神经退行性疾病的几种原因，并间接影响与神经系统疾病相关的肠道微生物组，CHP 有潜力缓解许多神经退行性疾病发病机制中的不同因素。

术语表

肌萎缩性脊髓侧索硬化症（ALS） 一种神经退行性疾病，其特征是肌肉痉挛、肌肉萎缩引起的快速进行性无力、言语障碍（构音障碍）、吞咽困难（吞咽障碍）以及由上下运动神经元变性引起的呼吸困难。受这种疾病影响的个人最终可能会失去控制所有自主运动的能力，尽管膀胱和肠道功能以及负责眼球运动的肌肉在疾病的最后阶段通常不会受到影响，且大多数患者的认知功能通常也不会受到影响。

生物膜　一种结构化的细菌细胞组成的群落,被封闭在自身产生的保护性聚合物基质中,附着在惰性或活表面上。

血脑屏障（BBB）　一种高度选择性渗透屏障,其能将循环血液与脑细胞外液（Brain Extracellular Fluid, BECF）分隔开。由紧密连接的内皮细胞形成,它允许对神经功能至关重要的分子通过,并防止潜在的神经毒素进入。

环二肽（CDPs）/2,5-二酮哌嗪　由二肽及其酰胺的非酶环化反应产生相对简单的化合物。它们是自然界中最常见的肽衍生物,由变形杆菌物种和人类合成。CDPs 具有蛋白水解的稳定性和促进与生物靶点相互作用的特点。

循环支架　一个六元环,由于其稳定的结构特征,药物化学中代表着一个重要的药效团。

人类微生物组　人体大约由 10 万亿个细胞组成,拥有 100 万亿个细菌,如在皮肤和肠道中。这是人类的"微生物组"并对人类的健康有着巨大的影响。然而,反过来,人类可通过影响居住在体内的细菌种类来影响微生物组。

炎症　先天免疫系统对有害刺激的反应,如病原体、受损细胞或刺激物等,是机体清除有害刺激并启动愈合过程的一种保护性尝试。典型的症状是疼痛、发热、发红、肿胀和功能丧失。

脂多糖（LPS）　革兰氏阴性细菌外膜上的一种糖脂。它被免疫细胞中的 Toll 样受体 4（Toll-like receptor 4, TLR4）识别,诱导促炎反应的激活。

小神经胶质　一类非神经细胞,构成大脑和脊髓的驻留巨噬细胞,作为中枢神经系统中主动免疫防御的第一和主要形式。

NF-κB（活化 B 细胞的核因子 κ-轻链增强子）　一个控制炎症反应的转录因子家族。研究最多的 NF-κB 家族成员是 p50-p65 异源二聚体,其在炎症刺激下诱导促炎介质的表达。

群体感应（QS）　一种通过分泌信号分子进行细胞间通信的机制。当细胞群体密度达到足以产生自诱导物的阈值时,分泌的自诱导物将调控一组特定基因的表达。

活性氧（ROS）　由分子氧衍生出来的一些活性分子和自由基,如单线态氧、超氧化物、过氧化物、羟基自由基和次氯酸。

促甲状腺释放激素（TRH）　由下丘脑产生的一种三肽激素,可刺激垂体前叶释放促甲状腺激素和催乳素。

未折叠蛋白反应（UPR）　一种与内质网应激反应相关的进化保守反应,称为 UPR。其最初目的是适应不断变化的环境,以重建正常的内质网功能。当适应失败时,内质网通过诱导编码宿主防御调控基因的表达来启动信号通路,并发出警报。过度和长期的内质网应激会触发细胞自杀,通常以细胞凋亡的形式出现,代表了多细胞生物摆脱功能失调细胞的最后手段。

毒力因子　病原体（细菌、病毒、真菌和原生动物）表达和分泌的分子,使它们能够通过破坏或逃避宿主防御而在宿主体内复制和传播。

缩写词

aaRS	氨基酸 tRNA 合成酶
AHSL	酰基高丝氨酸环内酯

AIP	自诱导肽
ALS	肌萎缩侧索硬化症
ARE	抗氧化反应元件
ATF6	激活转录因子6
CDO	环二肽氧化酶
CDP	环二肽
CDPSCDP	合成酶
CHOP	C/EBP 同源蛋白
CHP	环（组氨酸-脯氨酸）
ER	内质网
GRP78/Bip	结合免疫球蛋白/78 kDa 葡萄糖调节蛋白
GSH	谷胱甘肽
HO-1	血红素加氧酶-1
Hsp	热激蛋白
IKK	κB 激酶
IRE1	需肌醇酶-1
IκBα	κB 抑制因子
NF-κB	活化 B 细胞的核因子 κ-轻链增强子
Nrf2	NF-E2-调节因子-2
NRPS	非核糖体多肽合成酶
PCP	肽基载体蛋白
PEG	聚乙二醇
PERK	蛋白激酶 R（PKR）样内质网激酶
QS	群体感应
RAC1	Ras 相关 C3 肉毒杆菌毒素底物 1
RNS	活性氮
ROS	活性氧
SOD1	超氧化物歧化酶 1
TNFα	肿瘤坏死因子 α
TRH	促甲状腺释放激素
UPR	未折叠蛋白反应
WT	野生型

参考文献

ABBAMONDI G R, DE ROSA S, IODICE C, et al., 2014. Cyclic dipeptides produced by marine spongeassociated bacteria as quorum sensing signals. Nat. Prod. Commun., 9: 229-232.

BALIBAR C J, WALSH C T, 2006. GliP, a multimodular nonribosomal peptide synthetase in Aspergillus fumigatus, makes the diketopiperazine scaffold of gliotoxin. Biochemistry, 45: 15029-15038.

BASSLER B, VOGEL J, 2013. Bacterial regulatory mechanisms: the gene and beyond. Curr. Opin.

Microbiol., 16: 109-111.

BATTLE S E, MEYER F, RELLO J, et al., 2008. Hybrid pathogenicity island PAGI-5 contributes to the highly virulent phenotype of a *Pseudomonas aeruginosa* isolate in mammals. J. Bacteriol., 190: 7130-7140.

BECK J G, CHATTERJEE J, LAUFER B, et al., 2012. Intestinal permeability of cyclic peptides: common key backbone motifs identified. J. Am. Chem. Soc., 134: 12125-12133.

BELIN P, MOUTIEZ M, LAUTRU S, et al., 2012. The nonribosomal synthesis of diketopiperazines in tRNA-dependent cyclodipeptide synthase pathways. Nat. Prod. Rep., 29: 961.

BELLEZZA I, MIERLA A L, MINELLI A, 2010. Nrf2 and NF-κB and their concerted modulation in cancer pathogenesis and progression. Cancers (Basel), 2: 483-497.

BELLEZZA I, GROTTELLI S, MIERLA A L, et al., 2014a. Neuroinflammation and endoplasmic reticulum stress are coregulated by cyclo(His-Pro) to prevent LPS neurotoxicity. Int. J. Biochem. Cell Biol., 51: 159-169.

BELLEZZA I, PEIRCE M J, MINELLI A, 2014b. Cyclic dipeptides: from bugs to brain. Trends Mol. Med., 20: 551-558.

BIVAR X K, 2018. Bacterial interspecies quorum sensing in the mammalian gut microbiota. C. R. Biol., 341 (5): 300. pii: S1631-0691 (18): 30051-30059.

BLANA V A, NYCHAS G J, 2014. Presence of quorum sensing signal molecules in minced beef stored under various temperature and packaging conditions. Int. J. Food Microbiol., 173: 1-8.

BOCKUS A T, MCEWEN C M, LOKEY R S, 2013. Form and function in cyclic peptide natural products: a pharmacokinetic perspective. Curr. Top. Med. Chem. . 13: 821-836.

BONNEFOND L, ARAI T, SAKAGUCHI Y, et al., 2011. Structural basis for nonribosomal peptide synthesis by an aminoacyl-tRNAsynthetase paralog. Proc. Natl. Acad. Sci. U. S. A., 108: 3912-3917.

BORTHWICK A D, 2012. 2,5-Diketopiperazines: synthesis, reactions, medicinal chemistry, and bioactive natural products. Chem. Rev., 112: 3641-3716.

BORTHWICK A D, DA COSTA N C, 2017. 2,5-Diketopiperazines in food and beverages: taste and bioactivity. Crit. Rev. Food Sci. Nutr., 57: 718-742.

BORTHWICK A D, LIDDLE J, DAVIES D E, et al., 2012. Pyridyl-2,5-diketopiperazines as potent, selective, and orally bioavailable oxytocin antagonists: synthesis, pharmacokinetics, and in vivo potency. J. Med. Chem., 55: 783-796.

BOULAMERY A, DESPLAT-JÉGO S, 2017. Regulation of neuroinflammation: what role for the tumor necrosis factorlike weak inducer of apoptosis/Fn14 pathway? Front. Immunol., 8: 1534.

BRIGELIUS-FLOHE R, FLOHE L, 2011. Basic principles and emerging concepts in the redox control of transcription factors. Antioxid. Redox Signal., 15: 2335-2381.

BROWN R E, GHANNOUM M A, MUKHERJEE P K, et al., 2015. Quorum-sensing dysbiotic shifts in the HIV-infected oral metabiome. PLoS One, 10: e0123880.

BURTON N C, KENSLER T W, GUILARTE T R, 2006. In vivo modulation of the Parkinsonian phenotype by Nrf2. Neurotoxicology, 27: 1094-1100.

BYRD A L, BELKAID Y, SEGRE J A, 2018. The human skin microbiome. Nat. Rev. Microbiol., 16: 143-155.

CABALLERO-VILLARRASO J, GALVAN A, ESCRIBANO B M, et al., 2017. Interrelationships between gut microbiota and the host: paradigms, role in neurodegenerative diseases and future prospects. CNS Neurol. Disord. Drug Targets, 16: 945-964.

CAMPBELL J, LIN Q, GESKE G D, et al., 2009. New and unexpected insights into the modulation of Lux

R-type quorum sensing by cyclic dipeptides. ACS Chem. Biol., 4: 1051-1059.

CATORCE M N, GEVORKIAN G, 2016. LPS-induced murine neuroinflammation model: main features and suitability for pre-clinical assessment of nutraceuticals. Curr. Neuropharmacol., 14: 155-164.

CHEN M Z, DEWIS M L, KRAUT K, et al., 2009. 2,5 Diketopiperazines (cyclic dipeptides) in beef: identification, synthesis, and sensory evaluation. J. Food Sci., 74: C100-C105.

CHOI S A, SUH H J, YUN J W, et al., 2012. Differential gene expression in pancreatic tissues of streptozocininduced diabetic rats and genetically-diabetic mice in response to hypoglycemic dipeptide cyclo(His-Pro)treatment. Mol. Biol. Rep., 39: 8821-8835.

CLARDY J, WALSH C, 2004. Lessons from natural molecules. Nature, 432: 829-837.

CLARDY J, FISCHBACH M A, WALSH C T, 2006. New antibiotics from bacterial natural products. Nat. Biotechnol., 24: 1541-1550.

CLEMENT P, BERNABE J, COMPAGNIE S, et al., 2013. Inhibition of ejaculation by the non-peptide oxytocin receptor antagonist GSK557296: a multi-level site of action. Br. J. Pharmacol., 169: 1477-1485.

CONTESTABILE A, MONTI B, CONTESTABILE A, et al., 2003. Brain nitric oxide and its dual role in neurodegeneration/neuroprotection: understanding molecular mechanisms to devise drug approaches. Curr. Med. Chem., 10: 2147-2174.

COOK L C, FEDERLE M J, 2014. Peptide pheromone signaling in Streptococcus and Enterococcus. FEMS Microbiol. Rev., 38: 473-492.

CORNACCHIA C, CACCIATORE I, BALDASSARRE L, et al., 2012. 2,5-Diketopiperazines as neuroprotective agents. Mini Rev. Med. Chem., 12: 2-12.

CORNFORTH D M, POPAT R, MCNALLY L, et al., 2014. Combinatorial quorum sensing allows bacteria to resolve their social and physical environment. Proc. Natl. Acad. Sci. U. S. A., 111: 4280-4284.

CORREIA T, GRAMMEL N, ORTEL I, et al., 2003. Molecular cloning and analysis of the ergopeptine assembly system in the ergot fungus *Claviceps purpurea*. Chem. Biol., 10: 1281-1292.

COX L M, WEINER H L, 2018. Microbiota signaling pathways that influence neurologic disease. Neurotherapeutics, 15: 135-145.

CUADRADO A, MARTIN-MOLDES Z, YE J, et al., 2014. Transcription factors NRF2 and NF-kappaB are coordinated effectors of the Rho family, GTP-binding protein RAC1 during inflammation. J. Biol. Chem., 289: 15244-15258.

DANDEKAR A A, GREENBERG E P, 2013. Microbiology: plan B for quorum sensing. Nat. Chem. Biol., 9: 292-293.

DASURI K, ZHANG L, KELLER J N, 2013. Oxidative stress, neurodegeneration, and the balance of protein degradation and protein synthesis. Free Radic. Biol. Med., 62: 170-185.

DE ABREU P M, FARIAS P G, PAIVA G S, et al., 2014. Persistence of microbial communities including *Pseudomonas aeruginosa* in a hospital environment: a potential health hazard. BMC Microbiol., 14: 118.

DE KIEVIT T R, IGLEWSKI B H, 2000. Bacterial quorum sensing in pathogenic relationships. Infect. Immun., 68: 4839-4849.

DE ROSA S, MITOVA M, TOMMONARO G, 2003. Marine bacteria associated with sponge as source of cyclic peptides. Biomol. Eng., 20: 311-316.

DEGRASSI G, AGUILAR C, BOSCO M, et al., 2002. Plant growth-promoting *Pseudomonas putida* WCS358 produces and secretes four cyclic dipeptides: cross-talk with quorum sensing bacterial sensors. Curr. Microbiol., 45: 250-254.

DIGGLE S P, GARDNER A, WEST S, et al., 2007. Evolutionary theory about bacterial quorum sensing: when is a signal not a signal? Philos. Trans. R. Soc. Lond. Ser. B Biol. Sci., 362: 1241-1249.

DINKOVA-KOSTOVA A T, KOSTOV R V, KAZANTSEV A G, 2018. The role of Nrf2 signaling in counteracting neurodegenerative diseases. FEBS J., https://doi.org/10.1111/febs.14379.

EGUCHI C, KAKUTA A, 1974. Cyclic dipeptides, I. Thermodynamics of the cis – trans isomerization of the side chains in cyclic dipeptides. J. Am. Chem. Soc., 96: 3985-3989.

EINHOLM A P, PEDERSEN K E, WIND T, et al., 2003. Biochemical mechanism of action of a diketopiperazine inactivator of plasminogen activator inhibitor-1. Biochem. J., 373: 723-732.

ELLISON D W, MILLER V L, 2006. Regulation of virulence by members of the MarR/SlyA family. Curr. Opin. Microbiol., 9: 153-159.

ENGEL S, JENSEN P R, FENICAL W, 2002. Chemical ecology of marine microbial defense. J. Chem. Ecol., 28: 1971-1985.

FADEN A I, JACOBS T P, HOLADAY J W, 1981. Thyrotropin-releasing hormone improves neurologic recovery after spinal trauma in cats. N. Engl. J. Med., 305: 1063-1067.

FADEN A I, KNOBLACH S M, MOVSESYAN V A, et al., 2004. Novel small peptides with neuroprotective and nootropic properties. J. Alzheimers Dis., 6: S93-S97.

Faden A I, Movsesyan V A, Knoblach S M, et al., 2005. Neuroprotective effects of novel small peptides in vitro and after brain injury. Neuropharmacology, 49: 410-424.

FELNAGLE E A, JACKSON E E, CHAN Y A, et al., 2008. Nonribosomal peptide synthetases involved in the production of medically relevant natural products. Mol. Pharm., 5: 191-211.

FRIEDLAND R P, CHAPMAN M R, 2017. The role of microbial amyloid in neurodegeneration. PLoS Pathog., 13: e1006654.

FRIEDRICH A B, FISCHER I, PROKSCH P, et al., 2001. Temporal variation of the microbial community associated with the mediterranean sponge *Aplysina aerophoba*. FEMS Microbiol. Ecol., 38: 105-113.

FUQUA C, GREENBERG E P, 2002. Listening in on bacteria: acylhomoserine lactone signalling. Nat. Rev. Mol. Cell Biol., 3: 685-695.

FUSE Y, KOBAYASHI M, 2017. Conservation of the Keap1 – Nrf2 system: an evolutionary journey through stressful space and time. Molecules., 22: pii: E436.

FUSETANI N, MATSUNAGA S, 1993. Bioactive sponge peptides. Chem. Rev., 93: 1793-1806.

GALLAGHER L A, MCKNIGHT S L, KUZNETSOVA M S, et al., 2002. Functions required for extracellular quinolone signaling by *Pseudomonas aeruginosa*. J. Bacteriol., 184: 6472-6480.

GALLOWAY W R J D, HODGKINSON J T, BOWDEN S D, et al., 2011. Quorum sensing in gramnegative bacteria: small molecule modulation of AHL and AI – 2 quorum sensing pathways. Chem. Rev., 111: 28-67.

GARCÍA-ESTRADA C, ULLÁN R V, ALBILLOS S M, et al., 2011. A single cluster of coregulated genes encodes the biosynthesis of the mycotoxins roquefortine C and meleagrin in *Penicillium chrysogenum*. Chem. Biol., 18: 1499-1512.

GARDINER D M, COZIJNSEN A J, WILSON L M, et al., 2004. The sirodesmin biosynthetic gene cluster of the plant pathogenic fungus *Leptosphaeria maculans*. Mol. Microbiol., 53: 1307-1318.

GAUTSCHI M, SCHMID J P, PEPPARD T L, et al., 1997. Chemical characterization of diketopiperazines in beer. J. Agric. Food Chem., 45: 3183-3189.

GIESSEN T W, MARAHIEL M A, 2012. Ribosome-independent biosynthesis of biologically active peptides: application of synthetic biology to generate structural diversity. FEBS Lett., 586: 2065-2075.

GIESSEN T, MARAHIEL M, 2014. The tRNA-dependent biosynthesis of modified cyclic dipeptides. Int. J. Mol. Sci., 15: 14610-14631.

GIESSEN T W, VON TESMAR A M, MARAHIEL M A, 2013. Insights into the generation of structural diversity in a tRNA－dependent pathway for highly modified bioactive cyclic dipeptides. Chem. Biol., 20: 828-838.

GINZ M, ENGELHARDT U H, 2000. Identification of proline－based diketopiperazines in roasted coffee. J. Agric. Food Chem., 48: 3528-3532.

GONDRY M, LAUTRU S, FUSAI G, et al., 2001. Cyclic dipeptide oxidase from *Streptomyces noursei*. Eur. J. Biochem., 268: 1712-1721.

GONDRY M, SAUGUET L, BELIN P, et al., 2009. Cyclodipeptide synthases are a family of tRNA－dependent peptide bond-forming enzymes. Nat. Chem. Biol., 5: 414-420.

GONZALEZ J F, ORTIN I, DE LA CUESTA E, et al., 2012. Privileged scaffolds in synthesis: 2,5-piperazinediones as templates for the preparation of structurally diverse heterocycles. Chem. Soc. Rev., 41: 6902-6915.

GONZÁLEZ O, ORTÍZ-CASTRO R, DÍAZ-PEREZ C, et al., 2017. Non-ribosomal peptide synthases from *Pseudomonas aeruginosa* play a role in cyclodipeptide biosynthesis, quorum-sensing regulation, and root development in a plant host. Microb. Ecol., 73: 616-629.

GONZÁLEZ-REYES R E, NAVA-MESA M O, VARGAS-SÁNCHEZ K, et al., 2017. Involvement of astrocytes in Alzheimer's disease from a neuroinflammatory and oxidative stress perspective. Front. Mol. Neurosci., 10: 427.

GRAM L, DALGAARD P, 2002. Fish spoilage bacteria-problems and solutions. Curr. Opin. Biotechnol., 13: 262-266.

GROTTELLI S, BELLEZZA I, MOROZZI G, et al., 2015. Cyclo (His-Pro) protects SOD1G93A microglial cells from Paraquat-induced toxicity. J. Clin. Cell. Immunol., 6: 287.

GROTTELLI S, FERRARI I, PIETRINI G, et al., 2016. The role of cyclo (His-Pro) in neurodegeneration. Int. J. Mol. Sci., 17: pii: E1332.

GRUENEWALD S, MOOTZ H D, STEHMEIER P, et al., 2004. In vivo production of artificial nonribosomal peptide products in the heterologous host *Escherichia coli*. Appl. Environ. Microbiol., 70: 3282-3291.

GU Q, FU L, WANG Y, et al., 2013. Identification and characterization of extracellular cyclic dipeptides as quorum-sensing signal molecules from *Shewanella baltica*, the specific spoilage organism of *Pseudosciaena crocea* during 4℃ storage. J. Agric. Food Chem., 61: 11645-11652.

GUPTA S P, YADAV S, SINGHAL N K, et al., 2014. Does restraining nitric oxide biosynthesis rescue from toxins-induced parkinsonism and sporadic Parkinson's disease? Mol. Neurobiol., 49: 262-275.

HAHN M E, TIMME-LARAGY A R, KARCHNER S I, et al., 2015. Nrf2 and Nrf2-related proteins in developmental and developmental toxicity: insights from studies in zebrafish (*Danio rerio*). Free Radic. Biol. Med., 88: 275-289.

HALLMANN A, 2011. Evolution of reproductive development in the volvocine algae. Sex. Plant Reprod., 24: 97-112.

HAWVER L A, JUNG S A, NG W L, 2016. Specificity and complexity in bacterial quorum-sensing systems. FEMS Microbiol. Rev., 40: 738-752.

HEALY F G, KRASNOFF S B, WACH M, et al., 2002. Involvement of a cytochrome P450 monooxygenase in thaxtomin A biosynthesis by *Streptomyces acidiscabies*. J. Bacteriol., 184: 2019-2029.

HENSE B A, SCHUSTER M, 2015. Core principles of bacterial autoinducer systems. Microbiol. Mol. Biol. Rev., 79 (1): 153-169.

HENTSCHEL U, SCHMID M, WAGNER M, et al., 2001. Isolation and phylogenetic analysis of bacteria with antimicrobial activities from the Mediterranean sponges *Aplysina aerophoba* and *Aplysina cavernicola*. FEMS Microbiol. Ecol., 35: 305-312.

HO L, ONO K, TSUJI M, et al., 2018. Protective roles of intestinal microbiota derived short chain fatty acids in Alzheimer's disease-type beta-amyloid neuropathological mechanisms. Expert. Rev. Neurother., 18: 83-90.

HOGAN, D. A, 2006. Talking to themselves: autoregulation and quorum sensing in fungi. Eukaryot. Cell., 5: 613-619.

HOLDEN M T, RAM CHHABRA S, DE NYS R, et al., 1999. Quorum-sensing cross talk: isolation and chemical characterization of cyclic dipeptides from *Pseudomonas aeruginosa* and other gram-negative bacteria. Mol. Microbiol., 33: 1254-1266.

HSU M, SRINIVAS B, KUMAR J, et al., 2005. Glutathione depletion resulting in selective mitochondrial complex I inhibition in dopaminergic cells is via an NO-mediated pathway not involving peroxynitrite: implications for Parkinson's disease. J. Neurochem., 92: 1091-1103.

HUANG R, ZHOU X, XU T, et al., 2010. Diketopiperazines from marine organisms. Chem. Biodivers., 7: 2809-2829.

ITOH K, ISHII T, WAKABAYASHI N, et al., 1999. Regulatory mechanisms of cellular response to oxidative stress. Free Radic. Res., 31: 319-324.

JAYATILAKE G S, THORNTON M P, LEONARD A C, et al., 1996. Metabolites from an Antarctic sponge-associated bacterium, *Pseudomonas aeruginosa*. J. Nat. Prod., 59: 293-296.

JUNG E Y, LEE H S, CHOI J W, et al., 2011. Glucose tolerance and antioxidant activity of spent brewer's yeast hydrolysate with a high content of cyclo-His-Pro (CHP). J. Food Sci., 76: C272-C278.

JUNG E Y, HONG Y H, PARK C, et al., 2016. Effects of cyclo-His-Pro-enriched yeast hydrolysate on blood glucose levels and lipid metabolism in obese diabetic ob/ob mice. Nutr. Res. Pract., 10: 154-160.

KIM S W, LEE H K, SHIN J et al., 2013. Up-down regulation of HO-1 and iNOS gene expressions by ethyl pyruvate via recruiting p300 to Nrf2 and depriving it from p65. Free Radic. Biol. Med., 65: 468-476.

KITAI T, TANG W H W, 2018. Gut microbiota in cardiovascular disease and heart failure. Clin. Sci. (Lond.), 132: 85-91.

KOGLIN A, WALSH C T, 2009. Structural insights into nonribosomal peptide enzymatic assembly lines. Nat. Prod. Rep., 26: 987-1000.

KOO K B, SUH H J, RA K S, et al., 2011. Protective effect of cyclo(His-Pro) on streptozotocin-induced cytotoxicity and apoptosis in vitro. J. Microbiol. Biotechnol., 21: 218-227.

KUMAR N, GORANTLA J N, MOHANDAS C, et al., 2013. Isolation and antifungal properties of cyclo(D-Tyr-L-Leu) diketopiperazine isolated from *Bacillus* sp. associated with rhabditid entomopathogenic nematode. Nat. Prod. Res., 27: 2168-2172.

KWAK M K, LIU R, KWON J O, et al., 2013. Cyclic dipeptides from lactic acid bacteria inhibit proliferation of the influenza A virus. J. Microbiol., 51: 836-843.

KWAK M K, LIU R, KIM M K, et al., 2014. Cyclic dipeptides from lactic acid bacteria inhibit the proliferation of pathogenic fungi. J. Microbiol., 52: 64-70.

LAHOUD G, HOU Y M, 2010. Biosynthesis: a new (old) way of hijacking tRNA. Nat. Chem. Biol., 6: 795-796.

LAZOS O, TOSIN M, SLUSARCZYK A L, et al., 2010. Biosynthesis of the putative siderophore erythrochelin requires unprecedented crosstalk between separate nonribosomal peptide gene clusters. Chem. Biol., 17: 160-173.

LEE J, ZHANG L, 2015. The hierarchy quorum sensing network in *Pseudomonas aeruginosa*. Protein Cell, 6: 26-41.

LEE J, WU J, DENG Y, et al., 2013. A cell-cell communication signal integrates quorum sensing and stress response. Nat. Chem. Biol., 9: 339-343.

LEE H J, SON H S, PARK C, et al., 2015. Preparation of yeast hydrolysate enriched in cyclo-His-Pro (CHP) by enzymatic hydrolysis and evaluation of its functionality. Prev. Nutr. Food Sci., 20: 284-291.

LEHRER R I, COLE A M, SELSTED M E, 2012. θ-Defensins: cyclic peptides with endless potential. J. Biol. Chem., 287: 27014-27019.

LESZEK J, BARRETO G E, GASIOROWSKI K, et al., 2016. Inflammatory mechanisms and oxidative stress as key factors responsible for progression of neurodegeneration: role of brain innate immune system. CNS Neurol. Disord. Drug Targets, 15: 1-8.

LI W, KHOR T O, XU C, et al., 2008. Activation of Nrf2-antioxidant signaling attenuates NFkappaB-inflammatory response and elicits apoptosis. Biochem. Pharmacol., 76: 1485-1489.

LI J, WANG W, XU S X, et al., 2011. Lactobacillus reuteri-produced cyclic dipeptides quench agr-mediated expression of toxic shock syndrome toxin-1 in staphylococci. Proc. Natl. Acad. Sci. U. S. A., 108 (8): 3360-3365.

LISKAMP R M J, RIJKERS D T S, KRUIJTZER J A W, et al., 2011. Peptides and proteins as a continuing exciting source of inspiration for peptidomimetics. ChemBioChem, 12: 1626-1653.

LIU M, GRAY J M, GRIFFITHS M W, 2006. Occurrence of proteolytic activity an N-acylhomoserine lactone signals in the spoilage of aerobically chill-stored proteinaceous raw foods. J. Food Prot., 69: 2729-2737.

LIU R, KIM A H, KWAK M K, et al., 2017. Proline-based cyclic dipeptides from Korean fermented vegetable kimchi and from *Leuconostoc mesenteroides* LBP-K06 have activities against multidrug-resistant bacteria. Front. Microbiol., 8: 761.

LOZUPONE C A, STOMBAUGH J I, GORDON J I, et al., 2012. Diversity, stability and resilience of the human gut microbiota. Nature, 489: 220-230.

LYMAN M, LLOYD D G, JI X, et al., 2014. Neuroinflammation: the role and consequences. Neurosci. Res., 79: 1-12.

MAIYA S, GRUNDMANN A, LI S M, et al., 2006. The fumitremorgin gene cluster of *Aspergillus fumigatus*: identification of a gene encoding brevianamide F synthetase. ChemBioChem, 7: 1062-1069.

MARCHESAN J T, MORELLI T, MOSS K, et al., 2015. Association of synergistetes and cyclodipeptides with periodontitis. J. Dent. Res., 94 (10): 1425-1431.

MARIETTA E, HORWATH I, TANEJA V, 2018. Microbiome, immunomodulation, and the neuronal system. Neurotherapeutics, 15: 23-30.

MENEGATTI S, HUSSAIN M, NAIK A D, et al., 2013. mRNA display selection and solid-phase synthesis of Fc-binding cyclic peptide affinity ligands. Biotechnol. Bioeng., 110: 857-870.

MINELLI A, BELLEZZA I, GROTTELLI S, et al., 2006. Phosphoproteomic analysis of the effect of cyclo-(His-Pro)dipeptide on PC12 cells. Peptides, 27: 105-113.

MINELLI A, BELLEZZA I, GROTTELLI S, et al., 2008. Focus on cyclo(His-Pro): history and perspectives as antioxidant peptide. Amino Acids, 35: 283-289.

MINELLI A, CONTE C, GROTTELLI S, et al., 2009a. Cyclo(His-Pro)promotes cytoprotection by activating Nrf2-mediated up-regulation of antioxidant defence. J. Cell. Mol. Med., 13: 1149-1161.

MINELLI A, CONTE C, GROTTELLI S, et al., 2009b. Cyclo(His-Pro)up-regulates heme oxygenase 1 via activation of Nrf2-ARE signalling. J. Neurochem., 111: 956-966.

MINELLI A, GROTTELLI S, MIERLA A, et al., 2012. Cyclo (His-Pro) exerts antiinflammatory effects by modulating NF-κB and Nrf2 signalling. Int. J. Biochem. Cell Biol., 44: 525-535.

MISHRA A K, CHOI J, CHOI S J, et al., 2017. Cyclodipeptides: an overview of their biosynthesis and biological activity. Molecules, 22 (10): 1796.

MITOVA M, TOMMONARO G, HENTSCHEL U, et al., 2004. Exocellular cyclic dipeptides from a *Ruegeria* strain associated with cell cultures of *Suberites domuncula*. Mar. Biotechnol., 6: 95-103.

MONNET V, JUILLARD V, GARDAN R, 2016. Peptide conversations in gram-positive bacteria. Crit. Rev. Microbiol., 42: 339-351.

MORLEY J E, LEVINE A S, PRASAD C, 1981. Histidyl-proline diketopiperazine decreases food intake in rats. Brain Res., 210: 475-478.

MOUSSA S H, KUZNETSOV V, TRAN T A, et al., 2012. Protein determinants of phage T4 lysis inhibition. Protein Sci., 21: 571-582.

MOUTIEZ M, SCHMITT E, SEGUIN J, et al., 2014. Unravelling the mechanism of non-ribosomal peptide synthesis by cyclodipeptide synthases. Nat. Commun., 5: 5141.

NG W L, BASSLER B L, 2009. Bacterial quorum-sensing network architectures. Annu. Rev. Genet., 43: 197-222.

NICHOLSON B, LLOYD G K, MILLER B R, et al., 2006. NPI-2358 is a tubulin-depolymerizing agent: in vitro evidence for activity as a tumor vascular-disrupting agent. Anti-Cancer Drugs, 17: 25-31.

NOBLE E E, HSU T M, KANOSKI S E, 2017. Gut to brain dysbiosis: mechanisms linking western diet consumption, the microbiome, and cognitive impairment. Front. Behav. Neurosci., 11: 9.

ORTIZ-CASTRO R, DÍAZ-PEREZ C, MARTÍNEZ-TRUJILLO M, et al., 2011. Transkingdom signaling based on bacterial cyclodipeptides with auxin activity in plants. Proc. Natl. Acad. Sci. U. S. A., 108: 7253-7258.

PAN H, WANG H, WANG X, et al., 2012. The absence of Nrf2 enhances NF-kappaB-dependent inflammation following scratch injury in mouse primary cultured astrocytes. Mediat. Inflamm., 2012217580.

PARK D K, LEE K E, BAEK C H, et al., 2006. Cyclo(Phe-Pro) modulates the expression of ompU in *Vibrio* spp. J. Bacteriol., 188: 2214-2221.

PARK S W, CHOI S A, YUN J W, et al., 2012. Alterations in pancreatic protein expression in STZ-induced diabetic rats and genetically diabetic mice in response to treatment with hypoglycemic dipeptide cyclo (HisPro). Cell. Physiol. Biochem., 29: 603-616.

PEREZ-PARDO P, HARTOG M, GARSSEN J, et al., 2017. Microbes tickling your tummy: the importance of the gut-brain axis in Parkinson's disease. Curr. Behav. Neurosci. Rep., 4: 361-368.

PRASAD C, 1995. Bioactive cyclic dipeptides. Peptides, 16: 1511-1564.

PRASAD C, 2001. Role of endogenous cyclo(His-Pro) in voluntary alcohol consumption by alcohol-preferring C57Bl mice. Peptides, 22: 2113-2118.

PRASAD C, PETERKOFSKY A, 1976. Demonstration of pyroglutamyl peptidase and amidase activities toward thyrotropin-releasing hormone in hamster hypothalamic extracts. J. Biol. Chem., 251: 3229-3234.

PROKSCH P, EDRADA R A, EBEL R, 2002. Drugs from the seas—current status and microbiological implications. Appl. Microbiol. Biotechnol., 59 (2-3): 125-134.

PROSPERINI A, BERRADA H, RUIZ M J, et al., 2017. A review of the mycotoxin enniatin B. Front. Public Health, 5: 304.

RASH M, ANDERSEN J B, NIELSEN, K F, et al., 2005. Involvement of bacterial quorum-sensing signals in spoilage of bean sprouts. Appl. Environ. Microbiol., 71: 3321-3330.

REYNOLDS A, LAURIE C, MOSLEY R L, et al., 2007. Oxidative stress and the pathogenesis of neurodegenerative disorders. Int. Rev. Neurobiol., 82: 297-325.

RHEE K H, 2004. Cyclic dipeptides exhibit synergistic, broad spectrum antimicrobial effects and have antimutagenic properties. Int. J. Antimicrob. Agents, 24: 423-427.

ROCHA-ESTRADA J, ACEVES-DIEZ A E, GUARNEROS G, et al., 2010. The RNPP family of quorum-sensing proteins in Gram-positive bacteria. Appl. Microbiol. Biotechnol., 87: 913-923.

ROJAS MURCIA N, LEE X, WARIDEL P, et al., 2015. The *Pseudomonas aeruginosa* antimetabolite L-2-amino-4-methoxy-trans-3-butenoic acid (AMB) is made from glutamate and two alanine residues via a thiotemplate-linked tripeptide precursor. Front. Microbiol., 6: 170.

ROSIER A, BISHNOI U, LAKSHMANAN V, et al., 2016. A perspective on inter-kingdom signaling in plant-beneficial microbe interactions. Plant Mol. Biol., 90: 537-548.

ROSZYK E, PUSZCZEWICZ M, 2017. Role of human microbiome and selected bacterial infections in the pathogenesis of rheumatoid arthritis. Reumatologia, 55: 242-250.

RUSHWORTH S A, ZAITSEVA L, MURRAY M Y, et al., 2012. The high Nrf2 expression in human acute myeloid leukemia is driven by NF-kappaB and underlies its chemo-resistance. Blood, 120: 5188-5198.

RUSTENHOVEN J, JANSSON D, SMYTH L C, et al., 2017. Brain pericytes as mediators of neuroinflammation. Trends Pharmacol. Sci., 38: 291-304.

RYAN R P, AN S Q, ALLAN J H, et al., 2015. The DSF family of cell-cell signals: an expanding class of bacterial virulence regulators. PLoS Pathog., 11: e1004986.

SANDBERG M, PATIL J, D'ANGELO B, et al., 2014. NRF2-regulation in brain health and disease: implication of cerebral inflammation. Neuropharmacology, 79: 298-306.

SAUGUET L, MOUTIEZ M, LI Y, et al., 2011. Cyclodipeptide synthases, a family of class-I aminoacyltRNA synthetase-like enzymes involved in non-ribosomal peptide synthesis. Nucleic Acids Res., 39: 4475-4489.

SCHLUTER J, SCHOECH A P, FOSTER K R, et al., 2016. The evolution of quorum sensing as a mechanism to infer kinship. PLoS Comput. Biol., 12: e1004848.

SCHMIDT E W, OBRAZTSOVA A Y, DAVIDSON S K, et al., 2000. Identification of the antifungal peptide-containing symbiont of the marine sponge *Theonella swinhoei* as a novel δ-proteobacterium, "Candidatus Entotheonella palauensis" Mar. Biol., 136: 969-977.

SCHMITZ F J, VANDERAH D J, HOLLENBEAK K H, et al., 1983. Metabolites from the marine sponge *Tedania ignis*. A new atisanediol and several known diketopiperazines. J. Org. Chem., 48: 3941-3945.

SCHULTZ A W, OH D C, CARNEY J R, et al., 2008. Biosynthesis and structures of cyclomarins and cyclomarazines, prenylated cyclic peptides of marine actinobacterial origin. J. Am. Chem. Soc., 130: 4507-4516.

SCHWARZER D, FINKING R, MARAHIEL M A, 2003. Nonribosomal peptides: from genes to products. Nat. Prod. Rep., 20: 275-287.

SHERWIN E, DINAN T G, CRYAN J F, 2018. Recent developments in understanding the role of the gut microbiota in brain health and disease. Ann. N. Y. Acad. Sci., 1420 (1): 5-25.

SHIBOSKI C H, 2002. HIV-related oral disease epidemiology among women: year 2000 update. Oral Dis., 8 (Suppl 2): 44-48.

SHIBOSKI C H, WILSON C M, GREENSPAN D, et al., 2001. HIV-related oral manifestations among adolescents in a multicenter cohort study. J. Adolesc. Health, 29: 109-114.

SJÖGREN M, JONSSON P R, DAHLSTRÖM M, et al., 2011. Two brominated cyclic dipeptides released by the coldwater marine sponge *Geodia barretti* act in synergy as chemical defense. J. Nat. Prod., 74: 449-454.

SKANDAMIS P N, NYCHAS G J, 2012. Quorum sensing in the context of food microbiology. Appl. Environ. Microbiol., 78: 5473-5482.

SPRAGUE JR G F, WINANS S C, 2006. Eukaryotes learn to count: quorum sensing by yeast. Genes Dev., 20: 1045-1049.

STIERLE A C, CARDELLINA IIND J H, SINGLETON F L, 1988. A marine *Micrococcus* produces metabolites ascribed to the sponge *Tedania ignis*. Experientia, 44: 1021.

STRIEKER M, TANOVIC A, MARAHIEL M A, 2010. Nonribosomal peptide synthetases: structures and dynamics. Curr. Opin. Struct. Biol., 20: 234-240.

SUZUKI T, YAMAMOTO M, 2015. Molecular basis of the Keap1-Nrf2 system. Free Radic. Biol. Med., 88: 93-100.

SUZUKI T, MOTOHASH H, YAMAMOTO M, 2013. Toward clinical application of the Keap1-Nrf2 pathway. Trends Pharmacol. Sci., 34: 340-346.

TAPEINOU A, MATSOUKAS M T, SIMAL C, et al., 2015. Cyclic peptides on a merry-go-round: towards drug design. Biopolymers, 104: 453-461.

TAUBERT D, GRIMBERG G, STENZEL W, et al., 2007. Identification of the endogenous key substrates of the human organic cation transporter OCT2 and their implication in function of dopaminergic neurons. PLoS One, 2: e385.

TAYLOR M W, SCHUPP P J, BAILLIE H J, et al., 2004. Evidence for acyl homoserine lactone signal production in bacteria associated with marine sponges. Appl. Environ. Microbiol., 70: 4387-4389.

TEIXIDÓ M, ZURITA E, PRADES R, et al., 2009. A novel family of diketopiperazines as a tool for the study of transport across the blood-brain barrier (BBB) and their potential use as BBB-shuttles. Adv. Exp. Med. Biol., 611: 227-228.

THAKUR N L, HENTSCHEL U, KRASKO A, et al., 2003. Antibacterial activity of the sponge *Suberites domuncula* and its primmorphs: potential basis for epibacterial chemical defense. Aquat. Microb. Ecol., 31: 77-83.

THION M S, LOW D, SILVIN A, et al., 2018. Microbiome influences prenatal and adult microglia in a sex-specific manner. Cell, 172: 500-516.

THOMPSON IIIRD G R, PATEL P K, KIRKPATRICK W R, et al., 2010. Oropharyngeal candidiasis in the era of antiretroviral therapy. Oral Surg. Oral Med. Oral Pathol. Oral Radiol. Endod., 109: 488-495.

TOMMONARO G, ABBAMONDI G R, IODICE C, et al., 2012. Diketopiperazines produced by the halophilic archaeon, *Haloterrigena hispanica*, activate AHL bioreporters. Microb. Ecol., 63: 490-495.

TSANG A H, CHUNG K K, 2009. Oxidative and nitrosative stress in Parkinson's disease. Biochim. Biophys. Acta, 1792: 643-650.

UYEMURA K, DHANANI S, YAMAGUCHI D T, et al., 2010. Metabolism and toxicity of high doses of cyclo (HisPro) plus zinc in healthy human subjects. J. Drug Metab. Toxicol., 1: 105.

VALKO M, LEIBFRITZ D, MONCOL J, et al., 2007. Free radicals and antioxidants in normal physiological functions and human disease. Int. J. Biochem. Cell Biol., 39: 44-84.

VERDINE G L, HILINSKI G J, 2012. Stapled peptides for intracellular drug targets. Methods Enzymol., 503: 3-33.

WANG B, ZHU X L, KIM Y, et al., 2012. Histone deacetylase inhibition activates transcription factor Nrf2 and protects against cerebral ischemic damage. Free Radic. Biol. Med., 52: 928-936.

WEBSTER N S, HILL R T, 2001. The culturable microbial community of the great barrier reef sponge *Rhopaloeides odorabile* is dominated by an α-proteobacterium. Mar. Biol., 138: 843-851.

WEBSTER N S, WILSON K J, BLACKALL L L, et al., 2001. Phylogenetic diversity of bacteria associated with the marine sponge *Rhopaloeides odorabile*. Appl. Environ. Microbiol., 67: 434-444.

WEITZ J S, MILEYKO Y, JOH RI, et al., 2008. Collective decision making in bacterial viruses.

Biophys. J., 95: 2673-2680.

WYNENDAELE E, GEVAERT B, STALMANS S, et al., 2015. Exploring the chemical space of quorum sensing peptides. Biopolymers, 104: 544-551.

YANG H, DUAN Z, 2018. The local defender and functional mediator: Gut microbiome. Digestion, 97: 137-145.

YANG L, PALLIYAGURU D L, KENSLER T W, 2016. Frugal chemoprevention: targeting Nrf2 with foods rich in sulforaphane. Semin. Oncol., 43: 146-153.

YIN W B, GRUNDMANN A, CHENG J, et al., 2009. Acetylaszonalenin biosynthesis in *Neosartorya fischeri*: identification of the biosynthetic gene cluster by genomic mining and functional proof of the genes by biochemical investigation. J. Biol. Chem., 284: 100-109.

YU X, LIU Y, XIE X, et al., 2012. Biochemical characterization of indole prenyltransferases: filling the last gap of prenylation positions by a 5-dimethylallyltryptophan synthase from *Aspergillus clavatus*. J. Biol. Chem., 287: 1371-1380.

ZHU J, ZHAO A, FENG L, et al., 2016. Quorum sensing signals affect spoilage of refrigerated large yellow croaker (*Pseudosciaena crocea*) by *Shewanella baltica*. Int. J. Food Microbiol., 217: 146-155.

ZOUHIR S, PERCHAT S, NICAISE M, et al., 2013. Peptide-binding dependent conformational changes regulate the transcriptional activity of the quorum-sensor NprR. Nucleic Acids Res., 41: 7920-7933.

推荐阅读

CHEN J H, LAN X P, LIU Y, et al., 2012. The effects of diketopiperazines from *Callyspongia* sp. on release of cytokines and chemokines in cultured J774A. 1 macrophages. Bioorg. Med. Chem. Lett., 22, 3177-3180.

CHOI S A, YUN J W, PARK H S, et al., 2013. Hypoglycemic dipeptide cyclo(His-Pro) significantly altered plasma proteome in streptozocin-induced diabetic rats and genetically-diabetic (ob/ob) mice. Mol. Biol. Rep., 40: 1753-1765.

GIESSEN T W, VON TESMAR, A M MARAHIEL M A, 2010. A tRNA-dependent two-enzyme pathway for the generation of singly and doubly methylated ditryptophan 2,5-diketopiperazines. Biochemistry, 52: 4274-4283.

MANDER P, BROWN G C, 2005. Activation of microglial NADPH oxidase is synergistic with glial iNOS expression in inducing neuronal death: a dual-key mechanism of inflammatory neurodegeneration. J. Neuroinflammation, 2: 20.

MILLER B R, GULICK A M, 2016. Structural biology of nonribosomal peptide synthetases. Methods Mol. Biol., 1401: 3-29.

REIMER J M, ALOISE M N, HARRISON P M, et al., 2016. Synthetic cycle of the initiation module of a formylating nonribosomal peptide synthetase. Nature, 529: 239-242.

SCHMITZ F J, SCHULZ M M, SIRIPITAYANANON J, et al., 1993. New diterpenes from the gorgonian Solenopodium excavatum. J. Nat. Prod., 56: 1339-1349.

SEGUIN J, MOUTIEZ M, LI Y, et al., 2011. Nonribosomal peptide synthesis in animals: the cyclodipeptide synthase of *Nematostella*. Chem. Biol., 18: 1362-1368.

原著贡献者

Gennaro Roberto Abbamondi National Research Council of Italy—Institute of Biomolecular Chemistry, Pozzuoli, Italy

Fernando Jesus Ayala-Zavala Center for Research in Nutrition and Development, A.C (CIAD AC), Hermosillo, Mexico

Ilaria Bellezza Department of Experimental Medicine, University of Perugia, Perugia, Italy

Didem Berber Marmara University, Department of Biology, Faculty of Arts and Sciences, Istanbul, Turkey

Raffaele Coppola DiAAA, Department of Agricultural, Environmental and Food Sciences, University of Molise, Campobasso, Italy

Valeria Costantino TheBlueChemistryLab, Department of Pharmacy, University of Naples Federico II, Napoli, Italy

Adele Cutignano National Research Council of Italy—Institute of Biomolecular Chemistry, Pozzuoli, Italy

Antonio d'Acierno Institute of Food Science, ISA-CNR, Avellino, Italy

Vincenzo De Feo Department of Pharmacy, University of Salerno, Salerno, Italy

Gerardo Della Sala Laboratory of Pre-Clinical and Translational Research, IRCCS-CROB, Referral Cancer Center of Basilicata, Rionero in Vulture, Italy

Germana Esposito TheBlueChemistryLab, Department of Pharmacy, University of Naples Federico II, Napoli, Italy

Ilaria Finore National Research Council of Italy—Institute of Biomolecular Chemistry, Pozzuoli, Italy

Florinda Fratianni Institute of Food Science, ISA-CNR, Avellino, Italy

Barış Gökalsın Marmara University, Department of Biology, Institute of Pure and Applied Sciences, Istanbul, Turkey

Adriano Gomes-Cruz Federal Institute of Education, Science and Technology of Rio de Janeiro (IFRJ), Department of Food, Rio de Janeiro, Brazil

Daniel Granato State University of Ponta Grossa (UEPG), Department of Food Engineering, Ponta Grossa, Brazil

Angel G. Jimenez Department of Microbiology, University of Texas Southwestern Medical Center, Dallas, TX, United States

Margarita Kambourova Institute of Microbiology—Bulgarian Academy of Sciences, Sofia, Bulgaria

Onur Kırtel Industrial Biotechnology and Systems Biology Research Group, Marmara University,

Bioengineering Department, Istanbul, Turkey

Raphaël Lami Sorbonne Université, CNRS, Laboratoire de Biodiversite et Biotechnologies Microbiennes, LBBM, Banyuls-sur-Mer, France

Alba Minelli Department of Experimental Medicine, University of Perugia, Perugia, Italy

Filomena Nazzaro Institute of Food Science, ISA-CNR, Avellino, Italy

Barbara Nicolaus National Research Council of Italy—Institute of Biomolecular Chemistry, Pozzuoli, Italy

Ebru Toksoy Öner Industrial Biotechnology and Systems Biology Research Group, Marmara University, Bioengineering Department, Istanbul, Turkey

Matthew J. Peirce Department of Experimental Medicine, University of Perugia, Perugia, Italy

Annarita Poli National Research Council of Italy—Institute of Biomolecular Chemistry, Pozzuoli, Italy

Wim J. Quax Groningen Research Institute of Pharmacy, Department of Chemical and Pharmaceutical Biology, University of Groningen, Groningen, The Netherlands

Nüzhet Cenk Sesal Marmara University, Department of Biology, Faculty of Arts and Sciences, Istanbul, Turkey

Vanessa Sperandio Department of Biochemistry, University of Texas South western Medical Center, Dallas, TX, United States

Roberta Teta TheBlueChemistryLab, Department of Pharmacy, University of Naples Federico II, Napoli, Italy

Wim Van den Ende Laboratory of Molecular Plant Biology, KU Leuven, Leuven, Belgium

Maxime Versluys Laboratory of Molecular Plant Biology, KU Leuven, Leuven, Belgium

Jan Vogel Groningen Research Institute of Pharmacy, Department of Chemical and Pharmaceutical Biology, University of Groningen, Groningen, The Netherlands